D0385104

MANAGEMENT: A SYSTEMS APPROACH

McGRAW-HILL SERIES IN MANAGEMENT

KEITH DAVIS Consulting Editor

ALLEN	Management and Organization
ALLEN	The Management Profession
ARGYRIS	Management and Organizational Development: The Path from *XA* to *YB*
BECKETT	Management Dynamics: The New Synthesis
BENNIS	Changing Organizations
BENTON	Supervision and Management
BERGEN AND HENRY	Organizational Relations and Management Action
BLOUGH	International Business: Environment and Adaptation
BOWMAN	Management: Organization and Planning
BROWN	Judgment in Administration
CAMPBELL, DUNNETTE, LAWLER, AND WEICK	Managerial Behavior, Performance, and Effectiveness
CLELAND AND KING	Management: A Systems Approach
CLELAND AND KING	Systems Analysis and Project Management
CLELAND AND KING	Systems, Organizations, Analysis, Management: A Book of Readings
DALE	Management: Theory and Practice
DALE	Readings in Management: Landmarks and New Frontiers
DAVIS	Human Behavior at Work: Human Relations and Organizational Behavior
DAVIS AND BLOMSTROM	Business, Society, and Environment: Social Power and Social Response
DAVIS AND SCOTT	Human Relations and Organizational Behavior: Readings and Comments
DeGREENE	Systems Psychology
DUNN AND RACHEL	Wage and Salary Administration: Total Compensation Systems
DUNN AND STEPHENS	Management of Personnel: Manpower Management and Organizational Behavior
FIEDLER	A Theory of Leadership Effectiveness
FLIPPO	Principles of Personnel Management
GLUECK	Business Policy: Strategy Formation and Executive Action
GOLEMBIEWSKI	Men, Management, and Morality
HARBISON AND MYERS	Management in the Industrial World

HICKS	The Management of Organizations: A Systems and Human Resources Approach
JOHNSON, KAST, AND ROSENZWEIG	The Theory and Management of Systems
KAST AND ROSENZWEIG	Organization and Management: A Systems Approach
KOONTZ	Toward a Unified Theory of Management
KOONTZ AND O'DONNELL	Principles of Management: An Analysis of Managerial Functions
KOONTZ AND O'DONNELL	Management: A Book of Readings
LEVIN, McLAUGHLIN, LAMONE, AND KOTTAS	Production/Operations Management: Contemporary Policy for Managing Operating Systems
LUTHANS	Contemporary Readings in Organizational Behavior
McDONOUGH	Information Economics and Management Systems
McNICHOLS	Policy Making and Executive Action Cases on Business Policy
MAIER	Problem-solving Discussions and Conferences: Leadership Methods and Skills
MARGULIES AND RAIA	Organizational Development: Values, Process, and Technology
MAYER	Production Management
MUNDEL	A Conceptual Framework for the Management Sciences
PETIT	The Moral Crisis in Management
PETROF, CARUSONE, AND McDAVID	Small Business Management: Concepts and Techniques for Improving Decisions
PIGORS AND PIGORS	Case Method in Human Relations
PRASOW AND PETERS	Arbitration and Collective Bargaining: Conflict Resolution in Labor Relations
READY	The Administrator's Job
REDDIN	Managerial Effectiveness
SALTONSTALL	Human Relations in Administration
SARTAIN AND BAKER	The Supervisor and His Job
SCHRIEBER, JOHNSON, MEIER, FISCHER, AND NEWELL	Cases in Manufacturing Management
SHULL, DELBECQ, AND CUMMINGS	Organizational Decision Making
STEINER	Managerial Long-range Planning
SUTERMEISTER	People and Productivity
TANNENBAUM	Control in Organizations
TANNENBAUM, WESCHLER, AND MASSARIK	Leadership and Organization
VANCE	Industrial Administration
VANCE	Management Decision Simulation

MANAGEMENT: A SYSTEMS APPROACH

DAVID I. CLELAND
Professor of Systems Management Engineering
School of Engineering
University of Pittsburgh

WILLIAM R. KING
Professor of Business Administration
Graduate School of Business
University of Pittsburgh

McGRAW-HILL BOOK COMPANY

New York St. Louis San Francisco Düsseldorf Johannesburg Kuala Lumpur London Mexico Montreal New Delhi Panama Rio de Janeiro Singapore Sydney Toronto

MANAGEMENT: A SYSTEMS APPROACH

Library of Congress Catalog Card Number 74-39065

07-011314-9

4567890DODO798765

This book was set in Alpha Gothic by University Graphics, Inc., and printed and bound by R. R. Donnelley & Sons Company. The designer was Richard Paul Kluga; the drawings were done by F. W. Taylor Company. The editors were Richard F. Dojny and John M. Morriss. Peter D. Guilmette supervised production.

contents

preface

This book deals with the *management* process as viewed in *systems* terms. Management is not a new field, and the systems viewpoint is not a new concept. However, the application of systems concepts to the management of organizations is relatively new.

Of course, theorists and authors have expounded for some time on the systems view of organizational management. However, many of the applications discussed have been of the "toy illustration" variety. Certainly, no comprehensive theory of organizational management through systems concepts has been developed.

This book may be viewed as an attempt at moving toward such a theory. While it may not attain the objective, it does focus on the application of systems concepts to the management process, rather than to individual management problems. It also draws on systems ideas from many fields in an attempt at synthesis and generalization. In those senses, at least, it is an attempt at theory development.

Perhaps it is best to introduce such a book by making clear those objectives to which it does *not* aspire and those approaches and principles that it does *not* adopt.

1. The book avoids a descriptive institutional framework. It is *not* a book about business management, or industrial management, or educational management. Rather, it deals with the generic management process common to all modern organizations. Illustrations are drawn from such diverse organizations as consumer-oriented business firms, industrial-oriented business, education, government, the military, churches, and hospitals. In no case is more descriptive material presented concerning these institutions than is necessary to understand the issue at hand. Thus,

the reader will be prepared to *begin* developing an understanding of the institution in which he finds himself after he completes the book, and, it is hoped, he will be able to do this better and more efficiently because he understands the process required for managing that organization.

2. The book does not subscribe to a particular point of view or school of thought other than that broad one which can be described as the "systems view." Quantitative analysis is made use of, but so, too, are behavioral science and a variety of other fields of thought which themselves may be thought of as comprehensive approaches to management. Thus, the "behavioral school" is drawn on, as well as the "management science" school and others, in developing a view of the overall management process.

The level of disciplinary sophistication—be it mathematical, behavioral, or otherwise—required for an understanding of the book is not great. Quantitative approaches are described from a point of view of the concepts involved rather than in terms of the detailed mathematics. Thus, high school algebra will suffice for complete understanding.

3. The book does not adopt the traditional "principles" approach to management. However, it does not seek to avoid principles, per se. Indeed, one might characterize the approach as an attempt at blending principles with the more recent systems approaches. When one does this, he finds that they are not so disparate as many believe.

Thus, it is easier to describe what the book is *not* in a concise way than it is to describe what it is. Management is a complex process, and the process of understanding it is therefore not simple. Moreover, the systems approach is inherently broader and more complex than other viewpoints which tend to artificially compartmentalize the various aspects of management.

Perhaps then the best way of explaining what the book is, as opposed to what it is not, is to ask the reader to browse through the table of contents and the book itself. Here, the most enlightening thing which might be said concerns the organization of the book rather than its specific content.

In dealing with a subject which is multidimensional in nature in the context of a single-dimensional "linear" medium, one is necessarily restricted. Often, we feel, this basic difficulty leads to unnatural partitionings on the part of authors. Classification schemes are invented to reduce multidimensional concepts to a single dimension. While this is perfect for a chapter-by-chapter book format, it often leads to a deemphasis of the interdependencies among major topics and among the subtleties and nuances of individual topics.

While we have no special ability to deal with this problem within the context of the traditional written-word book format, we have not attempted to "linearize" the presentation in order to make the book appear to be exhaustive or to make the various chapters and sections appear to be disjoint. In fact, as with any complex system, the system chosen to describe management in this book is a series of interrelated and interdependent concepts and approaches. Thus, each chapter builds on materials in other chapters, and each chapter leads to various other chapters. The book can therefore be fully understood only by proceeding from beginning to end, since each chapter intrinsically depends on all of those coming before it.

Indeed, we do not even promise that the logical relationship between elements of the book is entirely unidirectional. There are some concepts introduced in later chapters which are needed for the full understanding of earlier chapters. In such cases we have attempted to introduce the concepts more than once, usually at different levels of sophistication. On first introduction the concept is given in an intuitive explanation and is illustrated by examples. The objective of the first introduction is to provide an intuitively appealing approach to the concept which is as close as possible to that view held by the layman. Subsequent introductions of the concept tend to refine the definition and to emphasize the differences between the layman's view and that of the sophisticated management practitioner.

The exercises at the end of each chapter are generally based on the material in the chapter. However, the student will quickly find that some exercises are cumulative in nature and that a few may require some additional research and work to answer properly. Both of these features are purposefully included to give some semblance of the flavor of real-world problem solving to the student.

This book is directed to integrating the most recent management thought and theory into the systems view of organizations. It represents our view of the best of systems theory, quantitative analysis, behavioral science, and qualitative management theory, all woven into an integrated whole. It is not a book that can be read and then parroted back in the form of rote answers. The good student will find that the book raises more questions than it answers, but, if we have accomplished our goal, he will also find that he is better able to answer those questions than he would be without the framework developed here.

There are numerous people and organizations deserving of thanks for their aid in making this book possible. Any listing would omit some; so we shall choose the easy way out by thanking only those who labored on the many versions through which the manuscript proceeded—Cathy Brown, Patricia Colin, Ruth Enright, and Barbara Wilson.

David I. Cleland
William R. King

section one

INTRODUCTION

1

modern management

Management is an ancient art—practiced in the distant past principally by kings, princes, prime ministers, clergymen, and generals, and in the more recent past by business executives and government bureaucrats. Today we recognize that management pervades virtually every aspect of our society—the individual, the family unit, the corporation, the government agency, the national government, the United Nations, and other international organizations. Virtually everyone is a manager in some phase of his daily life. Each of us must manage our time and other resources, so that everyone knows "something" about management. This widespread knowledge about management is both a blessing and a curse to the management educator, for he deals with a topic that is already familiar to his audience, but he must deal with many preconceived, and often misguided, notions about the field.

That we are not all management "experts" is attested to by the undesirable results which many of us produce in our personal lives and by the problems of multimillion-dollar corporations which are discussed in our newspapers. If the Penn Central Railroad bankruptcy of 1970, the Lockheed Aircraft Corporation's cash flow problems of the early 1970s, and the Department of Defense's continuing problems with cost overruns on weapons system projects do not serve to remind us of the need for good management, most of us will reach that conclusion by comparing our optimistic plans for today's work with our actual accomplishments.

MANAGEMENT AND ORGANIZATIONS

To begin a discussion of management—a field in which each of us is involved at least as an amateur practitioner—one should define the term "management." To do so is to answer questions such as: What is management? How does it differ from other fields? What skills does it require?

Although beginning the study of management with a definition of the term has a certain logical appeal, we must guard against providing a definition that has little intuitive appeal or operational significance. Instead of such a dictionary-style definition, we shall offer a tentative *operational definition* of the concept called "management." Like any definition of a complex concept, it will be subject to revision as we learn more about the idea in question. Indeed, one might think of this book as being focused toward developing, for the reader, a personalized, intuitively appealing, and meaningful definition of management.

An Operational Definition of "Management"

A standard dictionary-style definition presumes that the concept that is being defined is explainable in terms of a number of other already understood concepts. With "management," as well as with other complex concepts, such an approach is not very useful. More useful is an *operational definition*[1]—one that relates the management concept to a number of *observable* criteria which, if satisfied, indicate that a management environment exists. Such definitions are not so glib as are dictionary definitions, but they are more useful in that they can serve as guides to thought and action.

ORGANIZED ACTIVITY The first observable criterion having to do with management is that it is a *process of organized activity.* Groups of people are involved in working toward a common purpose. The organized activity may take a variety of forms, ranging from a tightly structured organization, such as a military combat unit, to the very loosely knit organization typically found in informal social groups. General Motors, with its plethora of vice-presidents and division heads, is obviously engaged in organized activity, but so, too, are the college students who form a commune. Indeed, even the four couples who join together to take in a rock concert are "organized," since they undoubtedly assign responsibility for ticket purchase, auto driving, etc.

The work group[2] consists of those people who have an interest in the

[1] Operational definitions are important in science, where everything must be defined in terms of what would be observed if certain operations were performed. The use of the term here is not precisely that familiar to scientists, but it is closely related to it. See R. L. Ackoff, *Scientific Method,* John Wiley & Sons, Inc., New York, 1962, chap. 5.

[2] For simplicity's sake, we shall use such highly connotative terms as "work group" throughout the book even though it is apparent that the activity in question may not be thought of by anyone as "work." Thus, for the sake of semantic convenience, we shall force ourselves into such untenable situations as referring to the participants in an orgy as a "work group."

work activity. Some may play the role of superiors, some may be subordinates, peers, or associates; but in all cases, each has a stake in the outcome or product of the activity. *The management process in such a situation serves to bring about a coordinated effort of the many individuals and subgroups.*

OBJECTIVES The second operational criterion for management is that *there is an objective toward which the organized group activity is directed.* In highly structured military organization, the objective may be explicitly stated; however, the objective of a social organization such as a Kiwanis club or a parent-teacher's association might be much less specific. But, whether it is explicit or implicit, the objective is the goal toward which organizational activities are directed.

In some business firms, objectives may be rather specifically stated in terms of a desired return on investment or a specified increase in per-share earnings. However, consistent with society's emphasis on the social consequences of capitalistic activity, most modern businesses are stating their objectives in much more sophisticated fashion. It is no longer surprising to see a firm's annual report dwell on organizational objectives regarding ecology, job training, and a variety of other "social responsibilities" that did not often appear in such publications only a decade ago.

RELATIONSHIPS AMONG RESOURCES The third operational criterion for management is a requisite that the goal-oriented organized activity be brought about by *establishing certain relationships among the available resources.* "Resources" is a general term which includes material, supplies, equipment, funds, and, of course, people.

In the case of material resources, the relationships which are established tend to be physical in nature, for example, in the layout of a production line. In the case of monetary resources, the relationship would involve allocation of available funds to such purposes as holding inventory, acquiring and maintaining plant and equipment, and specifications as to how and where the funds are to be obtained.

The greatest challenge to the manager is frequently the establishment of relationship among *human resources.* He must deal with both *formal relationships,* such as those which are established by the organizational charter (as in the appointment of officers), and the *informal relationships* that arise as people attempt to relate to each other in their daily lives. These relationships evolve in a constantly changing pattern which reflects how people work with each other and are influenced by each other in their organizational lives. The formal names for these personal interrelationships are *authority* and *responsibility patterns.*

Authority and responsibility are complex forces found in any organization. For the present, let us be content with simplistic definitions: *Authority is the legal right to command or direct the efforts of others; responsibility is the obligation to respond to directions from other people.*

One of the challenges in today's society is to understand the changing patterns of authority and responsibility among the participants in virtually every organization. The drive of people at all levels to participate in the destinies of their organizations has become so pervasive that the "prerogatives of rank" that once formed basic ties of the social structure have largely disappeared.

Nowhere are these changes so apparent as in our universities, where many students have been appointed to such "establishment" bodies as the board of trustees. For instance, the *Wall Street Journal* reports that such prestigious schools as Harvard, Princeton, Stanford, and Columbia are among those which have taken major steps toward increasing student involvement in those decisions which determine their futures.[3]

In the place of the "old order" based on rank and authority has come a new set of human relationships and a new breed of manager whose authority is based on his personal effectiveness, his expertise, and his knowledge of the world around him. This new variety of manager needs skills to perform his job that are different from those that were essential to the traditional manager who was "in command" of those around him.[4]

WORKING THROUGH OTHERS The fourth critical element of management involves *working through others to accomplish organizational objectives.* Since the manager exists in a group environment, he must take advantage of the various diverse talents of the group. He must avoid the natural human tendency to do the job himself and must assign the doing of various tasks to others. Thus, through *division of labor*—the assignment of limited areas of responsibility to individuals or groups—the manager can take maximum advantage of both his own talents and those of others.

Typically, the idea of working *through* others is interpreted in the traditional sense of assigning tasks to subordinates. But we shall see that the "new breed" of manager may find it more challenging and important to work through peers, associates, and even superiors to get the job done.

DECISIONS The final critical criterion in the definition of "management" is an active involvement with *decisions*—the evaluation and selection of alternatives in a complex environment which is often fraught with risk and

[3] *Wall Street Journal,* Jan. 4, 1971, p. 1.

[4] Cf. Max Ways, "More Power to Everybody," *Fortune,* May 1970, pp 173–299.

uncertainty. Whether the decision be what new product to introduce, how much to charge for one's product, whom to hire, where to locate a plant, or which small company to consider as an acquisition candidate, the manager is faced with an unending sequence of choices among alternative courses of action, which we refer to as *managerial decision problems.* The quality of the alternative which he selects in each case determines the organization's performance in terms of profit, per-share earnings, or whatever measure is considered applicable. In the long run, the entire future of the organization rests on the degree to which the "right" decisions are made by managers.

DEFINITIONAL SUMMARY Thus, the five operational criteria which we shall use as a first approximation to "management" are:

1. Organized activity
2. Objectives
3. Relationships among resources
4. Working through others
5. Decisions

If we can observe and assess these factors in a particular situation, we shall say that management is being practiced; if not, we shall conclude that "true" (meaningful) management is absent.

Generality of the Definition

As we conclude our definition of management, we should be cognizant of its generality and wide range of applicability. Each of these criteria can exist in a corporation, government agency, proprietorship, church, family, or virtually any other kind of human activity. In many of these contexts, the practice of management is not usually thought of as management or expressed in management terms, but parents are managers in the family just as surely as the company's president is a manager of U.S. Steel, and the housewife who organizes a bridge party is as much a manager as is that individual who holds the title with the New York Mets.

There is a secondary benefit which one derives from the pervasiveness of management in our society. If we are truly to study a field which delves into just about every aspect of our lives, at least our study must be widely useful. Thus, we can approach the remainder of the book with the confidence of a man who is about to embark on an enlightening and useful "trip." If the book fails to provide such stimulation to the reader, it is the fault of the authors—for, indeed, management itself is a topic which has unique stimulations for the receptive mind.

The Pervasiveness of Organizational Management

Organizations pervade our society to such an extent that management, which is inherently involved with organized activity, is necessarily important to everyone.

Even the simplest tasks performed by an individual are "involved with" management even though they might not meet our definitional criteria. For instance, the construction laborer makes decisions involving the allocation of his limited energy and operational decisions about how the job is to be done. He may informally perform other management tasks when he works alongside novice workers. He also "manages" when he acts to implement these decisions—i.e., to actually *do* that which he has *decided.* Thus, even the elementary job of a construction laborer has *some* (but not all) of the essential elements of management.

Indeed, since everyone must at some time make decisions, allocate his time and energy, and carry out these actions to "get things done," everyone does indeed perform certain essential management elements. Hence, each of us is likely to be able to use profitably any information or knowledge which would enhance the effectiveness of our practice of these elements of management.

However, the study of management and its practice as it applies to such elementary situations is rather sterile. One of the reasons for this sterility is that little is achieved in our complex society through the actions of a single person. The strain of individual achievement is strong in our folklore; but the vast proportion of the "things" which "happen" in our modern world do so because of *groups* of people involved in *joint efforts.*

The real payoff from "good management" is not therefore with the individual, who must manage his own life, but rather with the *organizations* —business, governmental, ecclesiastical, etc.—which so importantly affect the cultures and economies of nations and the daily lives of virtually every human being.

Today's world is truly one in which organizations, and increasingly *large* organizations, are the prime movers. While our society strives to recognize the worth and uniqueness of the individual, at the same time most individuals have little power except in their roles as components of organizations. For example, the individual employee has little power to influence the actions of his employer. If the employer chooses to act solely in his self-interest, he can use his economic power to the disadvantage of individual employees. How this can occur was vividly displayed in the employer-employee relations which existed between the time of the Industrial Revolution and the advent of labor unions. However, the concerted action of a group, such as a labor union, can, through collective bargaining, have great influence on the employer. So, too, can the individual consumer have little impact on the producer of defective merchandise or the individual voter have little effect on the political system unless he combines with others to take joint action.[5]

Of course, organizations and their management have long been important to society. The Industrial Revolution resulted in a production system which necessitated specialization, and the shoemaker who was at one time purchasing agent, production worker, and salesman was replaced by a shoe company in which the labor was divided among specialists. With this change came the need for *managers* who could get things done through others by overseeing their work, integrating their efforts, and planning for the future of the organization. Even prior to that, the management practices of such organizations as churches and military forces were important to the daily lives of most people.

THE IMPORTANCE OF MODERN MANAGEMENT

If the management of organizations has been important to the daily lives of individuals since long before the Industrial Revolution, how can one reasonably claim that it is so much more important today than ever before? Is this perhaps not a simple case of authors who are specialists in the field, and therefore convinced that their own area of interest is more important than all others, even those that they know nothing about?

It is probable that such an accusation is in part true. All of us are subject to the danger of working in an area because we believe it to be important and then reinforcing our perception of its importance precisely because we are working in it. Such "tunnel vision" allows us to see nothing but that which we know and to draw conclusions and make evaluations on the basis of partial information.

However the greatly increased importance of management in the modern world is, fortunately, demonstrable. It emanates from two characteristics of our society: *(a)* its size and complexity and *(b)* its increasing rate of change.

Our Complex Society

The very magnitude of the organizations with which all of us deal in our daily lives, be they public utilities, government, or manufacturers, implies that each has great economic, political, and social power. Corporations

[5]The existence of "consumer-protection" agencies at various levels of government is another form of organized activity designed to meet these ends. So, too, are such organizations as "Common Cause"—a loosely knit confederation of people who recognize that their impact on the political system must be organized and based on sound information. See James Reston, "Washington: The Paradox of American Politics," *New York Times*, Sept. 9, 1970, p. 48.

which were considered to be huge prior to World War II have increased in size and power manifold. In 1939, General Motors Corporation, perhaps the archtypical giant corporation, had *assets* of about $1 billion. Less than twenty years later, its annual *after-tax profits* reached this amount.[6]

Economic magnitude automatically implies potential power. Such concentrations of power mean that *the impact of individual organizational actions is great.* And, of course, it is the management of these organizations which determines which actions will be taken and how this power will be wielded.

Thus, when the management of one auto manufacturer decides to incorporate additional safety devices on its new models, this action may motivate other manufacturers to follow suit, and the result may be the saving of thousands of lives annually. So, too, when the U.S. Department of Defense awards a *single* major contract for the development of a new weapons system, the impact on employment in an entire region may be immense. Indeed, the aggregate actions of the defense establishment support about one-fifth of all United States citizens through the military services, the federal department itself, and the defense industry.[7] Even a single program of a large agency may be of immense size and complexity. For example, at the peak of the Apollo moon shot program of the late 1960s, about 20,000 contractors and subcontractors were involved.[8]

The great potential impact of individual management actions has never been so pervasive as it is today. In effect, the cost of individual managerial errors is much larger than what it has ever been before, and the potential for such error is greater. A slight miscalculation on the part of a corporate executive in deciding on a new plant location can result in great cost to his firm and to those people in the region in which the plant is built. A new product which fails in the marketplace affects not only the corporation but thousands of workers who produce it and the thousands of "support personnel"—grocers, plumbers, etc.—with whom they deal. When a new federal program is introduced, literally tens of millions of people may have their day-to-day lives seriously affected for better or worse.

For all of the impact which a single organization can have, the aspect of our complex society which necessitates good management is the *interdependence of organizations.* Modern organizations are linked together in complex ways. This means that not only are individual organizations large and complex, but the "systems" formed by the interlinked organizations, which in turn form our society, are even more so.

[6] John Brooks, "A Clear Break with the Past," *American Heritage,* August, 1970, p. 6.

[7] "The War Business," *Wall Street Journal,* May 28, 1969, p. 1.

[8] "Corporate Contributions to Lunar Landing," *Investor's Reader,* published by Merrill Lynch, Pierce, Fenner, & Smith, vol. 52, no. 12, June 18, 1968, pp. 2–5.

If, for example, a large corporation were to propose that their employees work a six-hour instead of an eight-hour work day, the reverberations would be felt throughout the nation. *Competitors* would be placed in the position of studying the feasibility of doing the same thing; *local communities* would have to look at their recreational and business facilities to see if they were capable of handling the additional demand; *public officials* might consider legislation to change existing laws to reflect the six-hour work day; some zealous government official would undoubtedly call for an investigation; *employees* would have to determine how they could best utilize their additional free time; *companies* engaged in manufacturing recreational equipment might sense an increased demand for such equipment and set out to increase their production; *union leaders* would probably seize the opportunity for negotiating similar contracts with other companies; *stockholders* might very well feel that such action would reduce overall profitability, and thus they might protest at a stockholders' meeting; *suppliers* of goods and services would have to adjust their working hours to conform to the new practices of this company; *customers* might expect higher prices for the product or service they were buying from the company.

Thus, one of the ways in which we can demonstrate the increased importance of management is to reveal how the size, power, and interdependence of modern organizations can have serious, widespread, and lasting effect on society.

Change in Our Society

The increase in the size and complexity of organizations does not itself directly warrant the conclusion that management is vastly more important today than ever before. The other element which is essential to this conclusion is *change.*

It has become almost trite to expound the idea that we live in a period of accelerating change. This change is reflected in both tangible and intangible form. Institutions look vastly different today from what they did only a short time ago. Of the 20 largest companies in the United States 40 years ago, only 2 are still among the first 20 in size. Of the 100 largest companies 25 years ago, almost 50 percent either do not exist today or have substantially declined in size.[9] Computers now monitor the operations of all large organizations. Minority groups have demanded and achieved a greater degree of freedom and civil rights. An age of political turmoil, punctuated by the grim repeat of the assassin's rifle, is upon us.

[9] L. A. Allen, *The Management Profession,* McGraw-Hill Book Company, New York, 1964.

One could go on for hours in a seemingly endless listing of vast and obvious tangible changes in our society and its institutions.

Along with these tangible changes have come more dramatic intangible ones. Attitudes and values which have prevailed for generations have been overturned almost overnight—in many cases, so quickly that we all suffer to a degree from "future shock."[10] Sex, which was a forbidden topic of discussion for our parents, is now openly discussed and practiced outside the traditional confines of legal marriage. Religion—at least, of the formal, churchgoing variety—which was an intrinsic part of our heritage, appears to be in decline. Yet, before we are certain that this change has occurred, we are faced with a "Jesus revolution" of increased fervor. Again, one could make an endless list of the dramatic changes with which we are constantly faced.

Max Ways has suggested that the intangible changes are due to two "revolutions." According to Ways, the first change ". . . widely discussed in the 60's, is a sharp and insistent upsurge of the desire for material goods: food, houses, cars, clothing. The second revolution concerns the psyche. It pertains to the dignity, status, personality, significance, and power of individuals. Many of the aims of this second revolution can be summed up in one word: participation."[11]

In subsequent parts of this chapter and the book we shall deal with these revolutions and their impact on management. The term "participation" will be one that comes up again and again, so that we should remember it and the roots from which it springs. Now let us go on to conclude this opening chapter with a brief look at "modern" management as a changed version of "traditional" management.

MODERN AND TRADITIONAL MANAGEMENT

Change affects organizations and their managers with equal impact. In an era of change, few things are certain. One certainty is that while we will be unable to precisely predict the nature and magnitude of the continuing change process, it will indeed continue. Thus, organizations and managers must learn to cope with continuing change.

Organizations and Change

Organizations that do not adapt to change find it difficult or impossible to survive. A vivid illustration of organizations which were adaptive for

[10] The term "future shock" was coined by Alvin Toffler in his book of that name (Random House, Inc., New York, 1970) to describe the psychic and physical damage brought about by subjecting man to too much change in too little time.

[11] Max Ways, "More Power to Everybody," *Fortune*, May, 1970, p. 173.

centuries, and which may only now be finding that their rate of adaptability is inadequate, is the organized churches. Most churches evolved through constant interaction with society and by changing dogma to reflect evolving social and economic forces. Railroad companies found their business declining because they clung to outworn concepts of "the business they were in." Their view of themselves as providing railroad service came into conflict with changing modes of ground transportation; the result was a severe loss of revenues.[12]

Sometimes organizations are slow to change but finally do so before crises occur. Often, the failure of an organization to adapt can be rectified only by severe medicine, such as through "revolutionary" behavior. The failure of industrial leaders to fathom the intensity of the desire by the American working class for unionization contributed to costly strikes and plant shutdowns. Even the government did not understand the immensity of these forces until the early 1930s, when legislation guaranteeing the right of collective bargaining was finally passed. More recent industrial history contains many examples of companies whose failure to adapt to change caused economic and management difficulties that were resolved through painful liquidation proceedings, through merger and subsequent loss of organizational identity, or through management changes and reorganizations.

Managers and Change

Since organizations must cope with change in order to survive, so, too, must managers. Managers are surrounded by the manifestation of change in their organizations. Sometimes the manager may think of these changes as "progress"—such as most people did when man first set foot on the moon in 1969—and sometimes he may think of them as "turmoil"—such as many did when our universities were faced with student demands for greater participation in the academic processes. However, change, whatever its nature and the manager's perception of it, is inescapable for him, and he must learn to deal with it.

The "rules" under which the manager's milieu operates are themselves changing. Few, for instance—be they students, workers, or executives—unquestioningly accept traditional ways as they once might have. Authority and responsibility patterns are being upset, and it becomes increasingly apparent that *the process of change can no more be ignored than can the dramatic changes which are manifestations of this process.*

The modern manager recognizes that established ways of viewing the

[12] See Theodore Levitt, "Marketing Myopia," *Harvard Business Review,* July–August, 1960, p. 45, for a more extensive discussion of such instances together with an analysis of their causes.

world and of affecting it will no longer necessarily work or suffice. To cope with such an environment, he must establish innovative ways of thinking and doing and flexible attitudes and beliefs about the existing patterns of organized life.

That most of us are quite capable of this flexible attitude is quite apparent in temporal contrasts in our personal behavior and attitudes. The "sexual revolution" which has taken place in America is illustrative of this, as is the outlook of most Americans on our role as a world power, our goals with regard to the needy of our own country and around the world, and just about everything else.

The Evolution of Management

Lest anyone come to the conclusion that modern management is simply a different state of mind, let us consider how and why modern management has evolved from traditional management.

In effect, modern management is built on the best practices of traditional management augmented by new approaches, directions, technique, *and* attitudes. Even though the new management has evolved systematically, its *impact* has been revolutionary. Today's manager practices in a milieu which is quite different even from that which existed only a decade or so ago. The organization which he must manage is different from what it was then; new organizational forms have been developed to permit more flexible patterns of operation, so that in some of its salient features the modern organization looks little like its pre-World War II predecessor.

The changes which have occurred in organizations have in part been brought about by changes in technology. Continuous industrial processes, such as those in the chemical and petroleum industries, require different organizations from those required by batch processes, as in the steel industry. New organizational concepts, such as that of the conglomerate business firm, require new managerial tools.

Today, organizations which never before recognized a need for good management have become leaders in developing new management ideas. Universities, hospitals, and various levels of government must be organized and managed in the same sense as must business organizations, for they, too, have large and complex tasks to perform in a complex world. However, because their objectives and functions are quite different from those of the business organizations which served as the models for most of traditional management thought, these organizations cannot be carbon copies of businesses in either their organizational structure or their management practices.

Thus, much of the management theory which was developed for low-technology business firms is indeed inapplicable to modern organizations.

Today's high-technology organizations are not staffed with large numbers of unskilled or semiskilled workers, but rather with skilled workers and professionals who must be dealt with in ways different from those that were successful with the unskilled workers of the past.

Many modern organizations are loosely knit confederations of professionals who have neither much loyalty to the organization nor a great dependence on it. Illustrative of this are hospitals, which are largely staffed by physicians who operate as independent entrepreneurs, and universities, whose professors serve government and industry as consultants. In both of these illustrations, *the primary loyalty of the professional is to his profession*—not to the organization in which he happens to be practicing it. And, in both cases, the problems of managing the organization are much greater than and much different from the management problems of the past.

In business, much the same is true. High-technology business firms utilize people who are independent and mobile, and in an economy where skills are scarce, the firms must exercise a new kind of management over them. Gone forever as anything of significance to our economy are the organizations which operated with masses of semiskilled operatives supplemented by a few specialists to handle accounting, planning, etc. Now virtually everyone is a specialist, everyone can afford to behave like a valued asset, and everyone must be "managed" in a way which is consistent with this status.

The Foundations of Modern Management

The new management's focus on change has not led it to discard traditional management concepts and philosophies. Rather, new management has evolved from traditional management by making use of the best which traditional management had to offer and supplementing this with a number of new conceptual developments which amplify the scope and effectiveness of the manager who must live in a world of constant change. In subsequent chapters, we shall deal with both traditional and modern concepts. Here, we shall briefly review the primary new developments to provide a foundation for subsequent in-depth discussion.

The important conceptual developments include the *systems concept* and *formal decision analysis*. Other developments are an increased awareness of the *importance of the human element of organizations* and the organization's awareness of its *social responsibilities*.

SYSTEMS CONCEPT The *systems concept* is a simple idea which has only recently been extensively applied in organizations. The systems approach is the opposite of the often used problem-solving approach of "cutting the problem down to size." It recognizes that "things"—be they organizations, factories, highways, or whatever—are made up of components, each of which have unique properties, capabilities, and mutual relationships. The *overall output* of the system is the most significant thing to the systems designer and planner. Thus, he "looks at the big picture" (the overall system), rather than focusing on its constituent parts. For instance, the highway planner who views each segment of highway as a separate entity will probably find that the parts "fit together" very poorly and that eight-lane superhighways with high capacity are restricted in their usefulness by access to two-lane roads with low capacity. The overall ability of the highway system to fulfill transportation needs will, therefore, be poor even though the individual components are each good.

Figure 1-1 depicts a systems view of a school district as a combination of interacting parts (subsystems) in a larger environmental system. It shows how all of the participants in the "clientele subsystem," as well as administrators, teachers, and students, form a part of the overall system. Of course, the myriad of actual interrelationships of all of the groups shown on this diagram cannot be displayed in a single figure. But such a broad viewpoint as is displayed in the figure provides the opportunity for broader considerations than would be possible with a nonsystems view.

The systems view recognizes the significance of *synergism*—the fact that a whole composed of various parts may be quite different from the simple sum of its parts. The formation of conglomerate business firms composed of individual firms in a variety of industries or areas is illustrative of this. The conglomerate may be vastly more effective than the simple sum of its constituent firms because it is made up of a number of "producers" and "users" of similar products or technologies, or because the separate companies have different profit cycles, or because of the advantages of centralized planning and financing, or because of a number of these factors operating simultaneously.

The pervasiveness of the synergistic idea is indicated by its applicability in unconventional areas such as research, as revealed by the text of the Kaiser Aluminum and Chemical Corporation advertisement:

You can organize research in the regular way:
Along the tight lines of single and separate disciplines;
Or you can try to break down the mental boundaries and establish an intellectual melting pot.

At Pleasanton, California, we are opening a new center for Technology:
We call it interdisciplinary in the sense that scientists from many different fields will think and work together.

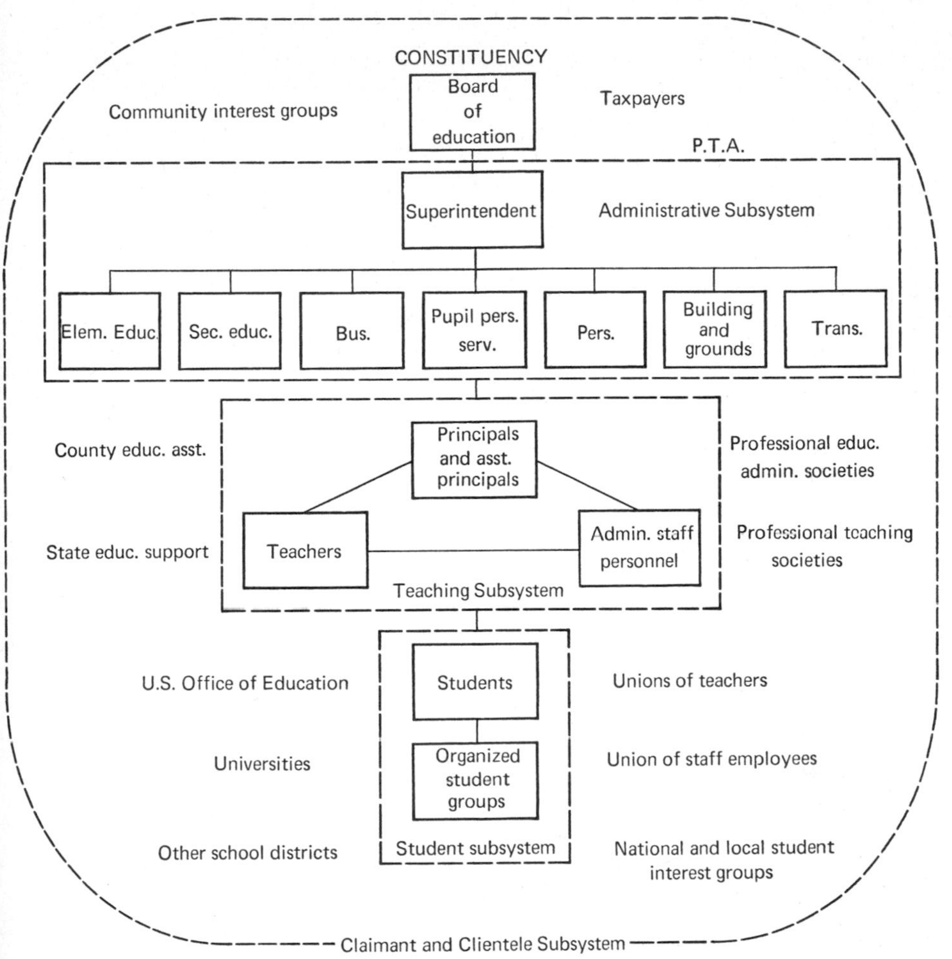

FIGURE 1-1 *A school district portrayed in systems terms.* SOURCE: *"Systems Management Applications to School District Administration," Research Report, Aug. 5, 1970, School of Engineering, University of Pittsburgh, Pittsburgh.*

In this atmosphere of synergism and mutual mind-rubbing, we hope to create an environment for the magic in research:
—when two seemingly unrelated facts are unexpectedly combined;
—when the thoughtstream of one area is imaginatively applied to another;
—when the need in one field is answered by an insight coming in from "left field."[13]

[13] *Fortune,* May 15, 1969, pp. 20–21.

Basic phenomena related to synergism and the systems idea are *interaction* and *interdependency.* The constituent elements of a "system" are linked together in such complex ways that actions taken by one produce far-reaching reverberations.

These interactions and interdependencies among the elements of a system are illustrated in the effects of a price rise in a basic commodity such as steel. What may have started out as a simple, sought-after wage increase by a union, when passed on to the customers in the form of a price increase, can affect prices in other commodities and products as well. When we see a price increase posted in basic steel products, we know that appliances and other products using steel will face price increases. An increase in wheat prices will be felt by the housewife, who has to pay more for the bread she buys. A reduction of the work force in a plant may cause a reduction in the income available to the workers who are laid off; in turn, this reduction in purchasing power will have some effect on business in the plant's locale.

Thus, the systems idea, with its emphasis on interaction and interdependence, fits naturally in a complex society whose elements themselves have these characteristics. However, the systems approach is more than a way of describing complexity; *it is a prescribed method for studying and changing systems.* What the systems approach requires the manager to do is to think in terms of the overall long-range effect produced by the interactions and interdependencies. This view is well stated in terms of a systems approach to building construction in the *Business Week* advertisement of Alpha Portland Cement Company:

> *The systems approach calls for advance planning of every detail in terms of user needs and circumstances rather than "what's available." Parts are designed to be integrated into multifunctioning components—programmed for compatibility. Means and methods are developed for these units (often prefabricated) to be readily joined.*
>
> *The aim is faster, more economical building—with equal or better quality.*[14]

In discussing the systems approach to building, Alpha Portland's ad emphasizes advance planning (the overall systems view), integration of system components and component compatibility (interaction and interdependence), and economical building with equal or better quality (overall system output).

[14] *Business Week,* Mar. 15, 1969, p. 99.

FORMAL DECISION ANALYSIS AND COMPUTERS Another major conceptual foundation of modern management is the formal analysis of decisions which is conducted by virtually all large organizations. Perhaps no aspect of modern government and business has undergone such radical change as has the decision-making process. Where once executive judgment and intuition reigned supreme, now formal decision analysis complements management judgment.

This complementarity of judgment and formal analysis amplifies the manager's ability to cope with complex decision situations. The formal analysis is based on decision theory—a theory regarding the best way to choose alternatives.[15] This theory is implemented by using mathematical and statistical procedures which are usually programmed into electronic computers.

The logic and consistency provided by decision theory, together with the power of mathematical and statistical solution techniques and the speed and accuracy of computers, combine to extend the scope of the analyses which can be performed on decision problems. No longer need the manager rely on trial-and-error approaches of limited scope. With formal decision analysis, the manager can explore a wide range of alternative actions which he might take in terms of their overall impact on the organization. He can do this because he has available the theory on which he can base such analyses, the logical procedures to perform them, and the computer facilities which enable him to do the complex calculations required in a short time.

All of these things provide the manager with a capability to ask "what if" questions which his management predecessor could only have dreamed of—*What* will happen to our profitability *if* we cut our price by 5 cents? *What* will be our unit cost *if* we spend $100,000 on new production equipment?

The computer's role in formal decision analysis should be placed in proper perspective; far too frequently it is either exaggerated beyond all reason or down-played to the degree that the computer appears to be nothing more than a fancy toy. In truth, the computer is an amazingly fast and accurate device for performing elementary operations (e.g., addition). Its speed and accuracy give it the capability to perform other operations (e.g., multiplication) by performing various combinations of the elementary ones.[16] The real power of computers lies in their speed and accuracy.

[15] For a more extensive discussion of elementary decision theory, see William R. King, *Probability for Management Decisions*, John Wiley & Sons, Inc., New York, 1968, chap. 3.

[16] For example, the product of 5 and 7 can be determined by "adding together seven 5s."

By using these traits to advantage, calculations can be routinely made that would be impractical to make without the computer.[17]

Thus, the computer extends the range and power of the analytical manager. He can perform analyses which he could never do without it, and he can investigate alternative courses of action which might have gone unexplored. Together with this, the computer and its associated data processing equipment bring into the hands of the manager a wealth of information which would be impossible for him to obtain otherwise.

A good illustration of the information and analytic power which the computer provides is a "simple" data processing operation which is done for the perusal of sales managers. By taking basic sales information from invoices and other original records and manipulating them in different ways, the computer system can provide the manager with sales data for individual salesmen, territories, and branch offices by product, size, and time period. The manager can use these breakdowns to identify weaknesses and strengths in his sales organization (e.g., a particular salesman may be performing poorly, or perhaps a branch office's sales are being affected by local competition) or in his product line (e.g., a particular package size may be selling poorly). Such elementary knowledge of "what is going on in the world" may not seem to be revolutionary to the uninitiated, but its value is apparent. And, prior to the advent of the computer—indeed, until quite recently in many large sales organizations—such information was unavailable without huge expenditures of time and funds to obtain it on a "one-shot," quickly outdated basis.

The revolutionary part which the computer has played in providing such capabilities is fully realized by those who have viewed the explosive growth of some organizations which historically showed little concern for the analysis of decisions. In a demand-oriented economy such as that which American business enjoyed in the past, built-in demands for a wide variety of products and services existed. Business firms naturally emphasized selling as a means to success. Formal decision analysis and the evaluation of potential profit opportunities were paid little heed. In effect, in such a situation, "doing something" (regardless of what it was) often had great payoff. The need to "do better" was therefore not always apparent.

Today's economy is a quite different one, in which demand must be "created" before products can be sold. In such an environment the significance of decision analysis is evident. The business firm must evaluate its markets in terms of the products that are likely to be of appeal, it must

[17] Virtually everyone has seen references to problems solved in a few hours or days on a computer which would have required many lifetimes for an individual to solve without the computer.

assess its capabilities for producing various products, and it must critically evaluate various profit opportunities.

Indeed, *the modern manager must seek the best use of all organizational resources—financial, physical, human, and intangible.* He can do this only if he takes advantage of techniques of formal decision analysis and the availability of information which is made possible by the computer.

THE HUMAN ELEMENT OF ORGANIZATIONS A major foundation of modern management is the recognition of the importance of human behavior in determining the effectiveness of organizations. Human beings are the one resource found in all human organizations—there are no "peopleless" organizations. Yet, if a Martian were to examine much of traditional management thought and practice, he might well conclude that some organizations are almost devoid of the human element.

The traditional view of the problem of motivating the people who make up organizations held that a system of rewards and punishments was adequate. The human element was looked upon as something which must be controlled via a system of rules, procedures, prescribed roles, and prescribed authority. This simple view, based on models of a man dominated by economic factors and lacking ambition and a sense of responsibility, has been largely discredited.[18]

The modern view of the human element is vastly different from this view; the multidimensional nature of human motivation is well recognized and accepted. Individuals are recognized as being motivated by many factors other than economic reward—social, psychological, and other motivations are important as well. For instance, the group in which the individual works is of critical importance to the satisfaction of his desires, for these satisfactions are in the form of self-fulfillment, self-esteem, and the approval and acceptance of the social group.

Today's accepted organizational value systems recognize the need for the industrial organization to assume more social responsibility as well as to contribute to the economic well-being of the individual. The values inherent in such a view include the integration of the needs of the individual with the needs of the social group to which he belongs. In turn, by meeting individual needs in terms of human dignity, recognition, and self-actualization, the goals of the organization are more readily achieved.

As we continue to become more complex and interdependent in society, the human element will play a greater role in the management of organizations. The drive for participation that we see today on the part

[18] See, for example, Douglas McGregor, *The Human Side of Enterprise,* McGraw-Hill Book Company, New York, 1960.

of employees, students, and other participating members of organizations is evidence of this.

> *In our society, where almost everybody has more power than almost anybody had a couple of hundred years ago, people get in one another's way. And nobody can have enough power to escape from dependencies on other people. "Crowding" and most other environmental ills are traceable to the explosive increase of per capita power. Political life and home life are full of conflicts because what A wants is not necessarily what B wants and both A and B demand that they be heeded.*[19]

SOCIAL RESPONSIBILITY OF ORGANIZATIONS One of the phenomena which have become of greatest importance to the management of modern organizations is the awareness that their responsibilities extend beyond the confines of the organization to larger systems of which the organization itself is a component. Thus, the business firm, which has traditionally believed (or acted as though it believed) that its sole duties were to its stockholders and management personnel, has come to recognize that its duties extend to its employees, its customers, and society in general.

As our society becomes larger and more complex, each individual and organization is increasingly dependent on other individuals and organizations. Carried to the ultimate logical conclusion, each of us affects all others in our society. In such a situation, each of us must indeed serve to some degree as our brother's keeper. This is especially true of the large organizations that wield great power. Since they can have far-reaching (and sometimes, unintended) effects, they have a responsibility to analyze these effects and to act responsibly.

Moreover, since organizations have much to gain in the long run from a stable society, it is in their best interest not only to behave responsibly, but *to take positive action to ensure the stability and preservation of society.*

Thus, the industrial company which has polluted the air or water is reasonably expected to develop plans for behavior which does not adversely affect the physical environment in which we all must live. And, since the cessation of one firm's contribution to our environmental pollution will not noticeably affect the quality of our lives, the firm has a responsibility to do more than to simply behave responsibly. It must act to encourage others to do likewise; it must share its knowledge and experience and offer aid and encouragement.

Perhaps more important than behaving responsibly in ways which

[19] Max Ways, *op. cit.*, pp. 173–299.

can be "pointed to with pride" in an annual report or publicity release is the responsibility of organizations to *make social consequences an intrinsic part of their decision-making considerations.* Thus, the modern business firm does not evaluate a new product's *potential* solely in terms of *market potential,* but also evaluates it in terms of the *social consequences.* What, for example, will be its impact on municipal solid waste disposal systems? What pollutants will its production process spew forth? How can these adverse consequences be avoided?

AN APPROACH TO THE STUDY OF MANAGEMENT

An exposition of traditional and modern management often leaves the management novice in a position similar to that of ". . . the donkey who starved to death between two bales of hay, unable to decide which to eat. We are posed in our times between reverence of the past and the acceptance of the future we see glittering there before us."[20]

In this book, we hope to make the best possible use of both the old and the new. Of course, this is a difficult task, and we shall leave it to the reader and the experienced manager to judge how well we have accomplished our purpose.

We adopt the *systems approach* or viewpoint as the central theme of our approach to management. In the next chapter, we will elaborate on this approach. Here, we shall contrast the as yet vaguely defined approach to more traditional views to at least make clear *what it is not.*

Traditionally management texts have employed a "principles" approach which seeks to make the best use of past experience through developing principles or guidelines for good management. While these principles are often useful and valid, they do not provide a full conceptual framework for management thought. At the worst, they can be *guides to action without thought.* If "principles" are taken as a checklist approach to management, they can be misleading and woefully inadequate.

The systems approach taken in this book seeks to develop a *way of thinking,* a *viewpoint,* a *conceptual framework,* together with a *methodology for implementation* rather than specific detailed guides to action. We attempt to be as specific as possible and to provide diverse real-world illustrations, but specific "answers" are not the primary objective.

As we have said that much of traditional management thought is worthy and useful, so, too, is much of the principles approach. However, a total reliance on principles is inadequate and, perhaps, counterproductive.

[20] D. Fabun, *The Dynamics of Change,* Prentice-Hall, Inc., Englewood Cliffs, N.J., 1968, p. 1.

Perhaps our attempt to blend the best of the old and the new in management thought is best described by Levitt in another context:

> *We must learn from the past in order to make any progress at all. Yet as we learn from and progress beyond the past, we run the risk of employing obsolete dogma for new times. The issue is not whether knowledge and know-how are transferable. It is whether they are applicable.*
>
> *This means that now even more than in the past the essential managerial skill is not a good memory for principles, as is often the case in law and medicine, but the ability to determine what the problem really is, and the ability to do that which distinguishes the chief executive from the janitor. Understanding is more important than dogma.*[21]

SUMMARY

Management is a field that everyone knows something about, but which many practice ineffectively.

An operational definition of management—one that relates the concept of management to a number of observable criteria—involves organized activity, objectives, relationships among resources, working through others, and decisions. Thus, the manager *works toward objectives, through others,* in an *organized environment* by *establishing relationships among resources and making decisions.* The degree to which the organization's goals are accomplished depends in great measure on the relationships which he establishes and the quality of the decisions which he makes.

This definition is widely applicable to various management environments—be they business firms, government agencies, hospitals, educational institutions, or families. Indeed, the concepts of management are applicable to one's own management of his personal resources—funds, energy, time, etc. However, the most significant area of application for management knowledge is the organization, for organizations, and increasingly large organizations, are the prime movers in our society.

"Good management" has always been important to organizations. However, it is more important today than ever before because of the size and complexity of our society and its increasing rate of change. The impact of individual management decisions, made in the context of a large organization such as General Motors or the Pittsburgh Bureau of Police, can be far-reaching and severe. Thus, *the cost of managerial error is very great,*

[21] Theodore Levitt, "The New Markets—Think Before You Leap," *Harvard Business Review,* May–June, 1969, p. 67.

and because of the high degree of interdependence among the elements of our social and economic systems, *the potential for managerial error is also great.* This, coupled with the ever-increasing rate of change which we are experiencing in virtually every aspect of our lives, makes "good management" more difficult to provide and more important to each of us.

Modern management is based on the foundations of the *systems concept,* formal *decision analysis,* and an awareness of the significance of the organization's *human element* and its *social responsibility.* It has emerged from traditional management by taking the best of traditional thought and marrying it with these new concepts and perspectives. Thus, "modern management" is evolutionary rather than revolutionary in concept, but it has had revolutionary impact on some organizations.

EXERCISES

1. How do dictionary definitions differ from operational definitions? Give examples by defining concepts such as a meter (unit of distance) in both ways.
2. Why is it desirable to have an operational definition of "management"?
3. Give, in your own words, an operational definition of "management."
4. Describe the significance of the five operational criteria used to define management in Chapter 1.
5. Why is the real payoff from good management considered to lie within the context of organizations?
6. Why is management more important to the daily lives of individuals today than it has ever been in the past?
7. Why did the Industrial Revolution enhance the importance of management?
8. What does the large *size* of modern organizations have to do with the importance of management, i.e., why is size important?
9. When a Wall Street brokerage firm was near financial failure in the early 1970s, it was supported by Ross Perot and associated interests. One of the reasons given by him for doing this was that he could forsee the failure of this firm leading to the collapse of the United States economy.[22] Can you envision how this might have occurred? What characteristic of modern organizational systems does such a potentiality describe?
10. Describe the possible impact on various interdependent organizations of a decision made by a large industrial firm to stop air and water pollution *immediately.*
11. What is "future shock"? How does it relate to change?
12. The text says (p. 14) that "Continuous industrial processes, such as those in the chemical and petroleum industries, require different organizations than do batch processes, such as the steel industry." Why?

[22] See "Ross Perot Moves in on Wall Street," *Fortune,* July 1971, pp. 90–115.

13. What are the salient differences between business firms and "public organizations" such as hospitals and universities with regard to their objectives? Are their objectives more similar or dissimilar now than they were in the last century?
14. How does the level of "technology" in an industry or organization affect its need for management? Give examples.
15. In your own words, summarize:
    a. the relationship of modern management to traditional management;
    b. the meaning of the conceptual developments which provide the foundation for modern management.
16. What is the systems approach? Give examples in terms of the way in which you might take the systems approach to the problems of:
    a. transportation planning for a city
    b. the evaluation of the worth of a harbor-dredging project
    c. the choice of a set of specifications for your new house
17. What are the disadvantages of the "nonsystems" view of problems?
18. What is synergism? How is it related to the systems concept?
19. What is the relationship of the concepts of *interaction* and *interdependence* to systems thinking?
20. What might be the impact of providing detailed historical data to consumer goods manufacturers on the consumer purchases of an individual household? Would it enhance their decision-making capabilities? How?
21. What is meant by the term "demand-creation economy"? How does this relate to the practice of management in such an economy?
22. Contrast the traditional management view of the individual and his motivations with the more modern view.
23. What is the organization's role in preserving society and improving the quality of life for individuals?
24. What are "management principles"? What is their role in management thought?
25. Develop an operational definition of management in the context of a specific organization by relating the operational criteria used in the text to the organization.
26. Explain the various sorts of relationships among resources and organizations which must be established by the manager.
27. Why do formal organizational relationships create so much difficulty in the daily work of the manager?
28. Select an organization which is known to you and prepare a diagram describing the interdependency and interconnection of the organization with other elements of the society.
29. Kenneth Keniston, a psychologist at Yale, writing in the *New York Times Magazine* (March 10, 1970) on "You Have to Grow Up in Scarsdale to Know How Bad Things Really Are," comments: "Even in the Scarsdales of America with their affluence, their upper-middleclass security and abundance, their well-fed, well-heeled children and their excellent schools, something is wrong. Economic affluence does not guarantee a

feeling of personal fulfillment; political freedom does not always yield an inner sense of liberation and cultural freedom. Social justice and equality may leave one with a feeling that something else is missing in life." How does this description of life in suburban America affect the manager's job (*a*) directly, and (*b*) through the analogy of affluent suburbia to affluent middle-management?

30. Management is often described as a "way of thinking" about organized activity. How would you explain this?

31. Describe a university or college in systems terms making use of interactions and interdependencies.

32. Much of the criticism of computerized decision analysis revolves around the accusation that the manager is relegating decision-making responsibility to computers. Is this a valid criticism?

33. Analyze the following statement: In our society individuals are motivated principally by the promise of economic reward; this explains why union officials seek higher wages for their members.

34. Do you believe that the single objective of United States business firms is the realization of a maximum return on stockholder's investments? Why or why not?

35. In Chapter 4 in the book of Genesis, verse 9, Cain responds to the question, "Where Is Abel, thy brother?" with the comment "I know not; Am I my brother's keeper?" How does this response relate to today's society and the role of the manager in an organization?

36. Interfaces—places where two bodies come together at a common boundary—are increasingly the subject of study by management theorists. Interfaces play an important role in change in the physical world. For instance, the boundary layers or interfaces between crystals and metals are the places where action takes place under force. Geologists say that the greatest changes in the earth occur at boundary lines such as the surf and the shore. Sculptors typically direct their chisels along faultlines of blocks of stone, just as diamond cutters cut along flaws. Is it possible that organizational change is also likely to take place along interfaces where different "organizational cultures" rub shoulders? If so, can you predict the direction of future organizational changes?

REFERENCES

"Aftermath of Price-fixing Case," *Business Week,* March 4, 1961.

Alexander, Tom, "The Unexpected Payoff of Project Apollo," *Fortune,* July, 1969.

Ansoff, H. Egor, and R. G. Brandenburg, "The General Manager of the Future," *California Management Review,* Spring, 1969.

Athos, Anthony G., "Is the Corporation Next to Fall?" *Harvard Business Review,* January–February, 1970.

Black, Guy, "Systems Analysis in Government Operations," *Management Science,* October, 1967.

Brandenburg, Richard G., "The Making of a Manager," *IEEE Student Journal,* January, 1969.

"(The) Campus as a Management Problem," *The Wall Street Journal,* December 9, 1970.

Cleland, David I., and William R. King, "Systems Management and Analysis," *Manage,* November–December, 1968.

Culbert, Samuel A., and James M. Elden, "An Anatomy of Activism for Executives," *Harvard Business Review,* November–December, 1970.

Fielden, John S., "Today the Campuses, Tomorrow the Corporations," *Business Horizons,* June, 1970.

Jasinski, Frank J., "Adapting Organizations to New Technology," *Harvard Business Review,* January–February, 1959.

Johnson, Richard H., et al., *The Theory and Management of Systems,* McGraw-Hill Book Company, New York, 1967, chap. 1.

King, William R., "The Systems Concept in Management," *The Journal of Industrial Engineering,* May, 1967.

Koontz, Harold, "The Management Theory Jungle," *Journal of the Academy of Management,* December, 1961.

Mackenzie, R. Alec, "The Management Process in 3-D" *Harvard Business Review,* November–December, 1969.

McDonald, John, "How Social Responsibility Fits the Game of Business," *Fortune,* December, 1970.

Mockler, Robert J., "The Systems Approach to Business Organization and Decision Making," *California Management Review,* Winter, 1968.

Morse, F. Bradford, "Private Responsibility for Public Management," *Harvard Business Review,* March–April, 1967.

Optner, Stanford L., *Systems Analysis for Business Management,* Prentice-Hall, Inc., Englewood Cliffs, N.J., 1968.

"Pupil Power," *The Wall Street Journal,* November 6, 1970.

"Ralph Nader Becomes an Organization," *Business Week,* November 28, 1970.

Samuelson, Paul A., *Economics,* 8th ed., McGraw-Hill Book Company, New York, 1970, chap. 1.

Thompson, Wilbur R., *A Preface to Urban Economics,* Johns Hopkins Press, Baltimore, 1965, introduction and chap. 7.

Votaw, Dow, and S. Prakash Sethi, "Do We Need a New Corporate Response to a Changing Social Environment," *California Management Review,* Fall, 1969.

section two

BASIC CONCEPTS

2

systems and models

The term "systems" has for some time been in vogue in learned discussions on topics ranging from philosophy to engineering. This book is no exception: here, we say that we shall take the "systems approach" to the study of management. However, our goal in adopting this approach is one of demonstrating that systems ideas offer the opportunity for a new kind of management—a kind which is effective, efficient, and, at the same time, considerate of personal and social values.

This demonstration is the task which we set for this book. It is also the goal which we establish for future managers, because the managers of the future must be different from the pattern laid out by managers of the past. If they are not, our society may crumble and our ideals may be lost. If "modern" management can be put to use in industry, government, churches, education, and other modern organizations, the worthwhile future envisioned by our forefathers will be a pale reflection of the real achievements of our progeny.

The future will indeed be molded by those of us who are willing and able to seize control of our destinies and act with vigor and aplomb. Of course, each of us can play a part. But, in an era of large, powerful organizations, it will be those in organizational management positions who will play the critical roles.

To develop an understanding of the way in which the manager can fruitfully take the systems approach to his job, let us begin by discussing some basic systems concepts.

SYSTEMS

A dictionary definition is a good point of departure for developing an understanding of systems and derivative concepts. A "system," says the dictionary, is "a regularly *interacting* or *interdependent* group of items forming a unified whole."

When one ponders this general definition for a moment, he recognizes that *just about anything is a system.* In particular, the everyday uses of the term indeed are valid according to this view. For example, we hear physicians discuss the "nervous system." Even if we do not possess a great deal of technical knowledge, we know that the nervous system is made up of a group of "items" (called "cells" and "neurons") which continually interact with one another to produce behavior.

Another everyday use of the word "system" is in the context of the "Mississippi River system"—the unified whole consisting of interacting rivers, streams, and small tributaries which combine to form the Mississippi River. In some sense, this is a simpler, or at least a more familiar, system; yet it satisfies all of the requirements of the dictionary definition. In fact, we can demonstrate the requisite interaction of the various constituent parts of this river system by considering what happens to the lower Mississippi when heavy rains occur in the region of its tributaries. Water is sent pouring down the tributaries into the Mississippi, and floods may be caused hundreds of miles from the rain source. Moreover, if one were to dam the Mississippi, the backup of water would eventually extend to the regions drained by each of its tributary streams. Thus, the parts clearly depend on one another (interdependence), and actions taken by, or on, one part clearly affect other parts (interaction).

The "solar system" is another that we commonly discuss. Here again, it is a unified whole made up of interacting parts. The moon, for example, affects the tides of the earth and provides reflected light for our nights. In turn, the earth's gravitational force keeps the moon in its orbit around the earth. Both earth and moon are warmed by the sun as they revolve about it.

Much of modern scientific research focuses on systems. For example, the *Wall Street Journal*[1] reported on research being done toward the construction of a "living system" in a test tube. Scientists engaged in this effort note that a single cell is a complex living system characterized by thousands of interrelated molecules, even though the cell may be less than 1/500 inch in diameter.

Subsystems

The idea of a single cell constituting a system is surprising to some, who think that the physical scope of a system must be large. Yet in fact, systems are composed of parts which are themselves systems. For instance, the human body is a system composed of various *subsystems* (nervous, cardiovascular, etc.). In turn, these subsystems are composed of cells, each

[1] *Wall Street Journal,* Oct. 30, 1967, p. 1.

of which is itself a system. Thus, systems typically exhibit a structure in which there are parts (sub-subsystems) imbedded within other parts (subsystems) within overall systems. And, of course, the imbedding of one system in another can go on through many stages; indeed, it can go on endlessly.

An *organizational chart* describing a business firm shows how subsystems can be imbedded in other subsystems to form a hierarchy of subsystems. Figure 2-1 is such a simplified chart for Company X, which is composed of three major departments—manufacturing, financial, and marketing. Each of these might be headed by a vice-president.

Now, in terms of Figure 2-1, let us illustrate some various imbedded subsystems which we might wish to consider. The sales subsystem is enclosed in dotted lines. It includes all activities related to sales and all personnel under the authority of the sales manager. The marketing subsystem is enclosed in a broken line. It encompasses the sales subsystem together with the advertising and marketing research subsystems. The company itself constitutes the next system level. It entails the marketing subsystem together with the other two major functional subsystems—manufacturing and financial. Thus, the various portions of Figure 2-1 depict the company system, a marketing subsystem, and a sales sub-subsystem, each one imbedded in the other.

Figure 2-2 shows how Company X can itself be considered a subsystem of larger systems. If, for instance, Company X is one of four producers of autos, these four firms collectively form an industry system, which may be termed the "primary auto subsystem." Other companies manufacture related items—tires, wheel covers, etc.—for new cars, so they collectively form the "secondary auto subsystem." The two subsystems together form a "new auto subsystem"—all of those firms which manufacture new autos or parts for new autos.

Other firms manufacture replacement parts for autos. Collectively, they form the "replacement auto subsystem." This, together with the "new auto subsystem," forms the "auto subsystem"—all of those firms which manufacture auto-related items.

Now, all of this may seem like only an interesting mental exercise until we recognize two things. First, all of these combinations which we have developed are indeed systems. The parts *interact*—by buying from and selling to one another, by adjusting products and designs to conform to those of other companies, etc.—and they are interdependent. For instance, the primary auto subsystem could not produce cars without the parts manufactured by the firms in the secondary auto subsystem, and they, in turn, could not exist without the primary system as a market. It is even more complex than that, since executives of the secondary auto subsystem firms drive cars produced by the primary auto manufacturers,

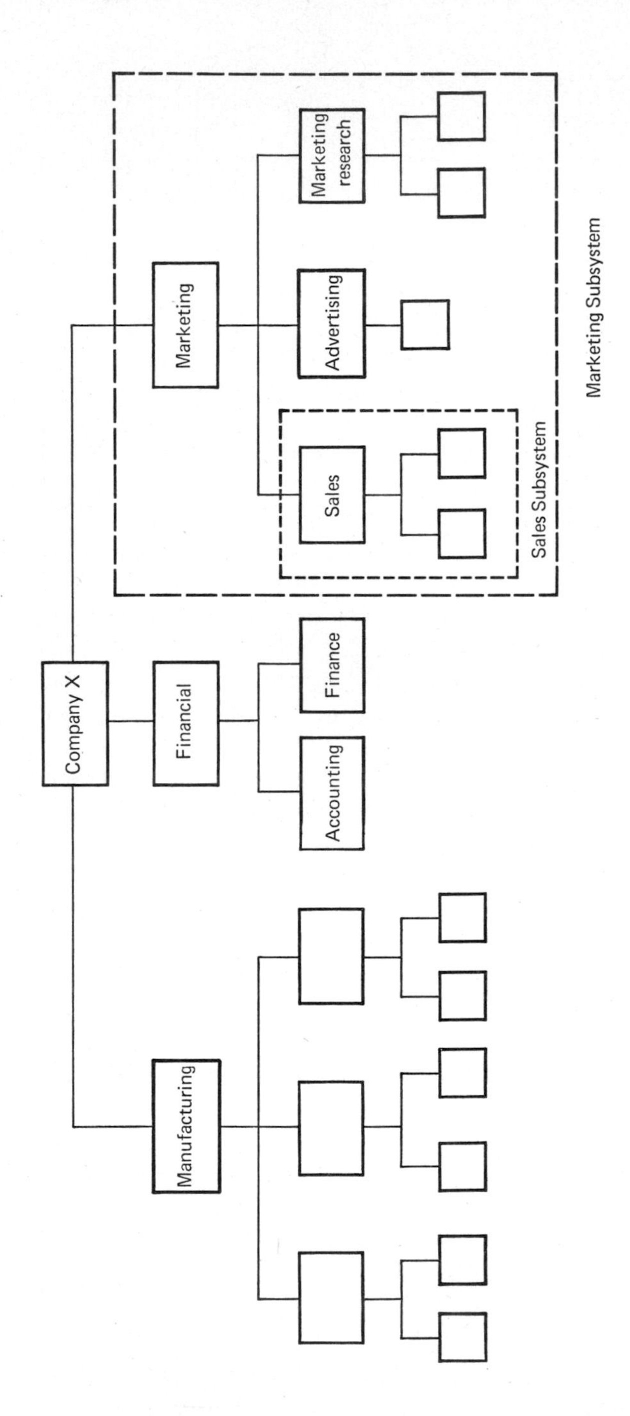

FIGURE 2-1 *Simplified organizational chart for Company X.*

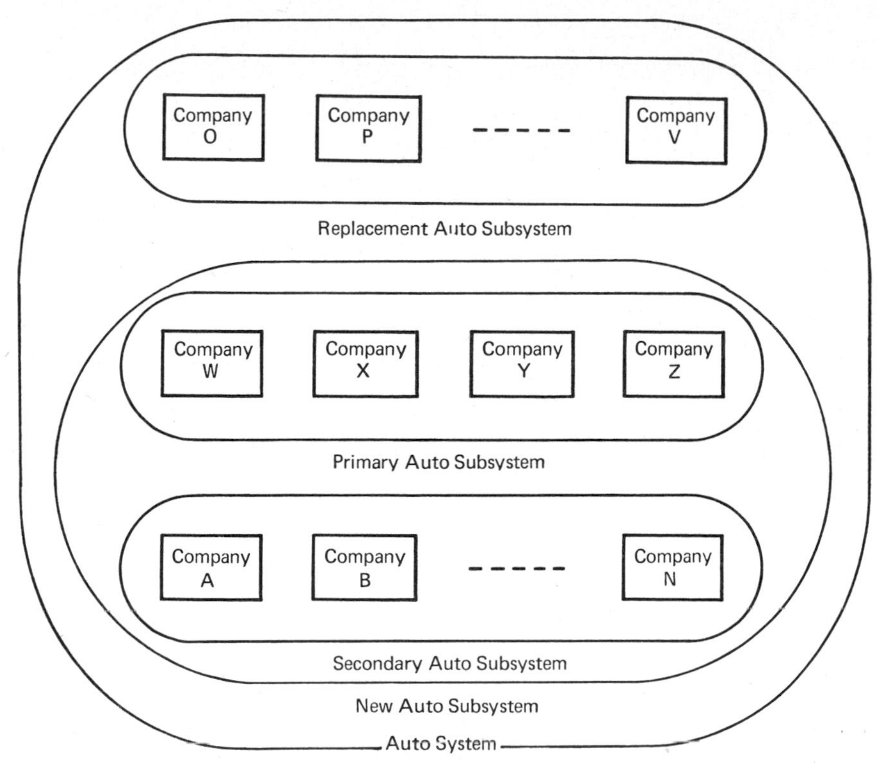

FIGURE 2-2 *Company X in a hierarchy of subsystems.*

and the secondary firms may also supply parts to the replacement subsystem firms.

The second important point which makes illustrations like Figures 2-1 and 2-2 of more than academic interest is that *the manager or analyst may define the various systems and subsystems in whatever way best suits his interest.* Thus, he may wish to consider the company as his system and to "divide" it into only the first level of subsystems. The detail involved in going to many derivative levels may not be worth his effort. Or, for another purpose, he may decide to consider the industry system and to decompose it only into individual firms as subsystems. In any case, the various systems and subsystems are always there; the question is how he may choose to represent them to suit a given purpose.

Input-Output Systems

The various parts of a system must be linked together in order for them to interact and be interdependent. Often, these linkages are adequately

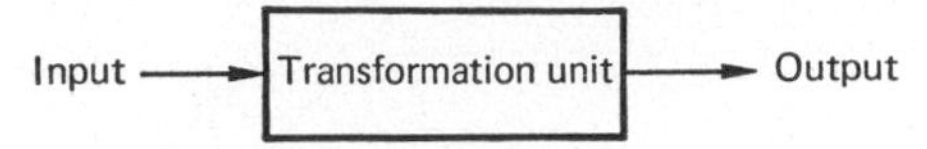

FIGURE 2-3 *Input-output system.*

described in terms of the *inputs* and *outputs* of each element. A diagram representing the simplest input-output system is shown as Figure 2-3. It shows "inputs" to a "transformation unit" which produces "outputs." The inputs and outputs may be physical in nature, as with materials or energy, or they may be informational.

For example, suppose that the system being described is a warehouse. Goods flow into it after they are manufactured and out of it after they are sold. The transformation is primarily an "aging" one, which is not significant except in cases of specific products (say, wine) or in cases where unintended transformations (say, spoilage or damage) occur. Information concerning company inventory policy also flows into the warehouse system and is transformed into action (in terms of which items are shipped out first and which are given priority on shipment) and into new information, such as data on stock levels, which then flows out.

A simple physical example of the input-output system conveys the "transformation" idea well. A gear train, which can be pictorially represented as shown in Figure 2-4, might well be thought of in terms of its salient process structure in the input-output fashion of Figure 2-3. Although the input-output diagram does not at all "look like" the gear train, as does the "picture" of Figure 2-4, it does describe these essential elements.

In the gear train of Figure 2-4, the input is the energy supplied by the lower rotating shaft. This input is transformed by the gear train into output energy, which has a different form. In this case, the gear ratio determines that the upper rotating shaft will rotate much more slowly than does the drive shaft. The function of the gear train illustrates that *an input-output system may be thought of as transforming inputs into outputs.* In this case, input energy is transformed into output energy which takes a different form from the input.

Among other systems which can be thought of in input-output terms might be:

1. A secretarial system which takes rough drafts as inputs and transforms them into finished manuscripts
2. A short-order-cook system which takes raw eggs, slab bacon, vegetables, and bread and transforms them into "BLT" sandwiches

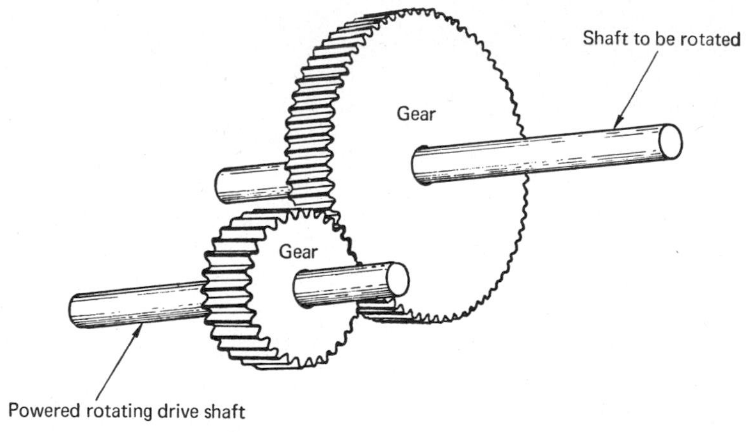

FIGURE 2-4 *Gear train.*

3. A stereo system which takes electronic signals and transforms them into audible music
4. A solid waste disposal system which takes garbage and transforms it into compost
5. An investment decision system which takes information concerning various corporations and the stock market and transforms it into buy and sell orders

Feedback Control Systems

Of course, it is easy to conceive of more complex graphical descriptions of the systems. A mechanical governor offers a physical illustration of a relatively complex and very useful system model called a *feedback control system.* A governor is a device designed to regulate the speed of an engine under varying conditions of load. If one wished to regulate the maximum speed with which an auto could be driven, he could put a governor on its engine.

The basic elements of a feedback control system, such as a governor, are that of a *goal*—e.g., a desired speed—a *monitoring and comparison device* which compares the actual speed with the goal, and a means for altering speed based on the comparison. Such a feedback control system is diagrammed in Figure 2-5. There, the speed goal which is input to the system is continually compared with the error. The transformation unit responds to reduce the deviation between the goal and the actual speed. Thus, if the engine is running too fast, the governor will detect this and act on the drive mechanism to reduce the speed.

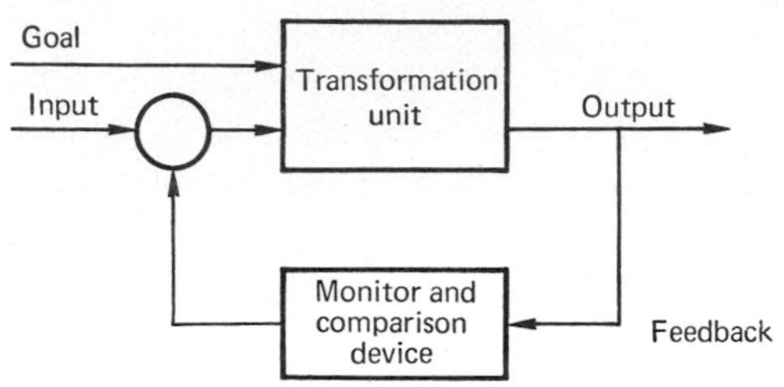

FIGURE 2-5 *Abstract diagram of feedback control system.*

We shall later find that the ideas associated with the feedback control system are of great importance in management. The significant aspect of the feedback control system at this point is the idea of a goal. None of the other simple systems which we previously considered possessed this feature. And we shall later find that it is the basic concept that distinguishes an organization from other systems.

Among other familiar systems which are of the feedback control variety are:

1. A home heating system in which the goal is set into the thermostat so that the system operates to maintain the desired temperature
2. An educational system in which examinations are conducted to provide the instructor feedback on student progress; this feedback enables him to take action to correct deviations from "normal" learning progress
3. An *exception reporting system,* which provides data on abnormal happenings to a business executive so that he can direct his attention to those aspects of his business which are most likely to require it

Feedback Systems with Memory

The simple feedback control system has a memory capability which is limited to the "standard," or goal, which is provided to the system from some outside source. For instance, the maximum permissible speed is the goal which is set into the engine governor mechanism. The system "remembers" this fixed goal, but it has no *selective* memory capability.

A feedback control system with selective memory might involve a *number of goals to be used under specified conditions.* Thus, not only is the system comparing existing conditions with a specified goal, but it is at the same time determining which goal to use.

A home thermostat with the setting determined by the time of day would be such a system, since it would at the same time compare the actual temperature with the standard and it would alter the standard periodically throughout the course of a day. One would need to connect the thermostat to a clock and to have a memory in which a *decision* rule was specified. The decision rule might be of the format of the table below.

TABLE 2-1

Time	*Thermostat setting*
12–8 A.M. . .	68°
8 A.M.–4 P.M. . .	72°
4 P.M.–12	70°

A diagram of such a second-order feedback system is shown as Figure 2-6. There, a memory device permits the changing of goals based on information conveyed through the outside feedback loop. The inside loop is a standard feedback loop which monitors the output, compares it with a fixed goal, and adjusts the transformation unit accordingly. The *decision rule* which permits the changing of goals is input directly into the memory and control device.

The "thermostat with memory" which we described earlier is a particularly good example of this sort of system because the nature of the information flowing through the two feedback loops is different. The reader should ensure that he understands that *time* information would flow through the outside loop in Figure 2-6 if it represented such a system, while *temperature* information would flow through the inside loop.

FIGURE 2-6 *Feedback system with memory.*

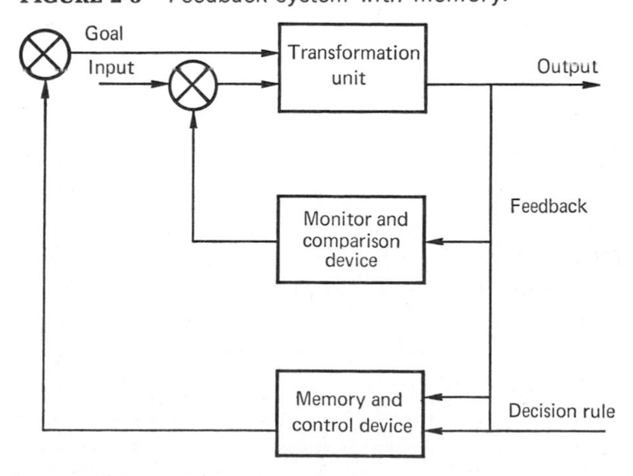

Some illustrations of a feedback control system with memory are:

1. A computer which monitors a production operation and orders changes in process conditions as prescribed amounts of various items are produced
2. A clerk who each month conducts an inventory count and places orders based on a chart which tells him the quantity to order in terms of the amount which is in stock
3. A physician who measures certain symptoms and then prescribes medication based on a comparison of those symptoms with sets of symptoms known to be associated with various ailments

Learning Systems

Higher-order systems may be referred to as *learning systems* or higher-order feedback systems. Basically, such systems involve at least one more feedback loop than described in Figure 2-6. These feedback loops permit the *changing of decision rules* based on measurements made of system outputs. Thus, the system is not simply told what to do; rather, it "learns" and changes its decision rules on the basis of some higher-order objective.

For our purposes, it is not useful to distinguish among various levels of higher-order learning systems. We should simply recognize that the people and organizations with whom we deal are (we hope) of this variety.

MODELS

The layman's idea of the meaning of the world "model" probably concentrates on that sort which are commonly found in *Playboy* magazine and in fashion shows. However, if pressed to consider other varieties, most of us would probably react to the idea by describing a model airplane. In doing so we would have brought to light the most important characteristics of models as they are used in management, in decision analysis, and in this book.

A *model* is a *representation of something else.* Usually, the "something else" is some observable system or phenomenon existing in the real world which is to be represented for purposes of display or analysis. For example, a child's model airplane is a representation of a real-world airplane. So, too, is a schematic diagram of the type typically used to represent electrical systems, such as that shown on the left side of Figure 2-7. This figure shows the equivalence between the elements of the actual system (on the right) and the elements of the model (on the left). A rectangle in the model represents an electronic device, and a line between two rec-

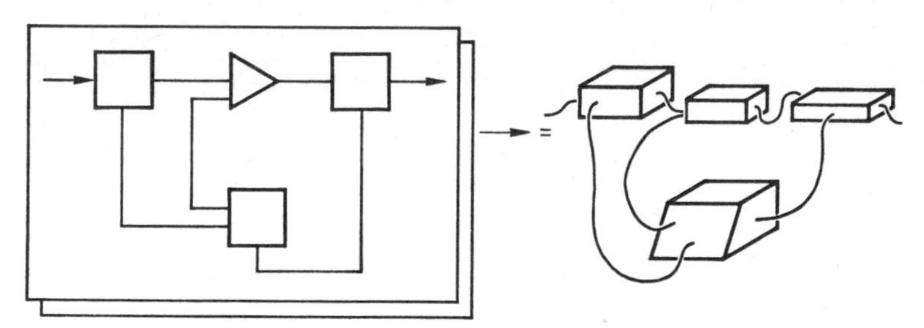

FIGURE 2-7 *Schematic diagram—physical system equivalency.*

tangles depicts a wire which serves as a linkage between two components within the electronic package.

A similar relationship is shown in Figure 2-8 between the elements of an organizational chart model and the human elements of an organizational system. There, blocks represent positions in the organization, while the lines connecting the blocks represent authority and responsibility linkages.

All of the diagrams used in earlier sections of this chapter represent models of systems. The organizational chart of Figure 2-1 is a model of an organization, since it represents the organizational elements and their interrelationships. An alternative model of the organization might focus on the way in which resources (say, cash) flow through the organization. Such a model, shown in Figure 2-9, emphasizes different elements of the organization from those of the model in Figure 2-1. Both are organizational models, but they are different in that they incorporate different elements of the real system into the model, and, correspondingly, they omit different aspects of the real system from the model. Similarly, Figure 2-4 is a model of a gear train because it is not the gear train itself, but rather a pictorial representation of it. As we noted earlier, so, too, can the input-output diagram of Figure 2-3 be thought of as a model of the same gear train if the input and output elements are appropriately defined.

FIGURE 2-8 *Organization chart and organization equivalency.*

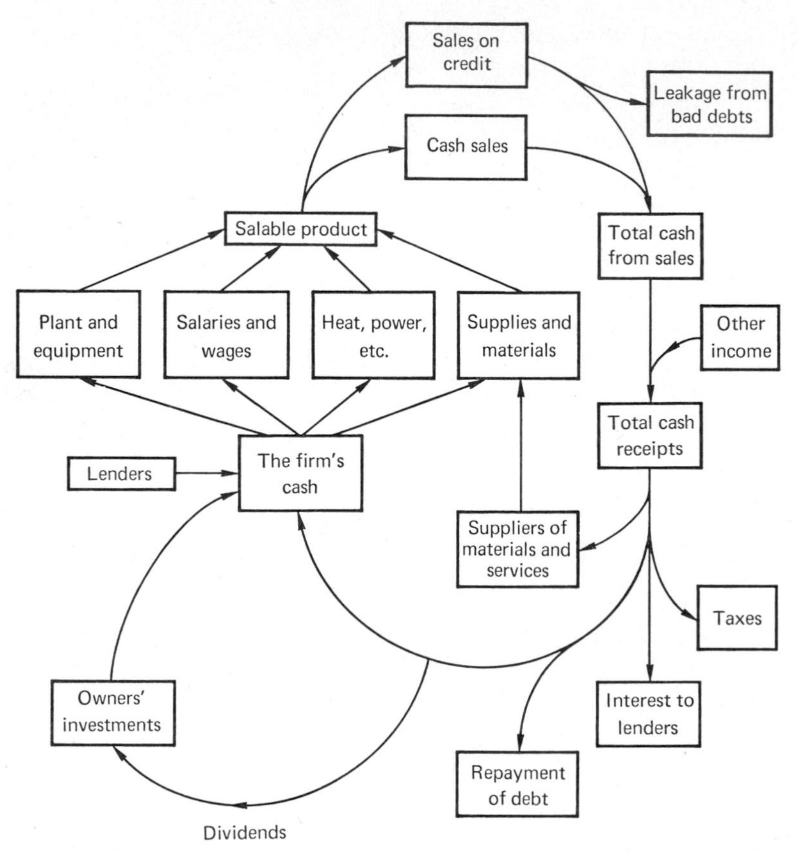

FIGURE 2-9 *Pictorial model of flow of cash through a business firm.* SOURCE: *A. M. Weimer, Business Administration: An Introductory Management Approach, Richard D. Irwin, Inc., Homewood, Ill., 1966, p. 426. Chart developed by Robert R. Milroy and Donald H. Soner. Reproduced with permission.*

The differences between these three entities—the gear train, the pictorial model, and the input-output model—offer interesting insights into the makeup of models. Most useful models are *abstractions of reality.* Thus, models incorporate certain aspects of the real system which they represent and simultaneously omit other features of the real system. For instance, the pictorial model of the gear train provides us with no information about the material of which the gears are made. Presumably, this information is a relatively unimportant aspect of the representation which was required in the illustrative discussion involving Figure 2-4. The utilization of the input-output model of Figure 2-3 to represent the gear train is even more abstract. Not only does this model omit the material, but also it does not even provide information on the number of gears or what general shape or configuration they have.

The differing degrees of abstraction in these models demonstrate that *one may have many different models of the same real-world system.* And one of these models is not intrinsically better than any other. *The description of a model as a "good" model is meaningful only in terms of the use to which the model is to be put.* A photograph of an airplane is a good model for providing magazine readers with some insight into its general "looks" and configuration; a plastic airplane model is good for a child's play or for decorating his room; a more substantial mock-up is good for testing purposes. Moreover, a training device whose exterior does not even "look like" the airplane itself is another model which is good for teaching pilots to respond to various wind conditions and to familiarize themselves with the responses of an aircraft's controls.

A Taxonomy of Models

There are many different kinds of models and there are many different kinds of classification schemes which have been applied to models. One of the taxonomies which is most useful in understanding the structural differences in models is that given by Churchman, Ackoff, and Arnoff.[2] They categorize models as either "iconic," "analog," or "symbolic."

ICONIC MODELS An *iconic* model is a simple scale transformation of the real-world system. There are only two ways in which an iconic model differs from the real system. First, as with all models, some aspects of the real system are omitted from the iconic model. Those aspects of the system which are incorporated into the model and which differ from the real system do so through a transformation of scale. Thus, the model airplane is an iconic model, since it omits some aspects of the real airplane (e.g., the interior wiring and electronics), it incorporates some aspects identically (e.g., color), and it incorporates some which are scaled-down versions of the corresponding feature of the real-world airplane (e.g., exterior dimensions). Iconic models and the real systems which they represent are therefore "look-alikes."

ANALOG MODELS A more abstract variety of model is the *analog.* In an analog model, properties are transformed—i.e., one property is used to

[2] C. W. Churchman, R. L. Ackoff, and E. L. Arnoff, *Introduction to Operations Research,* John Wiley & Sons, Inc., New York, 1957.

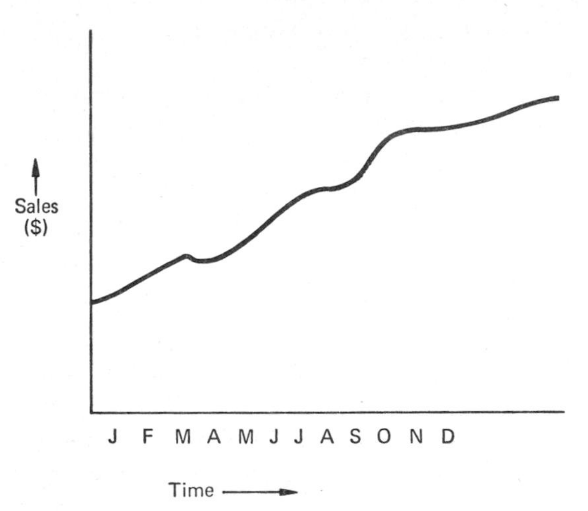

FIGURE 2-10 *Sales graph.*

represent another. A graph is the simplest illustration of an analog model. In the sales graph of Figure 2-10, for example, the property "distance along a horizontal line" is substituted for the property "time," and the property "distance along a vertical line" is used in lieu of "sales." Topographic maps in which various colors are substituted for "height above sea level" are also common analog models.

SYMBOLIC MODELS The most abstract variety of model is the *symbolic* model. In such a model, symbols are substituted for properties. For example, the equation

$$x = \frac{1}{2} gt^2$$

is a simple physical model if x is interpreted to be the distance traveled by a body falling from rest, g is a constant describing the acceleration caused by the force resulting from gravity, and t is the duration of time which the body is allowed to fall.

In management, symbolic models have long been used to describe simple phenomena. For instance, the model

$$P = R - C$$

or, "profit equals revenue minus cost," has long been recognized and used by managers. Only recently, however, have managers begun to use symbolic models for more complex systems.

Constructing System Models

Any model, whatever its nature, must be constructed through a process of determining which elements of the system are sufficiently important to be incorporated into the model and which are not. The importance of the various elements is determined by the use to be made of the model.

The methods used in constructing models are often more artistic than they are scientific. There are, however, some basic concepts which can be applied to the modeling problem of the manager and analyst.

SUBSYSTEM INTEGRATION A systems model may be developed by integrating the functions and processes of various subsystems. The outputs from one subsystem are simply the inputs to another subsystem in such an overall systems model. For example, this would be the case if subsystems A and B, shown in Figure 2-11, were the packaging and mailing systems for a catalog mail-order merchandiser. The product to be shipped comes to the packaging subsystem A, is wrapped, and then is output from A to be sent on as an input to the mailing subsystem B, where it is addressed, stamped, and becomes output from the overall system.

Other *flows* occur between the subsystems and between the system and its environment. Orders flow from the customer to the merchandiser's subsystems. People expedite the filling of special orders by communicating with those in other departments. Supplies necessary to the efficient operation of the organization flow through the subsystems. Management directives on new policies, personnel transfers, promotions, and other organizational changes are processed through the "system."

In general, subsystems may be interconnected in a myriad of complex ways. For example, the marketing and production subsystems within a business firm interact on a continuing basis. From the time at which the idea for a new consumer product is conceived until it is being sold on a mass basis, the production and marketing subsystems must process information and communicate with each other and with the outside world. When a new-product idea is proposed, its technical feasibility must be evaluated by elements of the production subsystem. When this information is passed to the marketing subsystem, it is further evaluated in terms of information

FIGURE 2-11 *System diagram showing component subsystems.*

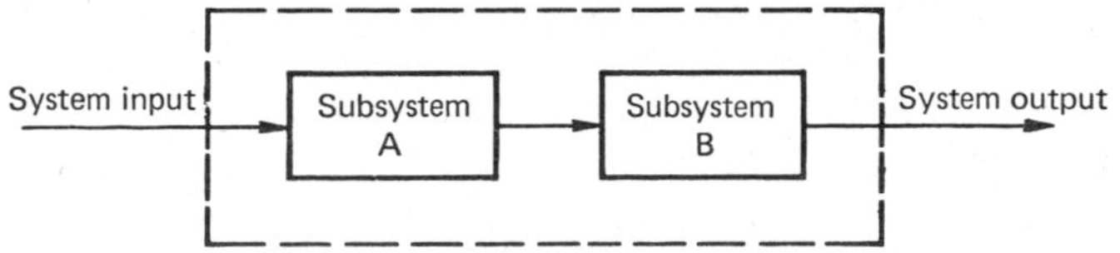

obtained in the marketplace concerning consumer desires, competitors' products, and the firm's ability to successfully market the product. It may be evaluated as being deficient in its market appeal, and recommendations may be made for design changes. The production system must then re-evaluate the product on the basis of these changes. When the new item goes into production, sales forecasts from the marketing subsystem are used to plan for the scheduling of production and for the quantities to be produced. When the product's appeal begins to wane, notification to the production subsystem must be given to avoid overstocking. Indeed, at every stage of a product's life cycle, there is need for an interplay and exchange between the production and marketing subsystems of the business firm as well as between the business system and the environment in which it operates.

In adequately describing such a complex system, a great deal depends on the use to which the description is to be put. It is not always necessary to give a detailed description of all subsystems of a system. Indeed, it is typically impossible to do so. An electronic computer is a system whose subsystems (registers, circuits, memory) would be of interest to the design engineer. On the other hand, the designer of a management information system would probably treat the computer itself as a subsystem without going into detailed description of the computer subsystems. The president of a business firm might be interested in the information system as a subsystem of the firm; he might then have little interest in the computer per se.

The idea of describing a system in terms of the use to which the description is to be put operates in both directions—in "building" systems from subsystems and in "decomposing" systems into their constituent parts. In the latter process, one can decompose systems into subsystems and subsystems into sub-subsystems almost indefinitely—perhaps to the level of the component atomic particles.

THE "BLACK BOX" APPROACH In the modeling of a system through the consideration of subsystems, it becomes obvious that at some point the effort required to describe the next subsystem level in detail is not worth the advantage gained in doing so. If this were not so, every model would involve detailed descriptions of atomic-particle behavior, and none would be very useful.

At some point of decomposition, one's interest in a subsystem is fully satisfied by a description of the transformation of inputs into outputs without regard to the internal structure of the subsystem or the details of the transformation. At this point, the subsystem may be described in simple input-output terms. In the language of the systems analyst, the subsystem is treated as a *"black box"*—i.e., an entity which does known things in unknown ways with unknown mechanisms. All that is known about a "black box" is that it operates as a transformation unit in changing measurable inputs into measurable output. Nothing is known about *how* it does this.

The "black box" approach may be applied to some very complex systems as well as to low-level subsystems of those systems with which one is primarily concerned. There are a variety of reasons for which one might do so. *The level of information which is needed concerning the system might be very gross,* so that no consideration would need be given to anything other than the input and output. Or, *one's understanding of the system in question may be so poor that he cannot describe it in any other terms.* Consider, for example, the research and development effort expended by government and industrial firms. One can hypothesize that such expenditures (the input) should have a great impact on the nation's economy (the output). The ways in which this impact occurs are so varied and complex, however, that it would be impractical to attempt to diagram the nation's "R&D system" in detail, since doing so would require the compilation of so vast a quantity of information that the results would be obsolete before they were organized. Consequently, one might view this complex R&D system in input-output terms and use measures such as the "total dollars expended by government and industry on research and development" in a particular year as an assessment of input and the "change in gross national product" in some subsequent year (say, five years later) as an assessment of output. If there is truly an effect of R&D on the economy, it might well be detectable at the national level in these gross input-output terms. Thus, even though the transformation of inputs (e.g., R&D expenditures) into outputs (e.g., GNP) may itself be exceedingly complex, it may not be necessary to understand and describe the transformation in detail in order to make use of it.

In any case, the black box approach to describe subsystems is always necessary at some level. One seldom even attempts to describe more than several levels of subsystems of a system, simply because of the scope of the task involved in doing so. In doing this, the task of describing the black boxes themselves may be completely ignored, if the system exists and the subsystems are adequately described in input-output terms, or it may be left to others. The leaving of subsystem description to others is typical in the design process, for example, where a general description of the transformation properties required of a subsystem (say, a computer which is part of a management information system) will be specified and the design of a subsystem (computer) which conforms to these specifications will be left to a computer manufacturer.

Using System Models

We have already indicated that the differences between various models of the same system are accounted for by the determination of which aspects of the real system are incorporated into the model and which are left out. The *primary value of a model is that it does leave things out.* If all models were "perfect" in the sense that they included *all* aspects of the real system, there would be no models but, rather, simply reproductions of real-world systems. As such, they would be useless for many purposes. A real DC-8 would have little utility for a child who wished to decorate his room with a representation of an airplane, for example. If he had an opportunity to obtain a "perfect" DC-8 model, and if his father could afford to do so, the child might well decide to park it in his back yard as a toy. And, even though his father might believe that it would be an interesting conversation piece for him, whatever use they eventually put it to, they could not use it for the stated purpose—that of decorating the child's room. Hence, it would be a bad model for that purpose.

Such a preposterous example demonstrates how important an aspect of a model is its omission of some aspects of the real system. Moreover, it again shows that *the worth of a model depends on the use to which the model is to be put.*

The primary value of a model lies in its simplicity relative to the real world. Because it omits some aspects of the real-world phenomenon or system which it describes, a model is necessarily simpler than that which it represents. This means that it is easier for us to understand the model. And understanding opens up the possibility of control—i.e., we must understand the system before we can control it and use it to our own ends.

The best illustration of the practical use of models is perhaps the space program. Models were used to *understand and predict* the implications of space flight before a man had ever left the earth's surface or ventured near the moon. Indeed, the models were not perfect, and our initial astronauts encountered some things in orbital flight and in the lunar environment that had not been anticipated, but the imperfect nature of the space models was well known in advance. The significant thing is that it would have been impossible for us to launch a man into space and recover him without the prior analysis of models which represented many of the phenomena which would be encountered. The models used were "good enough" to enable us to accomplish the basic objectives. The details which were left out were indeed sufficiently unimportant relative to those things which were included. The success of the space program bears witness to this.

All of science is heavily dependent on such models. *The impossibility or costliness of dealing with real-world systems leads the scientist to experiment on the model in lieu of experimenting on the system.* Thus, when a new ship design is proposed, the Navy does not simply build the full-size ship and try it out. Rather, it builds a model and tests it—i.e., it experiments on the model. Only when the model has been proven is the ship itself built.

The reason for experimenting on a ship model rather than building the ship itself is the relative cost efficiency. It would be impractically costly to build a ship and then discover that it had some major defect. The model helps the naval architect to discover flaws and to correct them before construction is begun.

The testing of models in wind tunnels is also an essential part of an aircraft development program. In the wind tunnel a replica of an aircraft is tested under varying conditions of air velocity, altitude, and climatic conditions. The data obtained from such tests have an important predictive value in the design of the actual aircraft itself.

In other instances, it is impossible to work on the real-world system. Consider the astronomer and the solar system, for example. He wishes to predict in advance when comets will appear and when eclipses will occur. Since he cannot manipulate the system itself, he instead manipulates a model and uses it as a basis for his predictions.

Management Models

Of course, the manager may be quite different from the scientist in his interests, abilities, attitudes, and needs. Why then should he use models? The simplest answer to this question is that we are not arguing that the manager should use models; in fact, all managers *do* use models. Indeed, they cannot avoid doing so.

Every complex problem or phenomenon must be thought of by the manager in terms of its salient parts. In all but the simplest instances, the manager must construct a "mental model"; he then thinks in terms of the mental model. This mental model has all of the appropriate characteristics of a more formal model—it is a *representation* in the mind of something that exists in the real world, it leaves some aspects of the real world out and includes others.

The need for even a model of the mental variety emanates from the complexity of the real world and the difficulty that the human mind has in handling vast complexity. To handle complexity and still maintain his sanity, the manager must construct a model and think in terms of it rather than in terms of the real world. Thus, the system model is substituted for the real-world system in the mind of the manager. This makes the situation comprehensible to him and enables him to do things which he could not do in the real world. He can, for example, ask "What if . . ." questions in terms of the model—i.e., he can manipulate the model whereas he may be precluded from manipulating the real system.

Of course, there are other kinds of models beyond mental models which are useful to the manager. In this book, we shall attempt to show the applicability and advantages of some of these. Basically, however, whatever sort of model the manager uses, his purpose is the same—*to predict.* A definition of a management model should make this purpose clear.

A management model is a representation of a system which is used to predict the effect of changes in certain aspects of the system on the performance of the system. Thus, a management model is simply a medium which enables the manager to ask his "What if . . ." questions: *What* would be the effect on profit *if* we increase advertising expenditures? *What* would be the effect on per-share earnings *if* we merge with XYZ Corporation? *What* would be the effect on children's health *if* a comprehensive child medical care program were financed with federal funds?

Any of the varieties of models previously discussed—iconic, analog, and symbolic—together with informal mental models, may be management models. The basic criterion is: "Are they used for predictive purposes?" Industrial engineers use iconic models of an industrial plant in planning the layout of the plant. In doing so they predict the effect of various alternative layouts on plant production, idle time, etc. Either analog or symbolic models (in the form of graphs or algebraic expressions) may be used to predict the performance of business systems. For example, the economist's *demand curve* is a simple model which relates price to quantity demand. The demand curve of Figure 2-12, if known, would enable us to predict that 100,000 units of our product will be sold if the price is established at $4 while 150,000 units will be sold if the price is set at $2.

Both the demand curve of Figure 2-12 (analog) and the algebraic expression which we might use to represent it (symbolic) can be management models, since they can be used to predict and because they are recognized to be imperfect descriptions. That is, we recognize that in any realization of these phenomena, the models may not perfectly describe what actually occurs in the real world. For example, these models do not consider the effect of monopolistic tendencies. What if the company to which the demand curve is related has some monopolistic power? Suppose customers are not completely free to go elsewhere to buy if the price is too high? These are things which are omitted from the simple demand curve model, presumably because they are relatively unimportant in the situation in which the model is to be used.

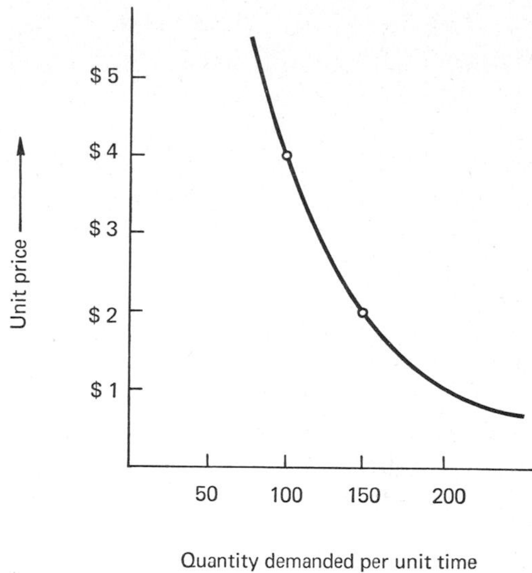

FIGURE 2-12 *Demand curve.*

SUMMARY

Just about everything that one can imagine can be described in systems terms—as collections of interacting and interdependent parts forming a unified whole. These systems and their constituent subsystems are defined by the manager or analyst in a way which best suits his needs.

At some point in the hierarchy of subsystems, a systems level which is best described in simple input-output terms will be reached. Such a system is completely described by inputs to, and outputs from, a transformational unit. In this simple system, the unit transforms inputs into outputs.

More complex systems incorporate feedback. In a *feedback control system,* actual performance data are "fed back" and compared with a pre-established goal. A *feedback system with memory* permits the selecting of various goals for specified conditions. For instance, a thermostatically-controlled heating system is a feedback control system. A heating system which automatically establishes desired temperature goals at various times of the day is a feedback control system with memory.

Learning systems—those in which decision rules can be varied according to the "learning" which occurs relative to a higher-order objective—theoretically describe the organizational systems with which the manager deals. However, it is frequently more useful, and often more accurate, to use one of the simpler system models to depict these systems.

Models of systems are an intrinsic part of the manager's life. Models represent systems, and as such, they can be manipulated, experimented upon, and used to predict in ways in which real systems cannot. For instance, an advertising response model can be used to predict the sales which will result from various levels of advertising expenditure. To obtain this same information in the real world would require much time and expense while perhaps doing irreparable harm to the firm's position in the market in question. The model makes it possible to avoid this, and if the model is "good enough" it permits the manager to make better decisions than he could without it.

A taxonomy of models includes iconic (scale-transformed), analog (property-transformed), and symbolic (symbol-transformed) types. All three varieties are useful to the manager in various decision situations. Scale models (iconic) can be used to lay out a plant. Graphs (analog) can be used to depict historical sales records. Mathematical models (symbolic) can be used to describe complex decision problems.

Models may be constructed using subsystem integration—by integrating the functions and processes of various subsystems. At some point, the "black box" approach—in which a system is described in input-output terms without regard for its internal workings—must be used, either because this level of the system is so little understood or because it does not require detailed understanding.

Any such models are only as good as the things that they omit. Any model is an abstraction, and their usefulness is primarily due to this abstraction. Management models are abstract representations of decision problems and processes which permit the manager to predict and assess the impact of possible courses of action on the performance of the overall system. In doing so, he should be able to select the best (or, at least, a good) course of action.

Thus, an understanding of systems and system models is very important to the manager. Such an understanding can lead him to better decisions and the organization to the better fulfillment of its goals.

EXERCISES

1. What is the difference between the idea of a "system" and the "systems concept"?
2. The systems approach in the final analysis involves essentially a "way of thinking" about an approach to complex problems. Describe what this way of thinking means to you.
3. In its broadest conception the systems approach represents a method of dealing with complex situations. In management theory, there seems to be some "conflict" between the systems approach to management and

earlier concepts of management. How does the systems approach relate to other views of management? Discuss.

4. Defend or refute this statement: Social systems within a given complex society have no influence on the business objectives of business organizations in that society.
5. The authors make the point that an organization from a systems viewpoint cannot be completely isolated from its environment. Can you identify any situations in history where an attempt was made to isolate organizations from their cultural, sociological, and economic environment? How successful have these organizations been?
6. One description of the general interaction model of a business organization identifies three major subsystems: the receptor system, the decision mechanism subsystem, and the processor subsystem. Identify the input and output stimuli to each of these systems. What role does feedback play?
7. The authors have provided a basic model which describes the business organization as a system. Take this basic model and apply it to the following type of organizations: (*a*) an ecclesiastical organization, (*b*) an educational organization, (*c*) a family unit, and (*d*) a social organization to which you belong.
8. Why do the authors encourage the use of models in understanding the making and execution of decisions?
9. The authors state, "The primary value of a model lies in its simplicity relative to the real world." Why do the authors make this statement?
10. The systems approach is neither new nor sophisticated. How would you defend or refute this statement? Give examples.
11. In the systems approach, an explicit recognition of the interdependencies—that "everything depends on everything else"—is a first step, if not a calming one, to the solution of organizational problems. How has this differed from our approach to organizational problems in the past?
12. In planning for an organizational system, many interrelated elements and subsystems must be taken into account. Planning does not simply involve the preparation of a document—a plan—at one organizational level and promulgation of this plan to subsequent lower levels. How would you describe the planning process from a systems approach?
13. Describe some of the subsystems that might be used in defining the following systems:

    a. A city
    b. An educational organization
    c. A business firm
    d. The production of a student research paper

14. Using one of the models of the business organization given in the chapter, identify the specific inputs and outputs to the system.
15. Subsystems may be interconnected in a variety of complex ways. De-

scribe how the marketing and production subsystems within a business firm interact on a continuing basis.

16. Describe how the teacher subsystem, administrative subsystem, and pupil subsystem interface within an educational system.

17. Why does the primary value of a model lie in its simplicity relative to the real world?

18. Describe how one could use a model in planning and developing a strategy for his own life.

19. Models have long been used in the physical sciences and engineering. However, they do not have a long history of usage in management. Can you explain why?

20. What are the various kinds of models? Give an example of each.

21. How can each of the various kinds of models described in the previous question be used as management models? Give specific illustrations. What characteristic in these models is essential if they are to be used as such?

22. A common system which is of particular significance to management is the organization. Such a system is often described as a purposeful system. What other systems are "purposeful"?

23. What is a learning system? How does it differ from a simple feedback system?

24. What is the "black box" approach? What are the various ways in which the black box approach can be used?

25. Give some illustrations of simple input-output systems, simple feedback systems, feedback systems with memory, and learning systems. Which of these models do you think best describes modern organizations?

26. Consider the possibility of modeling the organization of a university department using each of the models described in the previous question. What would be the different specifications of inputs and outputs which would be required for each kind of model? How would one go about determining which of these models is the best one to use?

27. Give some examples of some simple models with which each of us comes into contact in his everyday life.

28. Consider the manager's problem of describing an existing system which he wishes to better understand and the designer's problem of creating a new system from some specifications. How are the two problems alike and how are they different? What is the role of models in each?

29. It has been said that there is no reality other than the reality of models, since each of us thinks in terms of models rather than in terms of reality. What do you think of this viewpoint?

REFERENCES

Ackoff, R. L., "General Systems Theory and Systems Research: Contrasting Conceptions of Systems Science," *General Systems,* VIII, 1963.

———, "Systems, Organizations, and Interdisciplinary Research," in *Systems:*

Research and Design, Donald P. Eckman, ed., John Wiley and Sons, Inc., New York, 1961.

Ashby, W. Ross, *An Introduction to Cybernetics,* Chapman and Hall, London, 1956.

Bayliss, W. M., *Principles of General Psychology,* 4th ed., Longmans, Green & Co., New York, 1927.

Boulding, Kenneth E., "General Systems Theory—The Skeleton of Science," *Management Science,* April, 1956.

Bross, Irwin E. J., "Models," in Bross, Irwin E. J., *Design for Decision,* The Macmillan Company, New York, 1953.

Churchman, C. W., *The Systems Approach,* Dell Publishing Co., New York, 1968.

———, R. L. Ackoff, and E. L. Arnoff, *Introduction to Operations Research,* John Wiley and Sons, Inc., New York, 1957, chaps. 2 and 7.

Cleland, David I., and William R. King, *Systems Analysis and Project Management,* McGraw-Hill Book Company, New York, 1968.

DeGreene, Kenyon B., ed., *Systems Psychology,* McGraw-Hill Book Company, New York, 1970.

Farmer, Richard M., and William G. Ryan, "Management, Microeconomics, and Systems Theory," *MSU Business Topics,* Winter, 1969.

Glans, Thomas B., et al., *Management Systems,* Holt, Rinehart and Winston, Inc., New York, 1968.

Johnson, R. A., F. E. Kast, and J. E. Rosenzweig, *The Theory and Management of Systems,* McGraw-Hill Book Company, New York, 1967.

King, William R., *Quantitative Analysis for Marketing Management,* McGraw-Hill Book Company, New York, 1967, chaps. 1 and 4.

———, "Systems Analysis at the Public Private Marketing Frontier," *Journal of Marketing,* January, 1969.

———, "The Systems Concept in Management," *Journal of Industrial Engineering,* May, 1967.

Morrison, Edward J., "Defense Systems Management: The 375 Series," *California Management Review,* Summer, 1967.

Neuschel, R. F., *Management by System,* 2d ed., McGraw-Hill Book Company, New York, 1960.

Newell, A., J. C. Shaw, and H. A. Simon, *The Process of Creative Thinking,* RAND Document P-1320 (September, 1958; revised January, 1959), The Rand Corporation, Santa Monica, Calif.

Purcell, Edward, "Parts and Wholes in Physics," in *Parts and Wholes,* Daniel Lerner, ed., The Free Press, New York, 1962.

Schoderbek, Peter T., *Management Systems: A Book of Readings,* John Wiley and Sons, Inc., New York, 1967.

Simon, Herbert A., "The Architecture of Complexity," *Proceedings, American Philosophical Society,* 106, 6, December, 1962, 467–482.

Young, Stanley, *Management: A Systems Analysis,* Scott, Foresman and Company, Glenview, Ill., 1966.

———, "Organization as a Total System," *California Management Review,* Spring, 1968.

Zani, William M., "Blueprint for MIS," *Harvard Business Review,* November–December, 1970.

3

organizations

The word "organization," like the word "system," is a part of the vocabulary of virtually everyone. When one joins a fraternity, he becomes a member of an organization whose obvious manifestations are fraternity brothers, a fraternity house, dues, meetings, etc. And, when one accepts his first job, he generally becomes a member of a large corporate organization.

In fact, most of us become part of a basic social organization when we are born into the family unit. Indeed, many of the basic patterns of management and organization that are usually thought of in terms of government and industrial organizations can also be applied to this familiar organization. Certainly both formal organizations and families face similar problems of scarcity of resources, interpersonal relations, financial decision making, etc.

The parents in the family unit presumably have superior authority over the children, at least while the threat of physical punishment is viable. But, as the children grow in emotional and physical maturity, the patterns of authority and responsibility between the parents and the children take on a different tenor. The parents' effectiveness in the normal functions of "running the family" must be augmented by effectiveness in motivation, persuasion, and human relations. If the child feels acceptance in his family unit and is able to understand the obligations and responsibilities of membership in that basic social unit, his chances of being successful in the myriad of other organizations with which he will be affiliated in his lifetime are improved. On the other hand, if he is not accepted in the basic social unit and does not learn to accept responsibility, he may have serious problems with the many different forms of organizations which will affect his spiritual and physical well-being in later life.

Each of us is associated with many different organizations during a lifetime. Indeed, at any given point in one's life, he is simultaneously a member of many organizations. And, although our behavior at any given moment may be largely a function of our role in a single organization, it is likely to be influenced by the role which we play in others, also.

During any one day, every individual plays a variety of roles, many of them related to organizations. In the family unit, the father may be a leader, having *legal* authority over the other members of the family. On the way to his place of employment or school, he plays the role of a peer or associate in the car pool or on the bus. When he arrives at work, he is variously a subordinate, a peer, and a superior. In each of these roles he displays different patterns of behavior and takes actions which are consistent with different value systems.

For example, the college student's reaction to his courses is influenced by the "learning" which takes place in the environment of the family organization during the years of maturation. His reaction may be either positive or negative—e.g., formal learning may be accepted or rejected *because of* the informal learning from the family or *in reaction to* the mores of family and society which are being questioned by the student. In any case, the influence of nonacademic organizations on the formal learning process is significant.

So, too, with the businessman who must make decisions affecting the future of his organization. Even though a particular decision may involve strictly business considerations, it is the unusual executive who would not consider the implications of his decision on himself, his community, and his family. In doing so, he is allowing his membership in one organization to influence his actions in another.

ORGANIZATIONS AND MANAGEMENT

In Chapter 1, we gave an operational definition of "management" which included the elements organized activity, objectives, relationships among resources, working through others, and decisions. The element "organized activity" and the concept of an organization are obviously closely related.

When the layman uses the word "organization" as a noun, he is speaking of some entity, such as a corporation, club, or church. If he uses it in the sense of "being organized," he probably refers to a pattern of thinking and of utilizing resources in an effective fashion. For instance, if a suburban housewife maintains a checklist of things to do during the day and manages to accomplish several of these in a single trip with the family car, her husband would probably compliment her on being "organized."

Both of these common interpretations of "organization" are consistent with that used in management. In the broadest sense, an organization is simply *an agreement among people to cooperate in some endeavor.* Often, the agreement may be informal—as in the case of the "Park Street

Wednesday Nite Bridge Club"—or, of course, it may be a chartered corporation which has formally stated its objectives and formally established its officers in an application to the state government.

An Operational Definition of Organizations

Operationally, we can think of the elements which are common to organizations.

1. There is an explicit or implicit *objective* toward which the participants are working.
2. There exists some *pattern of authority and responsibility* between the participants.
3. There are usually some nonhuman elements involved.
4. There are many *decisions* (both routine and strategic) involved in the management of the human and nonhuman resources.

Thus, to operationally define an organization we need to use many of the elements which we used in our operational definition of management. This says that organizations and management must be intrinsically interlinked concepts. We shall concentrate on making this commonality more clear in the remainder of this chapter.

Organization Theory

When one begins a study of a concept, he should look to the theory which may exist concerning the concept. In the case of organizations, there is a body of knowledge known as *organization theory.*

There is much overlap between classical organization theory and traditional approaches to management. For instance, one of the mainstreams of classical theory is F. W. Taylor's "scientific management."[1] Taylor attacked the general problem of organizational theory—*"to analyze the interaction between the characteristics of humans and the social and task environments created by organizations"*[2]—in a specialized context. The context that he chose was that of specifying a method that would transform a general-purpose mechanism (a human being) into a special-purpose operative who could efficiently perform organizational tasks. Taylor therefore concentrated on low-level production operations and in

[1] F. W. Taylor, *Scientific Management,* Harper, New York, 1947.

[2] J. G. March and H. A. Simon, *Organizations,* John Wiley & Sons, Inc., New York, 1967, p. 13.

doing so accomplished two major ends—the introduction of objective measurement into organizational considerations and the defining of problem areas for others to further develop.

Other organizational theorists have focused on departmentalization, division of labor, and coordination.[3] Each of these concepts is closely related to management thought. Indeed, modern organizational theory is even more closely intertwined with management. Hence, in this chapter and the remainder of the book, we shall not attempt to offer a theory of organizations per se. Rather, our approach will be to use some of the approaches and conclusions of organizational theory in developing a conceptual framework for the practice of management.[4]

WHY HAVE ORGANIZATIONS?

There appears to be a natural tendency on the part of human beings to organize—if only in the loosest of fashion. The United States economy offers an illustration of this. Here, in a relatively unregulated environment, millions of people and thousands of companies produce and exchange goods to satisfy one another's needs. And, however inefficient the system may appear at times, it involves an amazing amount of inherent order and organization. This becomes especially vivid when one views "planned" economies which have not only failed to function efficiently but have failed to produce the goods and services desired by the populace. At lower levels, organizations also appear to be the natural thing: workmates develop value structures, standards of conduct, and status relationships among themselves; so, too, do schoolboys and "gangs" of school dropouts "organize" themselves.

Modern managers have recognized the existence of *informal organizations* within the structure of formal ones and have attempted to use these low-level informal organizations to the advantage of the overall formal organization. For example, *Fortune* reports on the significant results achieved from "job enrichment" programs which utilize the spirit of small work groups as a motivating factor to increase productivity.[5]

The need for organizations—both formal and informal—lies both in

[3] For instance, see L. H. Gulick and L. Urwick (eds.), *Papers on the Science of Administration,* Institute of Public Administration, Columbia University, New York, 1937.

[4] The reader who has a direct interest in organization theory is referred to the excellent book by J. G. March and H. A. Simon, *Organizations,* John Wiley & Sons, Inc., New York, 1957, as a starting point for his readings. A good collection of readings which can also be useful is J. A. Litterer (ed.), *Organizations: Structure and Behavior,* John Wiley & Sons, Inc., New York, 1963.

[5] J. Gooding, "It Pays to Wake Up the Blue-Collar Worker," *Fortune,* September, 1970, p. 133.

the psychological and social needs of human beings and in their desire to accomplish objectives. In a complex world, those significant things which can be accomplished by a single person become increasingly rare. Moreover, even those things which *could* be done by an individual cannot be *efficiently* done in such a fashion. This is reflected in our tendency toward increased specialization in virtually every field of human endeavor.

Specialization has characterized work activities since the Industrial Revolution. Prior to that time, the shopkeeper could be producer, salesman, and financial expert, all at the same time. The advent of mechanization meant that new markets had to be found and new expertise developed. This led naturally to the increased *division of labor*—the assignment of individuals or groups to limited areas of responsibility.

Thus, the primary rationales for the existence of formal organizations are *effectiveness* and *efficiency. An organization is differentiated from other systems by its purposeful behavior—its pursuit of objectives.* Some goals can be achieved only by the concerted action of a group of people who make use of nonhuman resources. When multiple goals are involved, problems of *resource allocation* come into play. For instance, the group must allocate their time, energies, and other resources to those activities where the greatest effect can be produced with the least expenditures, since a scarcity of resources is an economic fact of life.[6] No organization or individual possesses adequate resources to pursue all of its or his possible goals at the highest level of intensity; to assign an individual to one task is to make him unavailable for other tasks; to spend limited money on increased salaries is to preclude the possibility of using it in the hiring of additional workers. Thus, *the basic economic question of the allocation of scarce resources is at the heart of the need for formal organizations.* This is so because even though the natural tendency of human beings to organize might well lead to the same allocation of resources as is attained via a formal organizational structure, it would do so only after a series of trials and errors and it would do so very, very slowly. For an enterprise to function and prosper in an organized world, it must organize itself and seek efficiency.

THE ORGANIZATION AS A SYSTEM

An organization is a kind of system. At the superficial level this is made clear by one of the common system models which can be used to describe an organization—the organization chart.

[6]Of course, it is most frequently impractical to literally achieve the "greatest effect with the least expenditures." We shall explore this idea further in later chapters.

Organization Charts

An *organization chart* such as that shown in Figure 3-1 is a familiar model of the authority and responsibility relationships that exist between departments and people in a formal organization. An advantage of such a chart is that it helps to define organizational relationships; without such a chart it is difficult to determine who works with whom and how groups of people are related in the organization. The lines on the organizational chart are used to indicate formal authority and can provide insight into *formal* communications and structure in the organization. On the other hand, one of the major shortcomings of the traditional organization chart is its failure to depict how the organization operates in terms of *how* the people work together on a daily basis.

Cyert and March have given an interesting "plus and minus" appraisal of the utility of the organization chart model of the organization.

> *Traditionally, organizations are described by organization charts. An organization chart specifies the authority or reportorial structure of the system. Although it is subject to frequent private jokes, considerable scorn on the part of sophisticated observers, and dubious championing by archaic organizational architects, the organization chart communicates some of the most important attributes of the system. It usually errs by not reflecting the nuances of relationships within the organization; it usually deals poorly with informal control and informal authority, usually underestimates the significance of personality variables in molding the actual system, and usually*

FIGURE 3-1 *Simple organizational chart.*

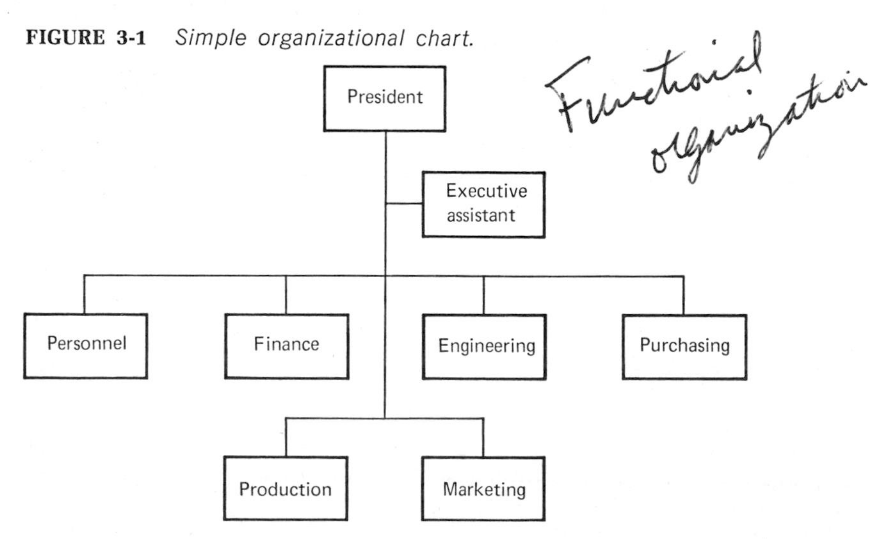

exaggerates the isomorphism between the authority system and the communication system. Nevertheless, the organization chart still provides a lot of information conveniently—partly because the organization usually has come to consider relationships in terms of the dimensions of the chart.[7]

Organizations—Purposeful Systems

While an organization chart is not a perfect model of an organization, it does show formal authority and responsibility structure, titles, planned tasks, and the anticipated relationships between tasks. In particular, the organization chart model of an organization emphasizes that an organization is a unified whole made up of regularly interacting parts. Hence, an organization is indeed a system. However, it is a special variety of system which we shall term a *purposeful system*—one that is seeking to attain a goal.

According to the idea of an organization as a goal-seeking system, the solar system is not an organization, since it exists without seeking any specific objective.[8] However, the mechanical governor does fall within the province of this definition, since it seeks a goal—the maintenance of a constant speed. So, too, is an ant colony an organization, since it has many interacting elements which constantly pursue goals of self-preservation and reproduction.

Most organizations which are of concern to us have human beings as elements of the system, however. Hence, it is apparent that General Motors is an organization. So, too, is our tennis club, the PTA, the University of Pittsburgh football team, and the respective families of the authors.

For some of these organizations it may be rather easy to spell out the goals. A university football team's primary purpose is to win games, thereby bringing recognition and credit to the university. Other secondary goals are apparent also—to provide healthy recreation for the players, jobs for the coaches, income via ticket sales to support other athletic functions, etc. For other organizations, it may be more difficult to spell out the goals, even if there is a written document which states them. In such cases, however, the existence of a goal is implicit in the existence of the system, since they are created by man.

The family offers a good illustration of a familiar organization. One

[7]R. M. Cyert and J. G. March, *A Behavioral Theory of the Firm,* Prentice-Hall, Inc., Englewood Cliffs, N.J., 1963, p. 289.

[8]Indeed, if there is such a goal, it is unknown to us mortals. We shall leave any further discussion of this to the philosophers.

important goal of most families is the reproduction of the human species. However, the family, as we know it, is the only one of many organizational forms which might be used to seek this goal. Communal "families" are an obvious alternative which has a great deal of appeal to some people.

Both varieties of families—traditional and communal—may have some common objectives, among them the reproduction of the species. However, both have other objectives also. For instance, the "quality" of the reproductions is of great importance to both kinds of family organizations. Neither wishes to simply reproduce; rather, they wish to raise "productive" members of society who will become "good citizens." Of course, the traditional family and the communal family may interpret the terms "productive" and "good citizen" differently, but they both are concerned with the quality of the offspring which they produce.

So, too, are nonfamily organizations concerned with the nature of the children being brought up in the society. The federal government has instituted many programs designed to enhance this quality, as have other governmental organizations, foundations, etc. Thus, even in this simple and familiar situation, it is easy to see the *interrelationship of organizations. Each organization has goals which are related to those of other organizations.* We shall see that this is an important and pervasive characteristic of organizational systems as subsystems of our society. As with all system views, this serves as an additional complicating factor, but, as we shall see, it also provides us with a point of view for analyzing organizational systems.

Traditional Models of Organizations

Systems ideas have produced radical changes in the nature of both the "mental models" which we use to think about organizations and the formal models which are used to describe and analyze organizations. Traditional organizational models have relied on two important organizational elements—*functional structure* and the *bureaucratic concept.*

FUNCTIONAL STRUCTURE OF ORGANIZATIONS The typical manufacturing organization chart (Figure 3-1) is a good model to use to describe the functional structure of the organization. Similar functional forms are found throughout government, industry, the military, ecclesiastical organizations, etc.

Functionalism is based on the concepts of *division of labor* and *specialization*—the assignment of subtasks to units of people who are "expert" in performing narrow, highly specialized functions. Thus, in forming functional departments, people of similar skill, training, occupational specialization, etc., are brought together. The assumptions are that the organization will be more efficient if people having one particular expertise are working together and that the strength of the organizational team provides an effective bond that brings about more efficient cooperative effort. Thus, the functional organization has engineers grouped into an engineering organization, production people working in a production organization, and financial people working in the financial department.

Figure 3-1 shows a simple, yet comprehensive, departmentalized structure oriented to those functions which are directly related to achieving the goals of the enterprise—production and marketing. Other elements—finance, personnel, engineering, and purchasing—support the production, marketing, and organization executive functions.

The *production* element of any organization—be it a manufacturing one or not—has responsibility for those *operations* which are necessary to produce the organization's "product." If the organization is a manufacturer, it has responsibility for fabrication, assembly, tooling, the purchasing of materials, and production control. If the organization's business is transportation, the operations function has responsibility for performing the transport function. If the business is research, the operations function is a research activity. Thus, something that may be a support activity in one organization may be the primary operational function of another. In any case, the production or operations function involves "doing the job" of the organization.

The *marketing* element of any organization complements the production function. An old marketing cliché succinctly describes this complementary relationship: "Everything is cost until the product is sold." Thus, even though the production function gets the job done, it does not really pay off until the product is delivered and paid for. Such is the role of marketing—to sell and distribute the product.

The marketing function involves sales, advertising, promotion, distribution, market research, and other activities. Although virtually every product or service must be "sold"—be it the selling to Congress of the recommended programs of a government department, the selling of a university's research capability to the government department which may fund the research effort, or the selling of paper towels to the university—as with various phases of production, the marketing activity or some marketing subactivity may be the "operations" function of some organizations. For instance, an advertising agency's primary "operation" is a subfunction of marketing.

Other functional elements of organizations have responsibilities which are generally described by their titles. "Personnel" deals with the human resources in the organization; "finance" deals with the financial resources; "engineering" is responsible for physical resources. Various organizations have their unique names for these functions; but, whatever the organization, they must all be performed. Hence, the functional model of the organization is both a natural and a useful one.

THE BUREAUCRATIC CONCEPT The traditional model of an organization emphasizes superior-subordinate relationships. Orders flowed from the top down a chain of command and individuals carried out orders. This, of course, was a vestige of the familiar military organization model. In effect, for some time, we used the military model to describe virtually all organizations, whether or not it applied well. In fact, this may explain the failure of the model to serve as a useful management model for *prediction* in organizations.

The *bureaucratic* model on which most traditional organizations are based emerged from the Industrial Revolution and saw its zenith in the 1940s. Bennis has described the key elements of the bureaucratic model as: ". . . a well-defined chain of command. A system of procedures and rules for dealing with all contingencies relating to work activities. A division of labor based on specialization. Promotion and selection based on technical competence. An impersonality in human relations."[9]

Systems Models of Organizations

In practice, the bureaucratic was not a very accurate description of most organizations. Many of the important activities of nonmilitary organizations are more heavily dependent for their success on relationships between people than on a rigid adherence to orders and directives. In business, decisions arrived at by a consensus negotiated among peers over a martini have always been important. So, too, have most successful managers learned to circumvent or "bend" rules and directives to "get things done."

For instance, one of President John F. Kennedy's close advisers is reported to have said that the primary asset of the "new generation" which took over the executive branch of the government in 1961 was their World War II experience, which had taught them to seek out the objective and to bend or break existing well-intentioned rules which prevented them from accomplishing it. Indeed, the experiences of the World War II generation may well have sown the seeds of the "situation ethics" which has come to pervade our society in recent years.

In all modern organizations, the individual interacts as much with

[9]Warren G. Bennis, "The Coming Death of Bureaucracy," *Think Magazine,* IBM Corporation, 1966, p. 30.

his peers and associates as with superiors and subordinates. Orders and directives usually do not cover such relationships, and although a formal bureaucracy does provide for the resolution of conflicts between peers, the tedium involved in such processes is so great as to preclude their use as anything but a last resort. Thus, informal relationships between people —both superior-subordinate and at the level of one's peers—are at the heart of modern organizations.

Modern organizations are also built upon a recognition of the need for *symbiosis*—the mutually beneficial living together of dissimilar organisms. This concept—the antithesis of functionalization—has been argued to be central to modern management. Tilles has defined the role of the general manager as one of presiding over these dissimilar systems as they seek to achieve symbiosis.[10]

Systems models of organizations recognize both the symbiotic concept and the importance of informal patterns of human behavior. They seek to describe the organization in terms of the *processes* and *flows* which contribute to the accomplishment of the organization's objectives.

Systems models provide a conceptual framework of an organization that is markedly different from our early concepts of an organization. In earlier times we would have limited our approach to the "components" of the organization, e.g., the people, equipment, etc., rather than considering *how the components relate to each other in the overall system.*

The concept of systems models also broadens one's perspective of the organization.

> [*It*] *goes beyond the traditional definition of organization. It views the enterprise as the central agency of an extended open system, encompassing a peripheral membership that interacts with, supports, and constrains the agency in its central membership. Failure to take this peripheral membership into account in modeling the significant pattern of energy exchange and transformation results in an oversimplification that renders the model useless for practical purposes. The systems concept is the only scheme that enables us to represent adequately the complexity of the interrelationships within a modern enterprise and establish or modify the understanding that must govern its performance.*[11]

GENERAL SYSTEMS MODELS A general systems model of an organiza-

[10] S. Tilles, "The Manager's Job: A Systems Approach," *Harvard Business Review,* January–February, 1963, p. 77.

[11] G. Gilmann, "The Manager and the Systems Concept," *Business Horizons,* August, 1969, p. 19.

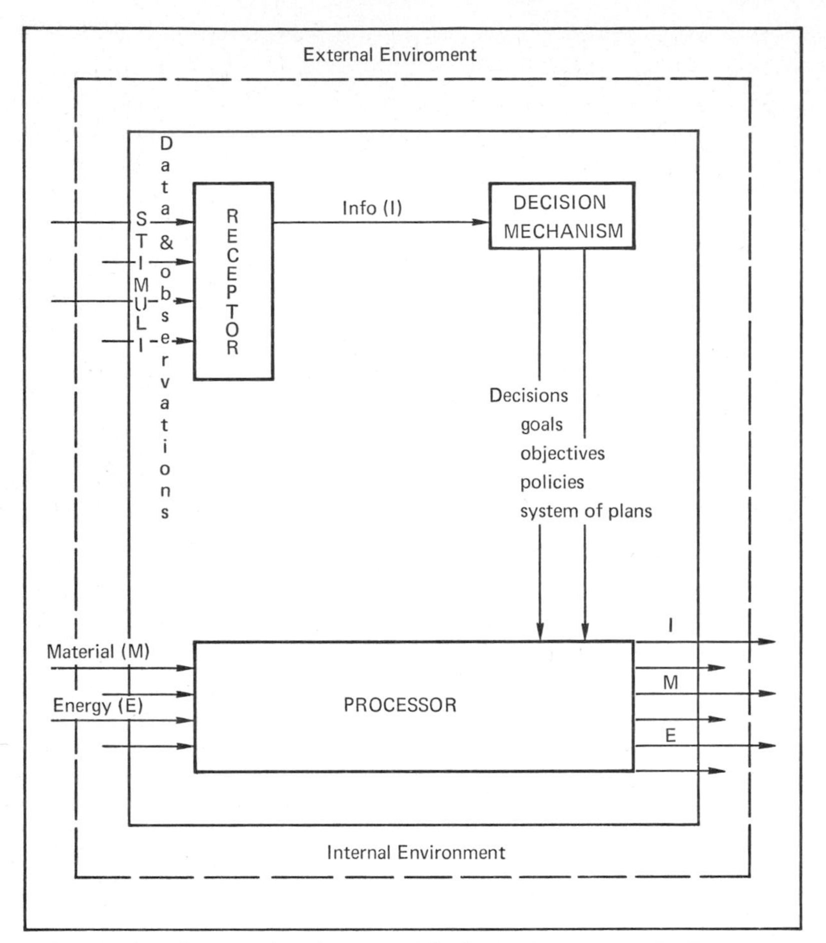

FIGURE 3-2 *A general systems model of a business organization.*

tion which is similar to ones presented by Johnson et al.,[12] Pask,[13] Katz and Kahn,[14] Young,[15] and others working in a variety of different fields of inquiry is shown as Figure 3-2. It emphasizes *flows* of information (I), material (M), and energy (E) through the organization, which is itself com-

[12] R. A. Johnson et al., *The Theory and Management of Systems,* McGraw-Hill Book Company, New York, 1967.

[13] G. Pask, "Comments on the Cybernetics of Ethical, Sociological, and Psychological Systems," in N. Weinger and J. P. Schade (eds.), *Progress in Cybernetics,* vol. 5, Elsevier Publishing Company, Amsterdam, The Netherlands, 1966.

[14] D. Katz and R. L. Kahn, *The Social Psychology of Organizations,* John Wiley & Sons, Inc., New York, 1966.

[15] S. Young, "Organization as a Total System," *California Management Review,* Spring, 1968, pp. 21–32.

posed of three entities—a *receptor,* a *decision mechanism,* and a *processor.* The receptor derives information from stimuli (data and observations) extracted from the internal and external environment. Information flows not only to points of decision, but also throughout the business organization in an elaborate network of communications. The decision mechanism uses information as the basis for deciding what should be done and for ordering the processor to do it.

Material flow is a primary concern of the processor activity. Material flow ranges from the acquisition of raw materials to the distribution of the finished product to the consumer. The decision mechanism plans the activities of the processor.

The energy flow consists of both human and nonhuman forms, described by Johnson et al., as:

> *Some source of energy is present in any operating system. It may be electricity obtained from available sources or generated by a firm-owned power plant. . . . Another obvious source of energy is people. Both physical and mental energy are required to operate business systems and people represent a renewable source, at least for the short run. People are quite variable as individuals. However,* in toto, *the group represents a reasonably stable source of energy for the system.*[16]

The various forms of energy are combined with materials and information by the processor as it executes the plans made by the decision mechanism.

The processor consists of all these management and operations functions relating to executing the decisions. Specifically, the processor includes not only the physical acquisition, conversion, and distribution activities, as in a manufacturing company, but also the management functions of organizing, controlling, and motivating. Also implied are all the informal, face-to-face, interpersonal relationships required in executing the decisions.

General systems models such as that of Figure 3-2 are useful for conceptualizing important facets of organizations which may be omitted from nonsystems approaches. For instance, they emphasize the significance of the decision-making–decision-execution feedback loop which operates via the environment. In addition they show that the systems view of organizations does not restrict itself to interrelationships of internal subsystems.

Of course, one can react to this general abstract systems model by saying that it is fine to pontificate that one should not restrict his

[16]Johnson et al., *op. cit.,* p. 343.

view to internal subsystems and that the environmental feedback loop should be considered. However, what are the practical implications of all this?

The answer to such a question is twofold. First, we can demonstrate that the practical consequences of a failure to take the systems view are great. Many of the inefficiencies that exist today in our society come about because of deficiencies in the ways in which organizations integrated their activities with those of others in the environment. For example, modern air transportation makes it possible to fly from one coast to another in a few hours so that one can become entangled in a traffic jam for several hours attempting to get to his final destination from the air terminal. Practically every air traveler has been delayed waiting for ticket processing, baggage handling, or other activities in air travel which involve queueing in one form or another. These delays have come about, not because of the inefficiency of a particular component of the total system, e.g., the aircraft flying from one point to another, *but because of the disparities among the components operating as a system.*

The second portion of the response to a question concerning the practical implications of systems models lies in the fact that alternative organizational forms, based on the systems approach, exist and are operational. In the next section, we give an introduction to such a systems application—project management.

THE PROJECT MANAGEMENT MODEL A "flows and processes" view, coupled with a recognition of the changes which are occurring in our society, leads one naturally to think about an objective-oriented transitory organizational setup. Such an approach would provide a way of organizing and applying resources to an ever-changing set of objectives and tasks. Bennis sees the evolution of such "temporary" organizations in the following fashion:

> *The social structure of organizations of the future will have some unique characteristics. The key word will be "temporary;" there will be adaptive rapidly changing temporary systems. These will be "task forces" organized around problems—to be solved by groups of relative strangers who represent a diverse set of professional skills. The groups will be arranged on an organic rather than program role expectations. The "executive" thus becomes a coordinator or "linking pin" between various task forces. He must be a man who can speak the diverse languages of research, with skills to relay information and to mediate between groups. People will be differentiated not*

vertically according to rank and status but flexibly and functionally according to skill and professional training.[17]

The *project organization* is illustrative of such a temporary organizational structure which is focused along systems lines. The Department of Defense has used project management effectively in pulling together the talents and resources required to accomplish a specific temporal objective such as the development of a weapons system. In the Manhattan Project's development and production of the first atomic bomb, the Department of Defense found that a *project* type of organization was necessary in order to provide a unity of purpose and to establish a focal point for pulling together the cooperative efforts of literally dozens of relatively autonomous organizations. Then in the 1950s, as the United States began the development of the ballistic missile system, the Department of Defense again turned to this form of organization and required contractors in the aerospace industry to do likewise.

Project management organizations take many different forms. In one form, an individual project is established as a distinct organizational element. Instead of one functional group, smaller functional counterparts are established within each project. Thus, rather than having one production organization for an entire plant, groups of production people are identified to support specific projects. Their functional units participate by providing members from the parent functional department to serve on the project teams. The focal point for all of the activity on a project going on within an organization is an individual designated as a "project manager." The sole task of this individual is to achieve the objective of the single project which is assigned to him.

A *project* may be defined as an ad hoc team of human and nonhuman resources pulled together in some authority and responsibility relationships to accomplish an end purpose. In today's complex society an organization typically faces a "stream of projects" that provide the basic work opportunity for the members of the organization. The management of these projects requires a different approach to the delegation of authority and the exacting of responsibility from what is traditional in management thought.

The application of the project idea brings about the emergence of two "complementary organizations" within a formal organization—the project organization and the traditional functional organization. The difference between these two organizations is illustrated by comparing Figure

[17]W. G. Bennis, "Organizational Revitalization," *California Management Review,* Fall, 1966, p. 59.

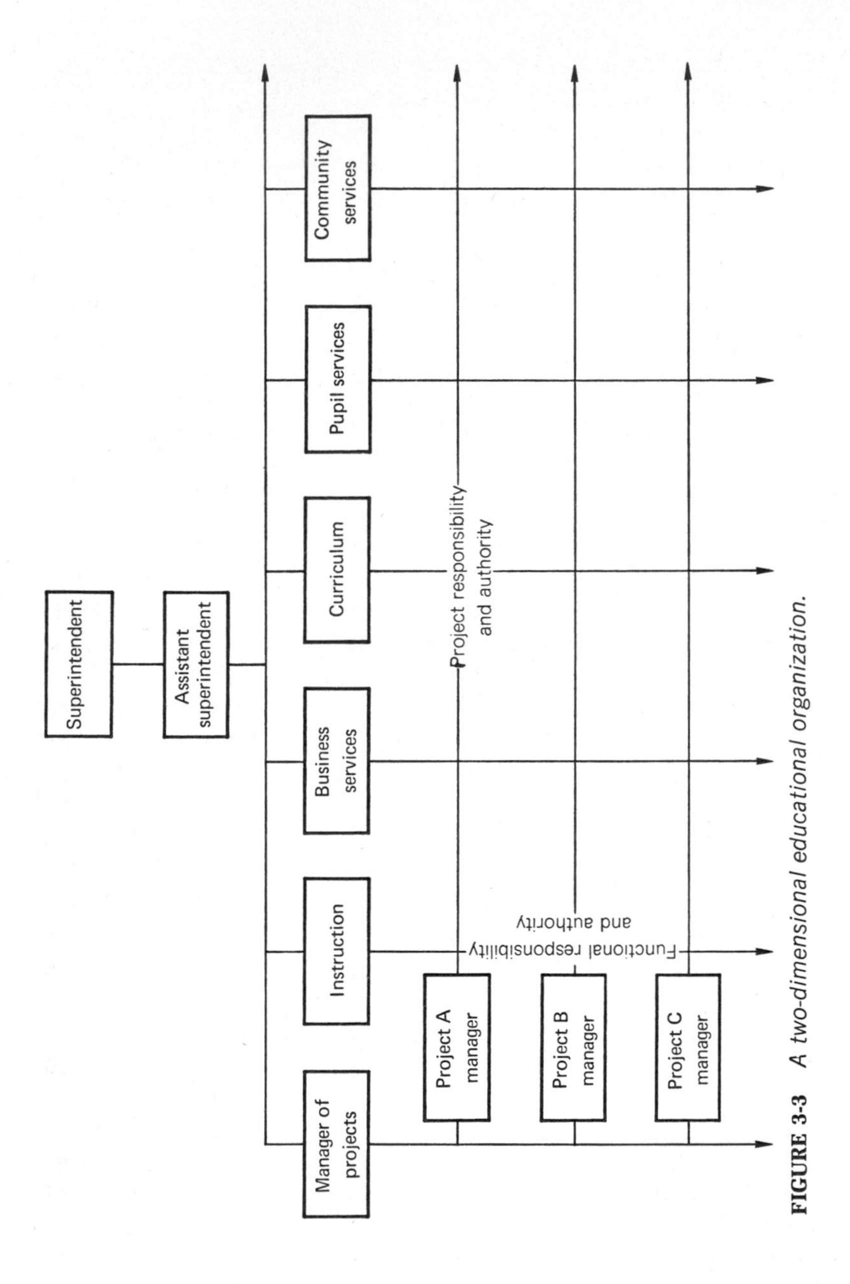

FIGURE 3-3 *A two-dimensional educational organization.*

3-1, which depicts a classical functional organization, with Figure 3-3, which shows an educational organization with these "two organizations" superimposed.

These figures (Figures 3-1 and 3-3) illustrate the essentially vertical nature of the traditional organization and the combined vertical-horizontal nature of the systems model. Figure 3-3 also illustrates how the effective accomplishment of project goals requires the crossing of traditional organizational lines. In this integration of the functional subsystems, and in the project goal viewpoint that focuses on the process of accomplishing the goal, the project management concept is indeed an operational systems model.

Evolving systems models such as this have already contributed greatly to management effectiveness; they are not simply textbook constructs. Evidence of this is seen in the *Fortune* comment, ". . . projectization has evolved in Apollo to the point of being a management revolution. It carries to its most elaborate development the task force concept now becoming the fashion in management doctrine."[18]

THE MANAGEMENT OF ORGANIZATIONS

An organization which is competing in an organized world must seek efficiency in order to survive. To be "organized" presumes that the achievement of efficiency is a *means* to the attainment of the basic objectives of the enterprise. To achieve efficiency requires that decisions be made and that actions be carried out. Moreover, it requires that control measures be taken to assure that decisions are implemented and that actions are properly executed.

Management is the process concerned with the selection of objectives and with the efficient achievement of those objectives. Hence, it involves two principal functions—*deciding* and *doing.* We shall later refer to these functions as the "planning" and "execution" phases of management. These functions are carried on respectively by the "decision mechanism" and "processor" elements of the general systems model in Figure 3-2.

The managers of any organization must decide what the organization's objectives are to be and what means are to be used to accomplish them. Thus, the owners and general manager of a baseball team must decide if they wish to win the championship at all cost, or to have the team make the greatest possible profit, or to seek some other objectives. The business manager must decide if the organization's goal is to be profit maximization, service to society, or something else. Then, both must decide on the actions which must be taken to achieve these objectives. For example, the baseball general manager may decide to buy new players for cash and the business manager may decide to introduce new products. In each instance, decisions regarding the objectives of the organization and their achievements are being made.

[18]"The Unexpected Payoff of Project Apollo," *Fortune,* July, 1969, pp. 114–156.

The "doing" function of management is a process of carrying out the actions indicated in the planning and decision phase. The manager who has decided to introduce new products must provide an environment within which new products can be developed, and the baseball general manager must go into the marketplace in pursuit of new personnel. Indeed, probably neither will actually perform all of the "doing" actions himself, so that both must instruct others and establish *control* procedures for ensuring that the instructions are carried out and that the chosen objectives are achieved.

In later chapters we shall deal with the two principal functions of the manager in greater detail. Here, two aspects of management deserve attention: the *objectives* which serve to distinguish the organization from other, non–goal-oriented systems and the role of those objectives in the *systems approach to management.*

Determining Organizational Objectives

The systems view of the organization necessitates the explicit consideration of the environmental system within which the organization exists. This becomes clear in a more penetrating fashion than nodding agreement and lip service when one recognizes the myriad of *claimants* to whom every organization is responsible. The business, for example, must be responsible to stockholders, customers, suppliers, unions, creditors, government, public organizations, etc., in order to survive.

To be responsible to these many claimants necessitates a multiplicity of organizational objectives. Stockholders might well be best served if the business were to concern itself only with attaining maximum profit. But what sort of profit do stockholders desire? Those interested in income through dividends would emphasize short-run profit, while those interested in the growth of the market value of their investment would be more interested in long-term profit. Thus, even the "simple" profit objective, which would appear to be paramount in the interest of one group of claimants —stockholders, is not so well defined at all.

Other claimants would be best served through other organizational objectives. Bondholders desire stability and safety, since their income from bonds is fixed. Customers might be best served if the business were to constantly seek maximum quality, the maintenance of excellent service, and low price. Unions desire the organization to emphasize the worker's "payoffs"—wages and working conditions. Of course, one could go on to an almost inexhaustible list of claimants and the objectives which they desire for the organization.

The most significant feature of such a listing of desired objectives would be their inevitably conflicting nature. One cannot simultaneously achieve maximum quality *and* minimum cost, or the highest possible profit *and* the best possible working conditions. *There is a conflict in such objectives* which is at the heart of the management process and the systems view of management.

We shall pursue this point further in the next section. Here, let us detour for a moment to consider the most important conflict of objectives facing the business organization today—that between responsibility to society and responsibility to stockholders and management, for it is on the resolution of this basic conflict that much of our society's future development may depend.

Both business and government managers have become increasingly aware of the interdependence of the two sectors in solving social problems. Business awareness of its major dual objectives—to shareholders and to society—has led naturally to an increased emphasis on sometimes neglected social objectives. Government, on the other hand, is led to look to business for help in solving social problems because of the specialized talent and resources which business can bring to bear.

Evidence of this trend is found in the study for the creation of a new agency in the Department of Housing and Urban Development of the United States government. The new office would attempt to promote private industry interest in social problems through "public interest partnerships—alliances between the Government and industry to solve problems in the public interest."[19]

An appreciation of the magnitude and potential of such an undertaking can be gained by realizing the extent of existing involvement of government scientists in experimental efforts enlisting industry aid. For example, government-industry groups have several projects in various stages of development:

1. *Housing: The HUD Department is asking industry to accept an expanding role in public housing. The agency's "Turnkey" program, which gives industry more initiative in constructing public housing and seeks private management of completed projects, now also contemplates an industry role in preparing tenants to own their own dwellings eventually. . . .*
2. *Neighborhoods: The Office of Economic Opportunity is asking industry to take on the problems not just of the jobless or the ill-housed, but of whole neighborhoods. Despite the national concern over slums, no one has ever "systematically analyzed the problems of an entire slum neighborhood. . . . Westinghouse*

[19] *Wall Street Journal,* Dec. 15, 1967, p. 1.

has received a special OEO grant to develop a comprehensive program to attack all the problems of a slum area in Baltimore. . . ."

3. *Public administration: The OEO may ask industry to take on a task that is astonishing in its implication. In California, a nonprofit corporation called Opportunity Industry Corporation has been formed to help cure the problems of the poor, partly with OEO money, by substituting private contractors for public agencies. . . .*[20]

Thus, the business is being considered as a societal subsystem (1) to produce a good or service and (2) to function as part of the overall economic and social system. Which role is most important? This is obviously not a question which can be simply answered. Yet, the business manager must constantly seek the answer to this question as he sets objectives and operates the firm.

The Systems Approach to Managing Organizations[21]

One of the conceptual foundations of modern management—the systems approach—provides a basis for viewing the manager's job in a new light. While systems ideas are neither complex nor new, their recognition and application in recent times has led to revolutionary changes in the way in which organizations are managed. Both the "deciding" and the "doing" phases of management have felt this impact.

A SYSTEMS ILLUSTRATION To illustrate the systems approach, we may consider the antithesis of the process of specialized knowledge through which problems are often handled in industry. Consider the mechanical engineer who is called on to study a piece of production machinery which operates in a production line. The inventive engineer may find that some simple changes in a machine can lead to increased production speeds. However, he cannot stop at that point. He must consider the *man-machine system* made up of the machine and its operator. If the operator cannot handle the increased potential of the machine, there may be no sense in making mechanical changes to permit increased output which cannot be utilized. Moreover, even if the operator can physically perform as is required, the psychological impact which increased speeds may have on him may be bad.

[20] *Ibid.*

[21] Parts of this section are adapted from William R. King, "The Systems Concept in Management," *Journal of Industrial Engineering,* May, 1967.

Even if these problems can be overcome, increased output at this machine may lead to in-process inventories building up between various work stations. For example, if the man-machine subsystem labeled "A" in the diagram of Figure 3-4 is the one whose output can be increased, and no change is made in subsystem B, an accumulation of partially completed items will build up between A and B. To obviate this requires the study of the overall production system encompassing man-machine systems A and B, which are *subsystems* of the production system. Perhaps another machine of the variety of B can be purchased and incorporated into subsystem B to handle the increased productivity of A, for example.

But then, what is to be done with the greater production quantities? Can they be sold? What will be the effect of this increased supply on the market price? Does the sales department have the staff to sell larger quantities? Are the financial resources available to purchase another machine B and to permit operating the enterprise at a higher output level?

THE SYSTEMS APPROACH Such an elementary series of successive generalizations and the asking of questions concerning the implications to other elements of an overall system are a part of the *systems approach. A system, by its very nature, is made up of interdependent elements. As such, actions which affect one element must affect others also. And actions of one element cause reactions on the part of others. The recognition of such interactions and interdependencies both within and without the organization is the essence of the systems viewpoint.*

Just as the systems viewpoint necessitates the consideration of the conflicting objectives of the claimants to the organization, so, too, is a recognition of conflict *within* the organization of central significance to systems ideas. Let us consider the idealized case of a firm whose leaders have already succeeded in resolving the claimants' conflicts and selected specific objectives for the company. The systems view at this stage re-

FIGURE 3-4 *A simple production system.*

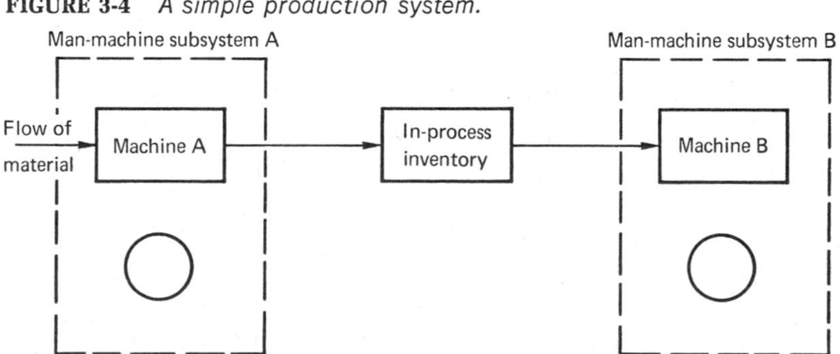

quires a recognition of the manager's desire to achieve these corporate objectives—to achieve maximum *overall* effectiveness. He wishes to avoid having the parochial interests of one organizational element or activity distort overall performance. And he must do this in an organizational environment which invariably involves *conflicting departmental organizational objectives and a scarcity of resources.*

To demonstrate this, consider the corporate viewpoint involved in the "simple" decision involving which products are to be produced and in what quantities. The production department of the enterprise would undoubtedly prefer that few products be produced in rather large quantities so that the number of costly machine setups which are necessary to convert from production of one product to production of another are minimized. Such a policy would lead to large inventories of a few products. Sales personnel, on the other hand, desire to have many different products in inventory so that they may promise early delivery on any product. Financial managers recognize that large inventories tie up money which could be invested elsewhere—hence, they want low total inventories. The personnel manager desires constant production levels so that he will not be constantly hiring new workers for short periods of peak production and laying them off in slack periods. One could go on to identify *objectives of almost every functional unit within an organization* relative to this simple tactical decision problem. As demonstrated, each of these objectives conflicts to some greater or lesser degree with some other objective —low inventory levels versus high inventories, many products versus few products, etc.

The same variety of situation can exist at every other level of the enterprise. The production department must constantly balance the speed of production with the proportion of rejects and the proportion of defective products which are not detected. The marketing function becomes involved when defective products cause complaints and lost sales. Indeed, wherever the "labor" has been divided in an organization, the management task of effectively integrating the various elements is paramount, and this can be effectively accomplished only by the manager adopting the systems approach to the "system" which is his domain.

The systems-concept viewpoint in this context is the simple recognition that any organization is a system which is made up of segments, each of which has its own parochial goals. Recognizing this, one can set out to achieve the overall goals of the organization only by viewing the entire system and then seeking to understand and measure the interrelationships and to integrate them in a fashion which enables the organization to efficiently pursue *its* goals.

Of course, this means that some functional unit within an organization may not achieve its parochial objectives—because what is best for the whole is not necessarily best for each component of the system. Thus, when a wide variety of products are produced in relatively small quantities, the apparent performance of the production department may suffer. Yet, if this leads to greater total revenues because no sales are lost, the overall impact may be positive.

One of the essential elements of the systems approach is an expanded perception that enables us to deal with complex situations. Using more traditional methods, the manager was forced to deal individually with the parts of the organization rather than with the whole. For instance, in a school system, the faculty was viewed as an entity; students were often looked upon as existing somewhat separately from the rest of the organization and were often treated as such, with little consideration as to how they fit into the larger whole. Alumni and administrators were often handled in much the same fashion without any recognition of the environmental system within which the school organization existed or of how the components interrelated.

A "COMMON SENSE" VIEW OF THE SYSTEMS APPROACH The systems approach has led to startling changes in the way in which some organizations are run. However great are the changes which it has produced, nevertheless, it is important to recognize that the concept is a simple and straightforward one. It really involves only the development of a new viewpoint on the part of the manager. Drucker agrees with this "point of view" philosophy of the systems approach in saying that ". . . the systems approach, which sees a host of formally unrelated activities and processes as all parts of a larger, integrated whole, is not something technological in itself. It is, rather, a way of looking at the world and at ourselves."[22]

Fricks emphasizes the "common sense" nature of the systems approach in saying that it is ". . . very little more than the application of common sense. For after all, as an idea or concept, it merely requires that the total problem be identified and attacked in a systematic manner —unquestionably a desirable approach to any problem."[23]

Paradoxically, while the systems approach has had great impact and has been paid even greater lip service in recent years, it has probably been too little applied. Conversely, many successful managers and entrepreneurs of the past recognized and used systems concepts, even though they did not call them that. For example, Robert Owen, a successful textile

[22] Peter F. Drucker, "Technological Trends in the Twentieth Century," in C. W. Persell, Jr., and M. Kranzberg (eds.), *Technology in Western Civilization*, vol. 2, chap. 2, Oxford University Press, New York, 1967.

[23] *Structural Problems in Organization Theory*, Systems Research Center, Case Western Reserve University, Cleveland, 1967, p. 2.

manufacturer in Scotland, made the following comments to a group of superintendents in 1813: ". . . from the commencement of my management, I have viewed the population with a mechanism and every part of the establishment as a system composed of many parts, and which it was my duty and interest to combine, so that every hand as well as every spring, lever, and wheel should effectively cooperate to produce the greatest pecuniary gain to the proprietors."[24]

SUMMARY

Each of us is associated with many different organizations at any given moment. The roles which we play in these organizations are varied and often interdependent, i.e., the role which we play in one organization may depend on that which we play in another.

An organization can be thought of as an agreement among people to cooperate in some endeavor. Its essential elements are objectives, patterns of authority and responsibility, and decisions. Since these are elements common to management, the two concepts are obviously closely interlinked.

Organizations exist because of the need for effectiveness and the search for efficiency. Thus, humans organize to accomplish some goal collectively that none can achieve alone. Because of the scarcity of resources, they seek to do this efficiently. Hence, the basic organization is created to do something, and then it is refined so that it can perform well, that is, meet the objective at low resource cost.

Organizations are purposeful systems—ones that are seeking goals. In modern society, the various goals sought by different organizations are interdependent, thus ensuring that the organizations are themselves interdependent.

Traditionally, organizations have been thought of in the terms of their *functional structure* and the *bureaucratic concept.* Functionalism in organization is based on division of labor and specialization—the decomposition of tasks into subtasks which can be delegated to work groups, each of which possesses a narrow expertise. Units such as the production, marketing, personnel, financial, and engineering departments illustrate this specialization at the first level in an organization.

The bureaucratic concept of the organization is another basis of traditional thinking about organizations. This model emphasizes superior-subordinate relationships and the chain-of-command.

[24] R. Owen, "An Address to the Superintendents of Manufacturies," in H. F. Merrill (ed.), *Classics in Management,* The Macmillan Company, New York, 1960.

More modern organizational models—*systems models*—depict the organization in terms of horizontal processes and flows rather than vertical hierarchy relationships. They focus on the interactions and interdependencies of the organizational components rather than the components themselves. This permits the analysis of large complex systems in terms of such things as the possible disparities occurring among the components operating as a system.

One illustration of the practical application of the systems model of organizations is the use of the *project* organization. Project organizations are temporary organizational structures established to pursue specific objectives. These objectives usually require activity of a finite duration and draw on the resources of a variety of functional departments. The project forms the focal point around which activity can be organized to seek the project objective.

An organization typically faces a stream of projects. Hence, it may have a myriad of temporary project organizations, each concerned with specific subobjectives of the overall organization, active at any given time.

The similarity of the operational definitions of "organizations" and "management" leads one naturally to consider the process of *managing organizations*—making and executing the variety of decisions which guide the organization toward its objectives. The manager's principal functions—*deciding and doing*—or the *planning and execution* phases of management, run the gamut of activity from the time at which general goals are established until they have been achieved. Of course, the process of management actually requires a continual reestablishment of objectives, redefinition of the ways in which the organization will seek the objectives, and reevaluation of how well it is doing. So literally, the manager's job is never done, since goals change as he seeks to achieve previously established ones and the alternatives available to him are in constant state of flux.

In determining overall organizational objectives, the manager must take account of the many *claimants* to whom the organization owes responsibility. Generally, these claimants have *conflicting objectives* in that the achievement of one precludes the achievement of others. For instance, business owes its stockholders a return; yet it also owes society an obligation. Often the two are in conflict; yet the conflicting objectives of these two general claimants must be resolved by the manager.

To do this, he must adopt the *system approach to management.* In the system approach, the manager takes as large a view of the system as is feasible, concentrates on assessing and controlling the interactions and interdependencies among the subsystems, and acts in the interest of overall systems effectiveness. To accomplish this, he must combine thoughtful analysis and common sense with a thorough understanding of the organization and its environment.

In later chapters we shall elaborate on the practical problems involved in doing this. The recommended approach will indeed revolve about systems ideals. Yet we should emphasize at this point that systems ideas do not provide a solution to the manager's problems. What they do provide is a *way of thinking about problems*—a model in terms of which these problems may be better perceived and better resolved.

EXERCISES

1. An organization may be simply defined as an agreement among people to work together in some cooperative effort. Show by example how organizations that you are familiar with meet this definition.
2. Organizations in our society have several common characteristics. Identify these characteristics and then show by example that you understand such characteristics.
3. In the Shakespearean play, *As You Like It,* the statement is made: "All the world's a stage, and all the men and women merely players. They have their exits and their entrances; and one man in his time plays many parts." How might this quote from Shakespeare be related to the organizational life of an individual?
4. It is impossible to separate an individual from his unit of organization and the greater environmental system in which that organization is found. Defend or refute this statement. Show some examples. How does this tend to confirm or deny the validity of the system approach?
5. What problems has the increasing division of labor in our society led to in contemporary organizations?
6. An organization can be described as a purposeful system. Explain.
7. Draw an organizational chart of your family unit.
8. There is some disagreement among management theorists as to the value of organization charts. What do you think?
9. Identify the characteristics of the bureaucratic form of organization.
10. Bureaucratic organization keynotes functionalism. What is meant by the term?
11. Contemporary management theory has been heavily influenced by military and church organizations. How would you identify this influence in today's management theory?
12. A modern systems model is not oriented toward the functional structure of an organization. Rather, greater emphasis is placed on the process and flows which contribute to the accomplishment of the organizational objectives. What are the processes and flows?
13. In determining objectives for a business organization, prime consideration must be given to the personal objectives of the stockholders, that is, the individuals who have provided the venture capital for the organization. Is this so? Why or why not?

14. The authors make the point that there is a conflict in organizational objectives which is at the heart of the management process and the systems view of management. Explain.
15. A system by its very nature is made of interdependent elements. Take a public school system as an example and demonstrate how you can identify these elements and how these elements are interdependent.
16. It has been said that the general manager constantly works in an environment in which there is conflict and the need for the maintaining of overall effectiveness of the organization. Show by example that you understand this concept.
17. How would you differentiate between the responsibilities of a functional manager and a project manager in an organization?
18. Relate "scientific management" to the "systems view of management." How are they similiar and dissimilar?
19. What is an "informal organization"? How does it relate to a formal one?
20. Discuss the general responsibility of the operations, marketing, and financial departments of a (*a*) business firm, (*b*) school district, (c) police department, and (*d*) hospital.
21. Discuss the general systems made of Figure 3-2 in *(a)* "black box" terms *(b)* input-output terms, *(c)* feedback terms.
22. What unique problems do you think might arise in an organization such as that described in Figure 3-3?
23. Who are the claimants to the following organizations: *(a)* hospital, *(b)* police department, *(c)* military unit, *(d)* fireman's drum and bugle corps?
24. Suppose a second man-machine subsystem B was being prepared to be added to the production system in Figure 3-4. What considerations would come into play in the system view?
25. Why *must* departmental objectives *necessarily* be in conflict?

REFERENCES

Carlisle, Howard M., "Are Functional Organizations Becoming Obsolete?" *Management Review,* January, 1969.

Cleland, David I., and David C. Dellinger, "Changing Patterns in Management Theory," *Defense Industry Bulletin,* January, 1966.

Ehrle, Raymond A., "Implications of a Systems Approach to Organization and Management," *Personnel Journal,* 44:2, February, 1965.

Gil, Peter P., "The Fundamental Organization Unit . . . The Superior/Subordinate Relationship," *Defense Management Journal,* Winter, 1967–68.

Gilmann, Glen, "The Manager and the Systems Concept," *Business Horizons,* August, 1969.

Greenwood, William T. (ed.), *Management and Organizational Behavior Series: An Interdisciplinary Approach,* South-Western Publishing Company, Incorporated, Cincinnati, 1965.

Kast, Freemont E., and James E. Rosenzweig, *Organization and Management—A Systems Approach,* McGraw-Hill Book Company, New York, 1970.

Koontz, Harold, and Cyril O'Donnell, *Principles of Management,* McGraw-Hill Book Company, New York, 1968.

Litterer, Joseph A., "Program Management: Organizing for Stability and Flexibility," *Personnel American Management Association,* September–October, 1963.

March, James G., and Herbert A. Simon, *Organizations,* John Wiley & Sons, Inc., New York, 1958.

Mee, John F., "Matrix Organizations," *Business Horizons,* Summer, 1964.

Reed, William H., "The Decline of the Hierarchy in Industrial Organization," *Business Horizons,* Fall, 1965.

Suojanen, Waino W., *The Dynamics of Management,* Holt, Rinehart and Winston, Inc., New York, 1966.

Walker, Arthur H., and J. W. Lorsch, "Organizational Choice: Product Versus Function," *Harvard Business Review,* November–December, 1968.

4

the organiztional environment

Every organization must exist in some context; no organization is an island unto itself. Each must continually interact with other organizations and individuals—the consumers of its products or services, its suppliers, unions, stockholders, governmental bodies, and many more. More significantly, however, each organization has goals and responsibilities related to others in its environment. Thus, not only must an organization deal with its environment in conducting its everyday affairs, but it must also give consideration to the goals of others as it establishes its own goals and conducts its operations.

The day is long past when a business firm or other organization had sufficient economic or political power to allow it to ignore the influences of its actions—both direct and indirect, intended and unintended—on others. A single organization can itself wield tremendous and far-reaching influence, but it is itself subject to the influence of diverse elements of its environment. Sometimes the manifestations of this condition are surprising—for instance, the influence of a single "crusader" (Ralph Nader) led to drastic changes in auto safety features during the late 1960s, while a relatively small group of civil rights protestors literally started the chain reaction which permanently altered the course of American social development earlier in that decade. In both instances, a small, vocal element of the environment had profound effects on powerful organizations—in the one case, the giant auto corporations and their product design, and in the other case, the personnel practices of virtually every American corporation.

THE ENVIRONMENTAL SYSTEM

The outer portion of the systems model in Figure 3-2 is labeled "environment." Information, energy, and material are shown to be "input-ed" to the organization from the environment and "output-ed" again to the environment. While this may seem apparent to the least perceptive observer of an organization, it is a fair judgment that most organizations know less about the environment in which they operate than they know about virtually any other aspect of their activities. Many organizations collect little information about their environment in either a formal or an informal way, and many organizations are inept at using the information which they do possess.

To begin a discussion of the environmental system of an organization, we abstract from various authors a listing of a number of "environmental subsystems."[1] This listing is obviously not exhaustive, nor are its elements disjoint. However, it provides a basic feeling for some important aspects of the environmental system.

Institutional	Physical	Technological
Cultural	Competitive	Sociological
Economic	Political	Educational
Ideological	Psychological	
Legal-Political		
Theological		

A Problem-oriented Approach

Perhaps the best way to understand the importance of environmental factors to an organization at a practical level is to deal with a specific situation involving that organization which has received the most attention regarding its handling of environmental problems—the business firm.

Let us consider the classic situation of the Ford Motor Company at Willow Run, Michigan, during World War II. The company began in April, 1941, to build the biggest mass production bomber plant in the world.[2] Within a year Ford had converted five or six square miles of woods and farmland into a great airport and had built a "three-quarter-mile factory." Ford recruited more than 42,000 workers, mostly people from the South. During the 43 months of actual production, the plant turned out 8,685 B-24 bombers. Inside the plant, production technology slashed man-hour bomber costs to less than one-tenth of the original cost. But technology

[1] S. Young, *Management: A System Analysis,* Scott, Foresman and Company, Glenview, Ill., 1966. H. Koontz and C. O'Donnell, *Principles of Management,* 4th ed., McGraw-Hill Book Company, New York, 1968. R. A. Johnson, et al., *The Theory and Management of Systems,* McGraw-Hill Book Company, New York, 1967. J. S. Bain, *Industrial Organization,* John Wiley & Sons, Inc., New York, 1968.

[2] The ensuing discussion of the Willow Run experience of the Ford Motor Company is adapted from L. J. Carr and J. E. Stermer, *Willow Run,* Harper & Brothers, New York, 1952.

stopped at the gate; nontechnical problems began on December 1, 1941, when Willow Run began to hire production workers.

> [*Willow Run was*] . . . *built out in the open country three miles from a little college-industrial town, Ypsilanti, in south-eastern Michigan and twenty-seven miles west of its natural labor pool, Detroit. . . . It began—believe it or not—with no provision whatever by anybody—by management or by state or by Federal governments —for living accommodations or community services for anybody anywhere.*[3]

War pressures soon made the "anywhere" the farming communities of Wastenaw County surrounding Willow Run.

> *The idea of 100,000 bomber workers driving the equivalent each year of 10 round trips to the sun on irreplaceable tires suddenly turned a seasick green. Somehow, somewhere, Willow Run was going to have to house thousands of bomber workers near the factory—and fast. So nobody with power and authority did anything about it for another full year.*[4]

Thus there existed unbelievably poor housing conditions and services for thousands of "newcomers" who were resented by the original residents of the community. For Ford, its responsibility for the workers stopped at the gate; the State and County resented the strain and overflow on all their facilities and denied any responsibility for the "migrants." The local businessmen and builders had their own parochial interests to protect and did so vigorously. Everybody (except the bomber workers) seems to have rejected the idea of federal government intervention, and since there was no overall pattern for cooperation, the result was a "social fiasco."

The effects, however, were not just social. Consider the problems that must have been caused by the following turnover and absentee rates.

> *During an active production life of 43 months, therefore, Willow Run used 4.16 workers for every job in the plant. . . . The number of daily stay-at-homes at the Ford bomber plant averaged as high as 17 per cent of the total working force. . . . It was the utter disregard by the Ford Motor Company and by the Federal Government of the necessary relations which exist in America*

[3] *Ibid.*

[4] *Ibid.*

between a factory and its serving community that sent absenteeism (and turnover) at Willow Run through the roof.[5]

The point in this illustration is, of course, that since the business firm is one of the powers in its community, its managers must make decisions concerning *both* the internal business system *and* the environmental system. The business firm cannot isolate itself from its environment, if only because of its awareness that individuals bring into their work situation the attitude and values which are derived from environmental subsystems such as the cultural, sociological, and theological. Likewise, their attitudes which are derived from their role in the company become a part of the environmental subsystems. If the business firm chooses to believe that it can exist in isolation, it has chosen one of the alternatives from those which are available to it—"noninvolvement." The selection of this alternative will have an effect in the same way as would the choice of a more "active" alternative.

Of course, this Ford situation existed in the distant 1940s. Today, companies are indeed generally more aware of environmental constraints. For instance, in a movement by firms away from the high prices and pressures of Manhattan office locations in the early 1970s, many firms took great pains to soften the impact of their arrival in a suburban setting. They did everything from designing buildings and preserving the natural atmosphere of their suburban office location to ensure that local property values would not be adversely affected to constructing their own sources of water and sewage treatment facilities to avoid overburdening local systems.[6]

Also illustrative of the concern of a large organization with the environment in which it must operate is the special-payment program instituted by the federal government to compensate local school districts for the additional burdens of educating large numbers of the children of government workers who may be assigned to a federal installation in the district. Such programs grew directly or indirectly from World War II experiences such as Willow Run.

Statement of the Organization's "Environmental Problem"

With these experiences in mind, one can begin to perceive the environmental problems faced by organizations. However, one of the difficulties in the environmental area is that an appreciation of some of the problems does not necessarily lead to precise statements of the problems or to solutions for them. Indeed, one of the great difficulties is that the environ-

[5] *Ibid.*

[6] "Offices in the Suburbs," *Time,* Feb. 8, 1971, pp. 53–54.

ment, by its very nature, is not precisely defined in terms of its constituent elements. Therefore, problems can continue to arise from previously unconsidered sources—a vexing situation for anyone who wishes to have both neat problems and neat answers.

Let us continue in the problem-oriented vein of the business firm to ask: If a business firm moves into a community which has poor schools, inadequate roads and services, insufficient housing, and a lethargic community spirit, what responsibility or obligation should the company take on? Certainly, there is no legal obligation to participate in these affairs of the community; but there are moral and practical obligations for it to participate in them. Going further, there are economic reasons for participating in the community affairs because the company and the community are inextricably interwoven—the fortunes of one will affect the fortunes of another. For example:

1. Inadequate housing or educational facilities may discourage potential employees from moving into the area where the company is located.
2. Inadequate utilities (fire protection, streets, highways, and police protection) may threaten the safety and efficiency of the company's resources.
3. Disruptions may occur on or near company property if the people of the local community do not accept the company's presence.
4. High turnover or absenteeism may occur if employees are not effectively integrated into the community.

Thus, one can take the viewpoint that environmental considerations are in the organization's enlightened self-interest. It is not necessary for an organization to take a "do-gooder" attitude for it to become involved with its environment. In many cases, it will find that its survival may depend on its doing so.

The "environmental problem" which is therefore posed for the manager is twofold:

1. How can the manager develop an appreciation of the need to assess environmental forces and a perception of the environmental forces which will significantly affect his organization?
2. How can the manager effectively integrate these forces into the organization's goals, objectives, and processes?

ENVIRONMENTAL FORCES

The first portion of the statement of the manager's environmental problem relates to the development of better appreciation and perceptions regarding environmental forces. To understand current environmental forces in turn requires an appreciation of some fundamental environmental concepts which underlie the activities and operations of every modern organization.

Fundamental Environmental Forces and Concepts

Few of the fundamental concepts of our environment are unknown to many of us. We travelers on this spaceship called "Earth" who are housed in the first-class section called "the United States" have all been schooled in the fundamentals of our political and social system. Indeed, on venturing into this section of the book, the reader may react by saying that the subject matter has become redundant with many other of his learning experiences.

However, the primary purpose in briefly noting a number of well-known environmental concepts is not to make the student or manager aware of them; rather, *often we are so familiar with them that we do not perceive how they affect our daily lives and the day-to-day problems which we face.* In the hope of developing the better appreciation and perception noted as desirous in the first portion of the environmental problem statement, we shall briefly discuss democracy, the political system, the economic system, and the various forms of business ownership. Subsequently, we shall deal with the role these can play both in creating the daily problems which managers must solve and in creating opportunities for greater managerial success.

DEMOCRACY The most basic element of our day-to-day existence—one that is so basic and taken for granted that we tend to think of it only on election day or when it is infrequently threatened in some small way—is our democratic form of government and the democratic concept.

The basis of democracy—the right of self-determination—is found in the ideas of Greek, Roman, and Christian philosophers. The Greeks practiced a pure form of democracy in the fifth and fourth centuries before Christ. They were probably able to practice pure democracy only because they lived in city-states, which were small geographical divisions. Today, even though our complex society precludes the practice of pure democracy, we consider the American way of life to be a democratic form.

The words of the Declaration of Independence established certain "self-evident truths." These include the equality of man and the inalienable rights such as life, liberty, and the pursuit of happiness. Securing these rights requires the establishment of governmental bodies which derive their powers from the consent of the governed.

The Constitution of the United States and its amendments establish limitations on the grants of power extended to the governing body. The Constitution establishes the legal framework from which self-respect is secured for the citizenry. Self-respect means the condition and result of man's self-realization—his self-direction and control of a free choice in doing those things in life which he thinks to be best. *At the same time "democracy" means respect for other men and their rights.* Thus, "democracy" means much more than just a way of governing a society; *it is an idea about how people should behave toward each other in an organized society.*

A democratic political system is built on a few guiding beliefs or assumptions concerning the nature of the citizen, society, and the political system. According to Irish and Prothro, "The democratic view of the nature of man is based on the principles of *equality* and *humanitarianism;* the democratic view of the nature of society is based on the principles of *individualism* and *progress.* The democratic view of the nature of politics is based on the principles of *majority rule* and *freedom of dissent.* Democratic theory leans on these beliefs—equality, humanitarianism, individualism, progress, majority rule, and freedom of dissent—and from them gains its enduring appeal."[7]

THE POLITICAL SYSTEM[8] The basic principle of our political system is the right of the people to govern themselves through chosen representatives. The people, therefore, have final authority through their right to choose senators, representatives, and the president and vice president of the United States. This right of choosing the members of government extends down through state and local governmental levels. All governments exist, according to a theory which has been subjected to stringent test in the United States only in recent years, through their ability to satisfy the majority of people in their needs and desires.

Some basic concepts which serve to define our political system are:

1. The federal system
2. Delegated powers
3. Divided authority
4. Independent legal system
5. Freedom and equality

[7] Marian P. Irish and James W. Prothro, "The Context of Ideas," in *The Politics of American Democracy,* 2d ed., Prentice-Hall, Inc., Englewood Cliffs, N.J., 1962, p. 45.

[8] The discussion of the political system is based in part on John C. Harvey, "Basic Principles of our Government," in *Federal Textbook on Citizenship: Our Constitution and Government,* U.S. Government Printing Office, Washington, 1955, pp. 172–179.

The *federal system* refers to the basic split between federal powers (such as to declare war and to regulate interstate commerce) and state powers (such as to collect taxes for state purposes). The Constitution provides a basic definition of federal and state powers, and if any change is desired in this relationship, the Constitution of the United States must be amended.

Our federal government has only those *powers delegated* to it by the citizens through the Constitution. All of the powers not delegated to the federal government or kept by the people are kept by the states. Theoretically, every act of the Congress and the executive branch has a foundation in the Constitution.

Authority is the legal and rightful power to act. At the national level of government, *authority is delegated to several branches.* Congress makes the laws; the executive branch sees that they are enforced; and the judicial branch interprets the meaning of the Constitution, laws, and treaties of the United States when questions concerning them arise in legal cases. This concept of divided authority is carried down through state governments. The division of powers of the national government provides for a system of "checks and balances" in government. Effectively, each branch is given some control over the other two; for instance, even the two branches of Congress have checks on one another. A bill must be passed by both houses before it can become a law. Even then, the President can veto a bill that has passed both houses of Congress, but in turn, Congress can pass it over the President's veto by a two-thirds vote in each house. The President appoints officers of the executive branch of the national government, but many of these appointments must be approved by the Senate. The President may make treaties, but they must be approved by a two-thirds majority of the Senate.

The *independent legal system* exists to ensure equal justice under the law. Judges of the federal court system are appointed to serve during good behavior, and their salary cannot be reduced while they are in office. Thus, the judge is free from influence from those who make or enforce the laws, and he can do his duty without fear of congressional or executive reaction. In the administration of justice, the Constitution, treaties, and the laws of the United States are the highest law. The officers of the government are expected to obey the Constitution and the laws of the nation. No officer can exercise authority except as the law gives him the legal right to act. Our system is, therefore, a government of laws—not a government of man.

Freedom and equality means the right to work, play, think, talk, and pursue one's destiny in life. Under our system, government has the duty of protecting these rights as they are set forth in the Bill of Rights of the Constitution. These rights have special implications with respect to our basic methods of speaking, writing, printing, and practicing our beliefs. The government cannot take away one's life, liberty, or property except in a manner prescribed by law. In a very real sense, our freedom and equality are guaranteed by convenants made between the people of our society.

Of course, the political system is designed to be flexible. As our society evolves and as changes in the basic structure of the system become necessary, our Constitution can be amended, and the laws can be interpreted by courts differently from the way they have been in the past. Thus, the system is theoretically able to meet the changing needs of our society. Whether it is sufficiently responsive to meet the fast-evolving needs of the modern world has been the subject of much debate in recent years. But we shall defer discussion of this problem until later.

THE ECONOMIC SYSTEM The general characteristics of the American economic system have been well summarized by Oxenfeldt:

1. *There is private ownership of the means of production, that is, individuals and not the government own most of the nation's factories, machinery, ships, railroads, natural-resource deposits and the like.*
2. *In their economic activities individuals strive primarily to obtain maximum money income. People ordinarily use their productive equipment, their money savings, and their labor skills in ways that give them the highest money return.*
3. *Most individuals are compelled to take outside employment in order to provide for even their minimum needs. Only a small proportion of the population finds support by working with self-owned means of production. The majority can earn a livelihood only by taking employment with persons who own means of production.*
4. *The basic economic "decisions" about what shall be produced, in what quantities, by what methods, and for whom are made through a "price system." There is no central group of planners that decides what goods shall be produced and specifies techniques of production and the like. These decisions are decentralized.*
5. *Production and sale of most goods and personal services occur under conditions of rivalry among buyers and sellers, each of whom is pursuing his own self-interest.*
6. *Individuals are compensated in accordance with their productivity, their rewards depending primarily upon their ability to*

produce the kinds of things that other people desire, their own diligence and the scarcity of the skills they possess.

7. *Compared with most other industrialized countries, the United States Government does not intervene to a large extent in business activity: Businessmen are relatively free to pursue any course of action they wish.*
8. *Industrial plants and equipment are available in very large quantities in relation to the size of the labor force. Production in the United States is very highly mechanized, partly because money capital is available to borrowers on relatively attractive terms.*[9]

BUSINESS OWNERSHIP Ownership of those organizations which are formed to provide goods and services in our economic system takes a variety of forms, depending on such factors as:

1. The resources available to the organization
2. Legal requirements of the states in which the organization is formed and will conduct business
3. The significance of such factors as permanency, transferability of ownership, and the control of the organization
4. The effect that the organizational form will have on other organizations with which the firm must interact

Thus, the forms of business organizations vary from free informal associations of persons to the legally defined entity called a "corporation," which must be created under the auspices of a government.

The *individual proprietorship* is the most commonly used form of business organization; it is essentially a one-person business. The individual proprietorship—often called the *sole proprietorship* or simply *proprietorship*—is that which results if an individual begins a business alone and takes no specific action to form another type of organization. In effect, there is no distinction made between the individual and his business. The shortcomings of this variety of business are the unlimited liabilities of the proprietor, the lack of permanency in that the business terminates if the proprietor dies, and the lack of legal separation between the owner and the business.

A *partnership* is an association of two or more persons who operate as co-owners of the business. A partnership comes into being through the execution of a verbal or written contract between the partners. The

[9] From *Economic Systems in Action,* revised edition by Alfred R. Oxenfeldt. Copyright 1952, © 1957 by Alfred R. Oxenfeldt. Reprinted by permission of Holt, Rinehart and Winston, Inc.

partnership agreement defines the conditions under which the partners will do business and how they divide up the duties and responsibilities of the partnership. Each partner is an agent for all other partners; i.e., the act of an individual partner binds the partnership. States have provided laws which stipulate the legal requirements governing the formation of partnerships. For business purposes, the partnership is considered to be a company; it lacks, however, the identifying earmark of a corporation—the legal separation between the owners and the business. As contrasted with the corporation, it also lacks stability, which makes it relatively more difficult for a partnership to enter into long-term contracts such as those involved in seeking capital.

A *limited partnership* is a special form of business association created by two or more persons under state statutes. The limited partnership must have one or more *general partners* who assume general liability for the partnership. The *limited partner* is not bound by the obligations of the partnership except in the specific contribution he makes to the partnership. The contribution of the limited party may be cash or other property but not services. In general, the limited partner has the same rights as the general partners except that he may not participate in the management of the firm.

The *corporation* is the dominant form of business ownership in the United States. Corporations exert some influence over most of us, at least in terms of providing goods and services which we use in our everyday lives. The corporate form of organization provides the means whereby capital (money, machine tools, and industrial facilities) can be combined with professional management in the production and marketing of goods and services. The corporation is essentially a fictitious being—described in Chief Justice Marshall's enduring words as ". . . an artificial being, invisible, intangible and existing only in contemplation of law."[10] The state brings the corporation into being through a multilateral contract between the state, the incorporators (those individuals desirous of forming the corporation), and the corporation itself (the legal entity or artificial being). The state which permits the formation of a corporation may withdraw the privilege, but this seldom happens, and then only for cause.

The only physical manifestations of the corporation are the assets which it owns. Evidence of the existence of the corporation is seen in viewing its property and through documentation such as stock certificates and annual reports.

The rights of the corporation include the purchase, ownership, and sale of property, contracting powers, the opportunity for legal redress in the courts, the carrying on of business affairs and producing of goods or

[10] *Dartmouth College v. Woodward H. Wheaton* 518 (1819).

services, participation in the economic, political, and social affairs of the day, and the ability to borrow money and pledge assets. Normally, three or more persons, a majority of whom are citizens of the United States, may form a corporation by filing with the responsible state official (normally the Secretary of State) articles of incorporation which describe such matters as the name of the corporation, the location of its principal offices, its officers, its purpose, the authorized number and par value of the shares of ownership, its capital, etc.

A number of essential features distinguish the corporation from other sorts of organizational makeup. In addition to those already mentioned, the corporation's distinguishing characteristics are:

1. The corporation has a *permanent life* beyond the death of any particular stockholder. This permanency is of great importance in such matters as issuing long-term bonds or mortgage notes and negotiating contracts whose performance extends years into the future.
2. There is a *free transferability of ownership.* The shareholders may sell or give away their shares of ownership without affecting the life of the corporation.
3. There is *limited liability* in that the owner of shares has no liability for the debts or legal damages incurred by the corporation. Once the stockholder has paid for his stock he cannot be held liable for any debt or damages sustained by the corporation. Of course, the stockholder is "liable" to lose his investment in the corporation if business is bad and the corporation a business failure.
4. The corporation is *managed through elected representatives*—directors, who in turn appoint the officers of the corporation. This form of representative management is necessary because of the virtual impossibility of having a large number of stockholders directly participate in the management of the corporation, e.g., to contract for goods and services.
5. The corporation is a free association of owners, managers, and employees. They are bound together by a contractual relationship and although they have different interests, they seek to accomplish their objectives by separate and harmonious means.
6. Typically, the corporation is *taxed separately* from the owners. After the expenses of the business have been paid, the residual is divided up between the stockholders and the government on the basis of applicable tax rates. Stockholders receive their share of the corporate earnings in the form of a dividend. Corporate profits not paid out in dividends are reinvested in the corporation. The stockholder must pay income tax on his dividends. (This can be contrasted with a general partnership, where each partner annually reports his portion of the

profits and pays taxes on them, whether such profits are withdrawn from the business or not.)

Thus, the corporation is an abstraction—a legal person clothed with many of the rights, privileges, and immunities that our society extends to natural persons.

There are a wide variety of other forms which have been created for the purpose of carrying on specialized activities. Among these are the nonprofit corporation, the joint venture, and the partnership association.

Cultural Forces

Our changing culture is a significant environmental force on every organization in the society. Cultural anthropologists have long known that cultural differences are reflected in different sets of values, which must in turn affect the ways in which organizations are run. Unfortunately, these value sets are not easy to define.

Perhaps we can illustrate different cultural value sets as they affect organizations by one author's observations regarding differences between United States and Latin American culture and "personality" as related to economic growth.

> *Comparatively the Latin American complex: (1) sacrifices rigorous economically directed effort, or profit maximization, to family interests; (2) places social and personal emotional interests ahead of business obligations; (3) impedes mergers and other changes in ownership desirable for higher levels of technological efficiency and better adjustments to markets; (4) fosters nepotism to a degree harmful to continuously able top-management; (5) hinders the building up of a supply of competent and cooperative middle managers; (6) makes managers and workers less amenable to constructive criticism; (7) creates barriers of disinterest in the flow of technological communication; and (8) lessens the urge for expansion and risk-taking. These Latin qualities are not necessarily detriments to the good life, perhaps just the opposite, but they are hindrances to material progress under the Anglo-American concepts of a market-oriented capitalist economy.*[11]

Our "market-oriented capitalist economy" has evolved a culture whose traditional value set is reasonably easy to characterize:

[11] T. C. Cochran, *The Inner Revolution,* Harper & Row, Publishers, Incorporated, New York, 1964, pp. 126–127.

> *. . . emphasis on the individual as the unit of social value; self-reliance; self-determination in forms of government and, concomitantly, resistance to other forms of public authority; the virtue of work; self-advancement as a moral obligation; demonstration of personal competence by material achievement; competition as a test of worth; objectivity and impersonality in economic decisions; regard for others as a matter of individual conscience; experimental activity and pragmatic solutions; compromise as a social principle; a view of change as progress.*[12]

However, cultures and their value sets are subject to change. Ours in the United States is changing at an accelerating pace. And it is not always easy to define in proper perspective the changes that are taking place. For instance, the competitive nature of our society has been described by many. Schools and other institutions which deemphasize competition have been developed. Yet, the assessment of whether this is an indication of a changing element of our value set or merely a temporary aberration created by a small minority is difficult.

So, too, it is difficult to determine at the time of this writing whether the "communal culture" which has been widely discussed and practiced is a serious cultural force or merely a temporary phenomenon. In either case, the implications to the management of organizations may be great. If it is temporary, it may merely provide a lucrative, short-term new market for the business firm's products and a new set of problems with which public agencies must deal. If it has a degree of permanence, it may foretell of basic organizational changes which will be necessary for survival.

Governmental Forces

Although the economic system is based on a profit objective and decisions are made privately by businesses, labor, and consumers, the role and significance of government in nongovernmental organizations are ever increasing.

Government and nongovernmental organizations interact most directly in terms of the legal rules and restrictions placed on business by law and through the actions of government agencies.

For instance, laws have been established to cover a wide range of business affairs. The Fair Labor Standards Act establishes minimum wages and hours. Other laws, such as the Sherman Anti-Trust Act, passed in

[12] N. W. Chamberlain, *Enterprise and Environment,* McGraw-Hill Book Company, New York, 1968, pp. 132–133.

1890, restrict noncompetitive behavior. The Robinson-Patman Act forbids price discrimination among purchasers where the effect would be to significantly lessen competition.[13]

There are also laws which serve to restrict competition. The subsidization of certain industries, such as the merchant marine and agriculture, clearly serves to reduce the degree of competitiveness that would otherwise exist. Other legislation has served to exempt special-interest groups from the provision of antitrust legislation. Illustrative of this are laws which have been passed permitting exporters to form associations and farmers and dairymen to form cooperative associations for purchasing and selling.

Thus, much of the formal legislation dealing with business is apparently contradictory. Some legislation is designed to encourage competition while other legislation discourages it. This is not so surprising in the light of the changing needs of a dynamic society and the fact that old laws are not necessarily wiped from the books when new ones are passed.

The role of government regulatory agencies with regard to business is significant and direct. The Civil Aeronautics Board (CAB) rules on rate change requests from major airlines, the Securities and Exchange Commission (SEC) registers firms as investment advisors, the Federal Communications Commission (FCC) allocates radio and television airwaves to stations, etc. The list of agencies and their powers is virtually endless.

Government directly affects nonbusiness organizations in a variety of ways. The taxing power of state and local government is a direct influence. Some nonprofit charitable and religious organizations are frequently exempted from certain taxes, for instance.

Governmental organizations are in turn affected by one another. Funds for certain federal programs are channeled through the states and counties. City officials are sometimes made responsible for the activities of certain "independent" federal programs operating in their area. For instance, a federal "Model Cities" program in Pittsburgh was carried out by an organization especially formed for the task, but city officials were made jointly responsible for the program.

The interlinkages between government—at any and all levels—and other organizations are numerous and diverse. The best forecasts indicate that these linkages will increase in both number and complexity in the future. Hence, their government will continue to represent one of the significant environmental forces which organizational managers must understand and use.

[13] M. C. Howard, *Legal Aspects of Marketing,* McGraw-Hill Book Company, New York, 1964.

International Forces

The growth of the "multinational firm" has led to an increased perception of the importance of international environmental forces which impact on business. Many American firms began to recognize the opportunities presented by international markets only late in the 1960s, and few have really exploited these opportunities effectively even now.

Conversely, many foreign companies have made significant progress in United States markets. Volkswagen can certainly be considered a major factor in the United States auto market, as can an increasing number of Japanese, French, Italian, English, and German firms. Japan has virtually taken over the United States' international role in shipbuilding, in addition to her "traditional" postwar eminence in photography, optics, and electronics. Imports of steel have for some time been of great concern to the domestic steel industry, and the steel industry, together with other threatened industries, has lobbied, directly or indirectly, for protective import regulations.

International organizations are becoming increasingly important to domestic organizations. The Organization of American States (OAS) and the European Economic Community (EEC), which is usually referred to as the "European Common Market," are illustrative of these. In most instances, these organizations involve governments which cooperate for some end. In the case of the Common Market, the purposes were to restrict the economic domination of Europe by the United States and the U.S.S.R. and to preserve the exploitation potential of the European market for the member states. Any United States firm which wishes to begin doing business in a nation which is a member of the Common Market would therefore be well advised to be aware of the EEC regulations regarding the firm's product.

Government forces and international forces can interact to jointly affect organizations. For example, the U.S. Department of Commerce has sponsored a series of "fairs" designed to bring small United States manufacturers into contact with potential foreign customers. The objectives of the United States government in this regard are to enhance foreign trade and, at the same time, to encourage small business enterprise.

International cultural differences are another interactive force on organizations. The ways in which one succeeds in negotiations or sales efforts in alien cultures may be very different from successful approaches used in one's own nation. "It works in the States" is not an indication that it will work in a foreign country, and the companies are legion which have discovered this in practical ways at great expense.

MANAGING ENVIRONMENTAL FORCES

In stating the manager's "environmental problem" we asked how he could develop an appreciation for general environmental forces, and then how he could integrate these forces into his organization and its development. The brief discussion in the previous section, "Environmental Forces," seeks to outline a conceptual framework for appreciation and understanding of the basic forces. In this section we go on to address the specific question of using environmental forces.

Developing an Organizational Model

In addition to a general understanding of the "environmental forces" which affect all organizations, the manager must be aware of those specific entities which affect his organization and the decisions which he must make. In effect, *he must determine which organizational model he is going to use.*

For instance, if the manager chooses to view his organization in terms of the organization chart model, he will undoubtedly deemphasize environmental considerations. They simply do not play an important role in such a model. The model abstracts from the real world by omitting these considerations.

In the process of selecting such a model, *the rational person should determine which aspects of the real world are sufficiently unimportant for the use to which the model is to be put, so that they may be omitted.* However, in reality a manager does not evaluate his model each time that he uses it. The model is literally the framework in which we think, and few of us question such basics frequently. Thus, in choosing a model which he will use under a variety of circumstances, he must be certain that he has not left out important considerations.

This is probably the case with the organization chart model. It is widely useful for identifying positions, people, and their formal relationships. Too often, however, it is used in circumstances which are inappropriate—for instance, in situations in which factors which it omits are of obvious importance.

The question of selecting an appropriate model to use may be thought of in terms of some of the alternative organizational models presented in Chapter 3. One model focuses on the internal system; another focuses on the industrial system; another focuses on the end-product system; still another gives consideration to the overall economic system. Each emphasizes different things and each is therefore useful for a different purpose.

The most significant element which distinguishes these various models is the organizational elements and components that are explicitly included in each. Some models abstract out all of the internal organizational elements to treat the organization as a single entity in an industrial system. Other models deal with the organization and those other organizations which directly interact with it. The question is: Which model to use?

An Organizational-claimant Model

An organizational model which is useful for the consideration of environmental forces is one which focuses on organizational claimants.

In the previous chapter, we introduced the idea of the multiple claimants and clientele of the organization. These clientele groups are the primary aspect of the environment with which the manager must deal—for their influence is direct and significant on a day-to-day basis.

CLAIMANTS *A claimant is an individual, a group of individuals, or some institution in the society that has a demand for something due.* Thus, an employee may be a claimant on a business firm under a workmen's compensation law. Of course, some claimants to a corporation have a more direct claim than others. The common stockholders as the residual owners have a more direct claim on the management of the business than do the creditors. On the other hand, the creditors may exercise contingency management rights if the interest or principal payments on a debt are in arrears. Preferred stockholders whose dividends are not paid because of economic difficulties that befall the firm may have the right to participate in the election of the board of directors.

Some claimants do not have legal contractual claims on the organization; yet, their claims are nonetheless real. Consider, for example, one firm's view of the importance of some of these noncontractual claims as illustrated by an advertisement in the *Wall Street Journal.*

> *"Help wanted: experienced manufacturing man capable of supervising integrated production facility—capable service club after-dinner speaker; must have proven record of effective plant personnel administration—and be good at running charity committees; profit-oriented—willing to serve without pay on hospital board or city council; must supervise recruiting and training procedures for plant personnel—and solicitors for community fund drives; the man we seek is a dynamic, aggressive producer—with the patience, tact, and understanding needed for leadership of church and civic groups."*

If we were to write a recruiting ad for a plant manager, that's how it might read. The underlined talents don't appear in any job description, but they are very real demands on the men who head up most of our 30 manufacturing plants.

With 14 of these plants located in towns of less than 25,000 people, our local managers are called upon for an unusual amount of community service. At any one time, we'll have two or three mayors or city councilmen, a dozen or so school-board members, and a brace of Rotary, Kiwanis, Lions, and other service club leaders among our plant managers.[14]

Table 4-1 on the next page reflects some of the multiple claimants to the business organization together with the nature of their claims. It represents a simple qualitative organizational-claimant model.

ENVIRONMENTAL FORCES AND CLAIMANTS Many of the environmental forces which were previously discussed are embodied in the claims shown in Table 4-1. Indeed, most of the so-called "fundamental forces" come into play in all of the claims. Stockholders' claims reflect not only the legal forces resulting from the nature of their ownership and the governmental forces resulting from laws and SEC regulations, but also concepts of "fairness" which grow from our democratic system. A protest by some General Motors stockholders regarding the absence of minority-group representation on the board of directors led to the appointment of the first black man to that body in the early 1970s. That protest reflected the stockholders' basic legal right, but it also reflected more complex environmental and cultural forces which were at work.

USING THE ORGANIZATIONAL-CLAIMANT MODEL A model such as that described in Table 4-1 is useful only if the relationships among its various components can be assessed and predicted. Thus, it is interesting to be aware of the fact that part of the government's claim on business is through taxes. It is at quite another level of utility to have *specific delineations* of the various taxes to which the organization is subject together with *predictions of the future patterns of taxation* and *evaluations of the impact of these patterns on organizational decisions.*

For instance, the manager who must make a recommendation concerning the location of a new plant must be aware of state and local taxes and their likely pattern of growth. He must be aware of the political climate which will affect taxation patterns and he must be able to evaluate

[14] Rockwell Manufacturing Company advertisement in the *Wall Street Journal*, Sept. 18, 1967.

TABLE 4-1 Claimants to the business firm

Claimant to the business firm	*General nature of the claim*
Stockholders	Participate in distribution of profits, additional stock offerings, assets on liquidation; vote of stock, inspection of company books, transfer of stock, election of board of directors, and such additional rights as established in the contract with corporation.
Creditors	Participate in legal proportion of interest payments due and return of principal from the investment. Security of pledged assets; relative priority in event of liquidation. Participate in certain management and owner prerogatives if certain conditions exist within the company (such as default of interest payments).
Employees	Economic, social, and psychological satisfaction in the place of employment. Freedom from arbitrary and capricious behavior on the part of company officials. Share in fringe benefits, freedom to join union and participate in collective bargaining, individual freedom in offering up their services through an employment contract. Adequate working conditions.
Customers	Service provided the product; technical data to use the product; suitable warranties; spare parts to support the product during customer use; R & D leading to product improvement; facilitation of consumer credit.
Supplier	Continuing source of business; timely consummation of trade credit obligations; professional relationship in contracting for, purchasing, and receiving goods and services.
Governments	Taxes (income, property, etc.), fair competition, and adherence to the letter and intent of public policy dealing with the requirements of "fair and free" competition. Legal obligation for businessmen (and business organizations) to obey antitrust laws.
Union	Recognition as the negotiating agent for the employees. Opportunity to perpetuate the union as a participant in the business organization.
Competitors	Norms established by society and the industry for competitive conduct. Business statesmanship on the part of contemporaries.
Local communities	Place of productive and healthful employment in the local community. Participation of the company officials in community affairs, regular employment, fair play, local purchase of reasonable portion of the products of the local community, interest in and support of local government, support of cultural and charity projects.
The general public	Participation in and contribution to the governmental process of society as a whole; creative communications between governmental and business units designed for reciprocal understanding; bear fair proportion of the burden of government and society. Fair price for products and advancement of the state-of-the-art in the technology which the product line offers.

alternative locations in the light of these predictions. To be aware of the fact that Pennsylvania's corporate income tax is the nation's highest is useful, but this fact must be complemented by information regarding other business taxes, sales taxes affecting business, local taxes, etc. Moreover, the likelihood of "tax reform" by a new state governor must be assessed and specific predictions must be made. Only then is rational decision making possible.

To put all of this together into useful predictions requires an understanding of *the objectives of the various claimants,* for it is only through an understanding of what they are trying to accomplish that useful predictions can be made. For instance, if a new governor's short-run objective is deemed to be the suppression of criticism of the state in order that he may have the time to make basic reforms, his campaign-oriented tax reform proposal may be a smoke screen to give him that time advantage. If so, its usefulness to business in predicting future taxation patterns may be minimal.

ORGANIZATIONAL-CLAIMANT MODEL SUMMARY The development and use of an organizational-claimant model require that:

1. Organizational claimants be identified.
2. The nature of each claim be specified.
3. Measurable elements be defined for each claim.
4. Predictions of the future pattern of these elements be made in the context of the claimant's objectives.
5. The impact of these predictions on organizational decisions be assessed.

Of course, the ability of any individual to do all of this informally is doubtful. We shall skip the practical "how-to-do-it" question for the moment. Now, we need only make clear that a mental model of the organization and its claimants usually need include only a specification of the claimants, the general nature of their claim, and an understanding of some of the assessable aspects of the claim. The making of measurements and predictions is a part of the task of the analyst or of the manager as he is faced with a particular problem. It is obviously not necessary that he carry all of these measurements and predictions around in his head; to ever attempt to do so would lead to unbounded confusion!

Environmental Factors in Planning and Decision Making

We have shown how the manager can be aware of and use environmental forces. But *where* and *when* can he use them? The answer is simple—in the organization's planning and decision making.

OBJECTIVES One of the basic sets of decisions which the organization must make is that concerning its objectives. What is it going to seek? Increasingly, objectives must be set in terms of the multiple organizational claimants.

For instance, business corporations have responsibilities both to shareholders and to the society in which they exist. The difficulty, of course, is that what is best for the shareholder might not be best for the society (and the converse). Robert J. Weston has identified modern managers as ones who can "identify needs in the future marketplace which offer opportunities for business profit, and identify needs in society which offer opportunities for service to society, and vigorously pursue *both* goals."[15]

Illustrations of the increasing involvement of business in the problems of society are legion. The National Alliance of Businessmen has placed thousands of hard-core unemployed in jobs. Owners and publishers of three newspapers in Charleston, West Virginia, offered free "situation wanted" advertisements to the unemployed. Western Electric Company donated the use of one of its tool and die shops and lined up volunteer instructors as part of a community effort to teach higher skills to the unemployed. U.S. Gypsum Corporation purchased slum buildings in East Harlem and rehabilitated them with completely new interiors. Then they returned the tenants to their apartments at higher rent, paid in part by rent supplements.[16]

The direct contribution of such apparently non-profit-oriented action to profit making was made apparent by the "spin-off" which was realized. For example, U.S. Gypsum developed three new products and found a use for an old one which had failed in the home building market during their East Harlem program.[17] The life insurance industry has recognized the direct influence of such efforts on their profit objectives. One illustration of this is the program under which they provided $1 billion to underwrite construction in slum areas and to create jobs for slum dwellers. *Business Week* has quoted an unidentified mortgage man from the insurance industry as stressing the importance of slum programs to insurers by saying, "If we kiss the cities good-bye, we kiss a hell of a lot of our conventional mortgages good-bye also. Our stake in them is too large."[18]

DECISION MAKING Environmental elements are an increasingly impor-

[15] Robert J. Weston, "Management's Next Generation: The New Service Commitment," *Saturday Review*, Jan. 13, 1968, p. 32.

[16] "Solving the Crisis in Our Cities," *Report to Business, no. 1,* American Business Press, Inc., 205 East 42nd Street, New York, N.Y. 10017.

[17] *Ibid.*

[18] "Life Insurers Push to Rebuild the Slums," *Business Week*, July 20, 1968, p. 52.

tant part of both short- and long-range decision making in organizations. We shall defer the treatment of the techniques of dealing with these factors until a later chapter. Here, the vehicle for demonstrating this point will be a simple one. Increasingly, organizations are finding themselves wrestling with *the same problems being dealt with by other organizations.* Of course, their objectives and viewpoints are different, but the simple fact that they are dealing with the same problem indicates that they must necessarily give consideration to one another's objectives and viewpoints. Thus, each must consider the other to be an element of its environment.

For instance, business and government increasingly become interested in the same problems. At the level of jobs and housing, their joint interests are fairly apparent. However, other problems in which their joint interests are not so apparent are increasingly being recognized. Consider, for example, the problem of establishing retail locations for chains of supermarkets. This is a problem which is a common one for supermarket chains. The systems implications of it are clear. Not only will each retail outlet compete with outlets of other chains and with small independent grocery stores, but each will interact with other outlets of the same chain. To determine where a retail supermarket should be located requires answers to such basic questions as, Where do people shop? When do people shop? Where do they live? For example, it is critical to know whether people tend to make shopping trips only to supermarkets, or whether they tend to visit a number of stores on a single trip. Do their trips emanate from home, or in a particular city; do they tend to shop on the way home from work? The answers to such questions are not at all obvious, and the choice of the best alternative location is not at all clear.

Now consider the government's role in the problem of locating retail businesses. Studies have shown[19] that purchases of food and housing account for about one-half the budget of poor families. Moreover, low-income neighborhoods have often been largely ignored by the large chain supermarkets. The result is that poor families are served by small independent retailers who must charge higher prices.[20] Indeed, if one views a map of existing chain store locations in most major cities, he finds that, aside from areas in which the poor have migrated to encompass existing stores, there is almost a total absence of chain stores in poor neighborhoods.

The implications of this are clear. The significant savings on food

[19] Sho Maruyama, "Cost-Effectiveness in a Poverty Program," unpublished paper at Economic and Youth Opportunities Agency of Greater Los Angeles, Sept. 30, 1966.

[20] Bureau of Labor Statistics, U.S. Department of Labor, "A Study of Prices Charged in Food Stores Located in Low- and Higher-income Areas of Six Large Cities," Summary, June, 1966.

made possible by economies of scale and the management practices of large supermarket chains would be equivalent to an increase in real income if supermarkets were located where the immobile poor could use them. Federal monies are being spent in other ways to achieve the same objective, so why should they not be spent to encourage supermarket chains to so locate? On the other side of the coin is the increasing social awareness and responsibility of business managers. Are supermarket chains pure profit seekers? Do they have other social objectives?

The answers to such questions are not at all clear. However, the simple posing of these questions clearly illustrates that both government and business have an intrinsic common involvement in such problems and that they must include one another in their respective environmental models.

PLANNING The management function termed "planning" entails all of the strategic decision making of the organization. However, the term "long-range planning" has come into vogue to indicate the function ". . . of selecting the enterprise objectives and the policies, programs, and procedures for achieving them."[21] The "long-range" modifier typically means that an attempt is made to direct the organization's activities for 2 to 10 years into the future.

The "quality" of any decision which has long-range implications depends on the worth of the predictions which play a role in the decision-making process. Thus, if the organization is considering new products which it will sell in the future, it needs to be aware of cultural trends and other environmental factors which will affect the demand for the product. Such predictions necessarily depend on a close study of the environment and its trends.

One important aspect of planning which we shall go into more deeply later is the idea of using planning *to influence the environment.* The manager need not simply accept his fate. Rather, he can take action to affect the future. For instance, he may begin an advertising program which is designed to forestall some aspect of the public's attitude which, if allowed to develop, will be negative for his organization. And, although the businessman should be concerned with keeping abreast of new government regulations, he should also endeavor to influence legislation through his legislators and lobbyists and, if necessary, to propose new legislation to his legislators.

The key to using environmental elements in this way is understanding. And, since all of us think in terms of models, understanding comes through the development and use of an environmental model.

[21] H. Koontz and C. J. O'Donnell, *Principles of Management,* 2d ed., McGraw-Hill Book Company, New York, 1959, p. 35.

ENVIRONMENT AND THE SYSTEMS MODEL

The general systems model of an organization as depicted in Figure 3-2[22] can serve to place this chapter's discussion of the environmental system in perspective. The reader will recall that the crucial elements of the organizational model were a *receptor,* a *decision mechanism,* and a *processor,* and that its emphasis was on the *flow* and *processing of information, materials,* and *energy.*

The organization's *receptor* mechanism extracts data and observations from both the internal and the external environment and converts it into information which is meaningful to the decision mechanism. Since stimuli are obtained from each environmental subsystem, the receptor must be sensitive to whatever specialized form the stimuli may have. For instance, the market demand aspect of the sociological subsystem is usually derived from personal observations, interviews, and questionnaires.

Efficiency of the system can be impaired by barriers which reduce the efficacy of the receptor. A company which does not encourage its technical personnel to keep abreast of technological advances is creating an awareness barrier. Psychological, sociological, cultural, educational, theological, and other subsystems of the environment create personal attitudes which determine how the human subsystem perceives the flow of stimuli in the environment. In the case of Willow Run, although the Ford Company was aware of the social upheaval, their predispositions led them to believe they had no responsibility to become involved. Such was the thinking of the times.

The organizational *decision mechanism* is really a large number of decision points distributed throughout the company at all management levels. The "decision maker" may be one man or a group of men. The decisions themselves are both programmed and nonprogrammed. The programmed decisions apply to routine or recurring situations that can be resolved through predetermined decision rules and procedures. The nonprogrammed decisions apply to a nonrecurring situation that must be dealt with in its own individual context.

Decision outputs are in the form of goals, objectives, and plans that govern the utilization of resources by the processor. What an organization is and what it does are largely dependent on the "personality" of the decision mechanism. The predispositions and biases brought in by the human decision makers from their environment pervade all decisions. For instance, a decision mechanism with the "personality" of an "economic man" would select organizational goals and objectives based primarily on a narrow criterion of self-interest such as return on investment, profit maximization, or sales maximization; at the other extreme would be

[22] See page 68 for Figure 3-2.

found the idealist, who would base his decisions on altruistic purposes and neglect the need for economic survival.

In responding to the changing environment, the decision mechanism and the receptor must be combined into an information processing system. According to Prince, the receptor and the decision mechanism cannot be isolated from each other in this role.[23] Indeed, the decision maker and the receptor may be the same individual. A supervisor, making decisions involving the morale of his workers, collects much of his data through personal contacts with his crew. From a general decision-making point of view, each decision area (financial, research, labor and public relations, production, etc.) generates specific information requirements. In this context, one of the operational functions of the receptor is to match sources of data with the information requirements.

This information may be as general as forecasts of the economic, technological, or sociological subsystems of the environment in the form of market studies, trends in basic research, or population trends. On the other hand, it may be a detailed technical analysis of a new product or an incremental cost analysis of a proposed numerically controlled production process. When integrated together, this general and detailed information becomes a basis for establishing goals, objectives, and plans.

The organization's processor performs all those activities associated with the operation and control of the sales, finance, production, personnel, and research and development functions. Again, it is not actually distinct from the receptor and decision mechanism in that it may be the instrument for carrying out some action, such as an internal company reorganization, involving activities in these components. Thus, the processor completes the feedback loop by creating the information, material, and energy flows back to the environment.

SUMMARY

Perhaps the least understood element of any organization is the environment in which it operates. Illustrations are legion of business firms, governmental agencies, and other organizations which failed in their purpose either because they did not appreciate the importance of environmental influences, they did not understand the structure and operation of their environment, or they did not have good information concerning their environment.

The manager's "environmental problem" therefore involves (*a*) the development of an appreciation of the need to assess environmental

[23] T. R. Prince, *Information Systems for Management Planning and Control,* Richard D. Irwin, Inc., Homewood, Ill., pp. 15–27.

forces, (*b*) the development of an understanding of the environmental forces which affect his organization, and (*c*) the development of methods for effectively integrating these forces with the organization's goals, objectives, and processes.

Among the various environmental forces with which the manager must deal are the concept of democracy, our political system, the concepts of freedom and equality, the economic system, and principles of business ownership. These fundamental concepts and forces, along with cultural forces, governmental forces, and international forces, form the environmental framework within which the manager must operate. For him to operate successfully, he must understand these forces and the ways in which they may affect his organizations.

To use his understanding of environmental forces, the manager must determine which organizational model he will require. One useful model which emphasizes environmental considerations is an *organizational claimant model.* An organizational claimant is an individual or institution which has a demand for something due from the organization. These demands embody many of the environmental forces affecting the organization. Hence, an enumeration of the various claimants together with an understanding of their claim, constitutes a simple organizational claimant model.

To use such a model, the manager must have specific *assessments* of the various claims and *predictions* of their future development. These, together with an understanding of the nature of the impact of the claims on the organization, enable him to assess the specific impact of hypothesized or predicted changes. With such assessments, the manager can proceed to establish objectives, make decisions, and plan with greater assurance than he could ever do in the vacuum which ignores environmental forces.

EXERCISES

1. "To decide not to decide is to decide." What is the significance of this statement to the attitude of the business firm toward its environment?
2. Why might a business firm choose to go beyond its legal responsibility to participate in community problems?
3. What is the manager's "environmental problem" in terms of (*a*) his perceptions and (*b*) his possible actions?
4. The authors seem apologetic in dealing with such basic environmental forces as "democracy." What is their argument for including these basic concepts in the text? Defend or refute it.
5. How would the convening of the United States into a "pure democracy" change the daily lives of (*a*) the "average man," (*b*) the business executive, (*c*) the congressional lobbyist?

6. Give a one-sentence definition of "democracy." (If you have difficulty, you may wish to return and review your response to exercise #4.)
7. How does the existence of the federal system affect the day-to-day operations of the business enterprise?
8. Answer exercise #7 for (*a*) the independent legal system, and (*b*) the concept of freedom and equality.
9. Discuss modern-day deviations from the characteristics of the American economic system as summarized on pages 93–94.
10. What are the physical manifestations of a corporation?
11. What are the advantages of the corporation over other forms of business ownership? How may some of these "advantages" work to the detriment of society?
12. How would the large-scale development of communes as a life style in the United States affect American business? What other organizations in our society might be greatly affected?
13. The novice corporate attorney might well feel that business-related laws are extremely contradictory. Would he be correct? Give an illustration. Can you explain (rationalize?) this?
14. How are various levels of government interdependent? Can "government" be viewed as a system? What advantages might this have?
15. What are some of the potential new problems which the business firm might encounter as it "goes international," i.e., as it begins to operate in many nations?
16. What is an organizational claimant? Give examples.
17. Who are the claimants to (*a*) a business firm, (*b*) a hospital, (*c*) a law enforcement agency?
18. How are "claimants" and "environmental forces" related?
19. Illustrate how one can use a simple organizational-claimant model in assessing and predicting the future of the organization. Do this for (*a*) a business, (*b*) a hospital, (*c*) an urban police department.
20. What are the ways in which environmental forces affect the objectives, planning, and decision making in an organization?
21. How do environmental forces relate to the information system in an organization? Describe this relationship in terms of the general systems model of Figure 3-2.
22. Suppose that you have decided to begin to sell a new product that you have invented. How would you decide on a form of ownership for your new business? (What considerations would be relevant?)
23. Profit is the single best criterion to use to assess the success of a business firm. Do you agree? Why or why not?
24. During the year the stock market will frequently respond to statements made by government officials. Explain why this occurs.
25. A United States auto, here called the "Arrow," was advertised in Spain using the phrase "Arrow is power." In Spanish, this phrase implied a lack of sexual vigor on the part of the purchaser. What can you infer from this concerning the potential problems in the operation of multinational firms?

26. When an American firm entered into an agreement with a foreign firm for a joint venture, the foreign firm's owner was quoted as saying, "There are no better business partners than the Americans. There is no limit to their drive, ingenuity, and imagination." What potential problems do you think that this "drive, ingenuity, and imagination" might pose for the foreign firm?
27. During Nelson Rockefeller's 1969 visit to Argentina, some supermarkets operated by Rockefeller interests in Buenos Aires were fire bombed. Later, certain Argentine politicians argued that insurance claims should not be honored since they did not cover damage due to political terrorism.[24] What potential difficulties with international business venture does this incident illustrate?

REFERENCES

Cleland, David I., and William R. King, "Adapting Management to the Modern World," Paper presented to the Eleventh Annual Academy of Management Conference, Midwest Division, Washington University, St. Louis, Apr. 6, 1968.

Culbert, Samuel A., and James M. Elden, "An Anatomy of Activism for Executives," *Harvard Business Review,* November–December, 1970.

Davis, Keith, and Robert L. Blomstrom, *Business, Society, and Environment,* 2d ed., McGraw-Hill Book Company, New York, 1971.

Drucker, Peter F., "Management's New Role," *Harvard Business Review,* November–December, 1969.

Eells, Richard, and Clarfefeence Walton, *Conceptual Foundations of Business,* Richard D. Irwin, Inc., Homewood, Ill., 1969.

Koontz, Harold, "Ad Model for Analyzing the Universality and Transferability of Management," *Academy of Management Journal,* December, 1969.

Lawrence, Paul R., and J. W. Lorsch, *Organization and Environment,* Richard D. Irwin, Inc., Homewood, Ill., 1969.

McDonald, John, "How Social Responsibility Fits the Game of Business," *Fortune,* December, 1970.

Miller, J. Irwin, "A New Partnership: Business Education, Society," *Business Horizons,* Spring, 1967.

———, "Business Has a War to Win," *Harvard Business Review,* March–April, 1969.

Taylor, Stuart A., "Is Management Truly Involved in the Urban Crisis?" *Business Horizons,* April, 1969.

Votaw, Dow, and S. Prakash Sethi, "Do We Need a New Corporate Response to a Changing Social Environment?" *California Management Review,* Fall, 1969.

[24] "Business Abroad: Hard Times Hit a Rockefeller Enterprise," *Business Week,* July 26, 1969, p. 60.

section three

ELEMENTS OF MODERN MANAGEMENT

5

the management of organizations

Organizations must be guided and directed toward the accomplishment of their objectives. It is the basic function of management to select appropriate organizational goals and to guide and direct the organization in the achievement of these goals.

MANAGEMENT FUNCTIONS

The management process has frequently been viewed in terms of various *managerial functions. A management function is a major activity of management.* There is a good deal of difference of opinion among management theorists as to the interpretation of the word "major" in this statement. In other words, the categorizations which are made and used for analytic purposes by various theorists differ. There is little difference of opinion as to either what managers ought to do or what they actually do. The main differences lie in the relative importance placed on the activities in various management models and, to some degree, in the terminology which is used.

Management Functional Models

Table 5-1 contains a comparison of the models used by various management authors. This table is not intended to be exhaustive—rather it is representative of the models and terminology used by selected authors of basic management texts.

In each of these instances, the author uses a qualitative model to describe management functions. Table 5-1 demonstrates that many different models are appropriate. Some of these differ only in semantics; others differ in the assessment which has been made of the relative impor-

TABLE 5-1 Major management functions as seen by various authors

	Planning	*Organizing*	*Control-controlling*	*Communication-communicating*	*Actuating*	*Directing-direction*	*Staffing*	*Innovation*	*Representation*	*Creating*	*Motivating*	*Directing and marketing*
Johnson et al.[1]	x	x	x	x								
Terry[2]	x	x	x		x							
Jucius & Schlender[3]	x	x	x			x						
Davis[4]	x	x	x									
Dale[5]	x	x	x			x	x	x	x			
Koontz & O'Donnell[6]	x	x	x			x	x					
Haimann[7]	x	x	x			x	x					
Hicks[8]	x	x	x	x						x	x	
Longenecker[9]	x	x	x									x

[1] R. A. Johnson et al., *The Theory and Management of Systems,* McGraw-Hill Book Company, New York, 1967.
[2] G. R. Terry, *Principles of Management,* Richard D. Irwin, Inc., Homewood, Ill. 1964.
[3] M. J. Jucius and W. E. Schlender, *Elements of Managerial Action,* Richard D. Irwin, Inc., Homewood, Ill., 1960.
[4] R. C. Davis, *The Fundamentals of Top Management,* Harper and Brothers, New York, 1951.
[5] E. Dale, *Management: Theory and Practice,* McGraw-Hill Book Company, New York, 1965.
[6] H. D. Koontz and C. O'Donnell, *Principles of Management,* McGraw-Hill Book Company, New York, 1955.
[7] T. Haimann, *Professional Management: Theory and Practice,* Houghton Mifflin Company, Boston, 1962.
[8] H. G. Hicks, *The Management of Organizations,* McGraw-Hill Book Company, New York, 1967.
[9] J. Longenecker, *Principles of Management and Organizational Behavior,* Charles E. Merrill Books, Inc., Columbus, Ohio, 1964.

tance of the elements to be included as "major" functions and those to be entirely omitted from the model.

Fayol's Functions

One of the earliest writings in management theory introduced the concept of dividing industrial undertakings into different categories. This categorization, developed by Henri Fayol, established a pattern for the study and practice of management. It grouped industrial undertakings into six groups:

1. *Technical (production, manufacture, adaptation).*
2. *Commercial (buying, selling, and exchange).*
3. *Financial activities (search for and optimum use of capital).*
4. *Security activities (protection of property and persons).*
5. *Accounting activities (stocktaking, balance sheet, costs, and statistics).*
6. *Management activities (planning, organization, command, coordination control).*[1]

The management function, according to Fayol, is not ". . . an exclusive privilege nor a particular responsibility of the head or senior member of the business; it is an activity spread, like all other activities, between heads and members of the body corporate."[2] Fayol felt that the first five categories of activities were well known, but that the management activity required considerable research; consequently, he devoted the most of his book to it. He described managerial activities as consisting of the functions of planning, organizing, commanding, coordinating, and controlling. In his words:

> *To manage is to forecast and plan, to organize, to command, to coordinate, and to control. To foresee and provide means for examining the future and drawing up the plan of action. To organize means building up the dual structure, material and human, of the undertaking. To command means maintaining activity among the personnel. To coordinate means binding together, unifying and harmonizing all activity and effort. To control means seeing that everything occurs in conformity with established rule and expressed command.*[3]

Management Subfunctions

An examination of management functions reveals that each management function can be visualized as composed of subfunctions. The planning function, as an illustration, can be said to consist of:

[1] Henri Fayol, *General and Industrial Management,* Sir Isaac Pitman and Sons, Ltd., London, 1949, p. 3.

[2] *Ibid.,* p. 6.

[3] *Ibid.,* pp. 5–6.

1. Strategic planning
2. Tactical planning

Strategic planning, in turn, involves:

1. Development of goals and objectives
2. Forecasting of an expected environment
3. Developing planning assumptions
4. Evaluating alternatives in the selection of strategic goals
5. Developing the plan of action

Other management functions might be broken into subfunctions of activities such as:

Organizing:

- Identifying resources
- Procuring resources
- Establishing relationships
  - Laying out resources
  - Developing authority and responsibility patterns

Motivating or directing:

- Sensing needs and wants
- Communicating purpose
- Counseling
- Providing suitable environment
- Individual follow-up

Controlling:

- Establishing standards
- Observing performance
- Comparing performance
- Taking corrective action

The complexity of the management process and the various subfunctions which can be considered in a model are well illustrated in the pictorial model of Figure 5-1. According to the developer of the model, Alec MacKenzie, ". . . the functions in the diagram . . . have been selected after a careful study of the works of many leading writers and teachers. While the authorities use different terms and widely varying classifications of functions, I find that there is far more agreement among them than the variations suggest."[4]

[4] R. Alec Mackenzie, "The Management Process in 3-D" *Harvard Business Review,* November–December, 1969, p. 87.

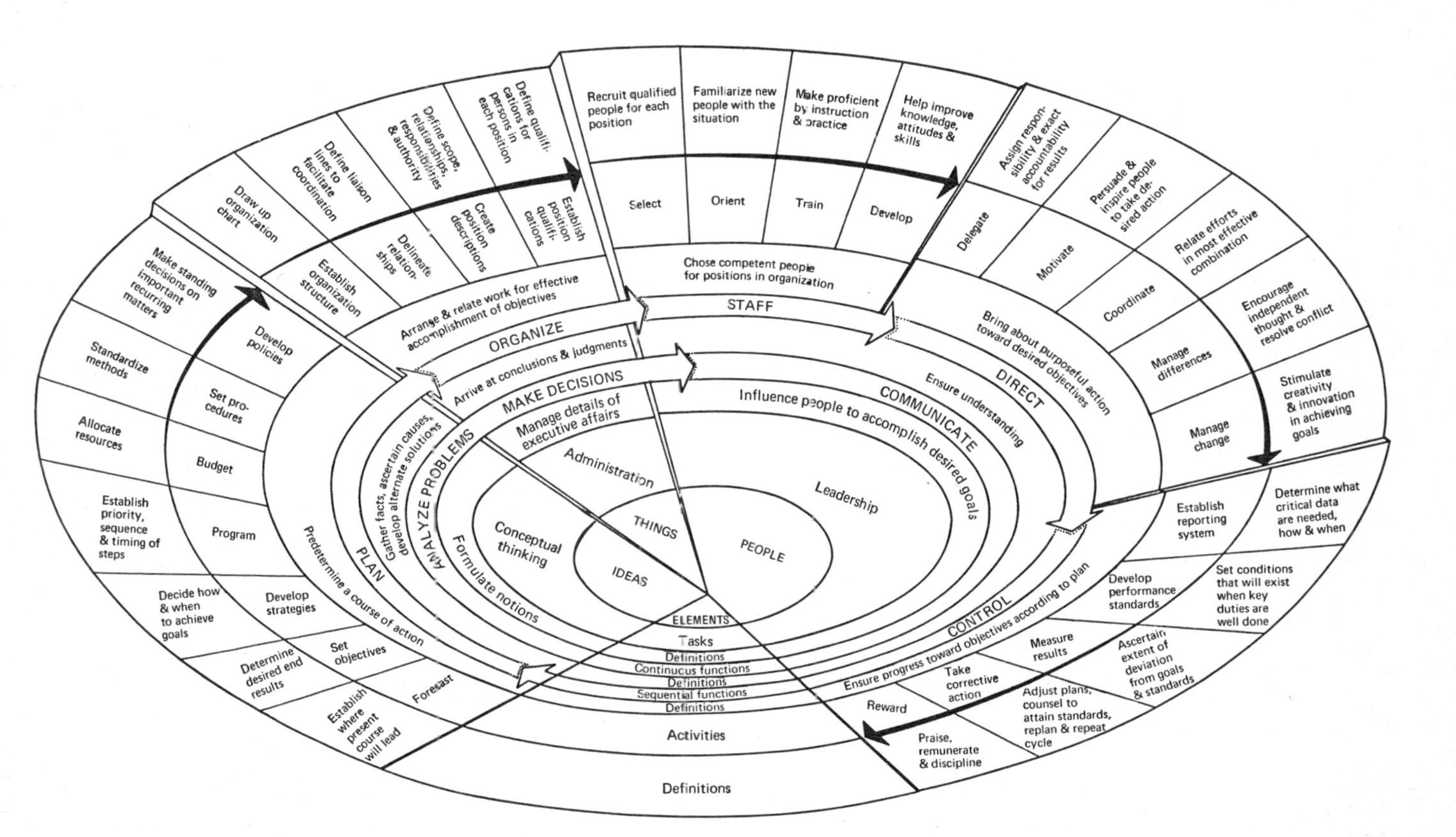

FIGURE 5-1 *A model of the management process.* SOURCE: *R. Alec MacKenzie, "The Management Process in 3-D,"* Harvard Business Review, *November–December, 1969, p. 87. (© 1969 by the President and Fellows of Harvard College; all rights reserved.)* Used by permission.

Management Functional Models Compared

Of all the qualitative models which may be used to describe management functions, the one introduced in Chapter 1 is undoubtedly the most abstract of all. There, we considered only two management activities—*deciding and doing.* In some instances, it is useful to operate at this high level of abstraction. In other cases, other models are more descriptive and conducive to learning. In this text, we shall not feel tied to any single model, since, like all models, each is useful for a particular purpose. We shall, however, attempt to place in context the various models which are used.

For example, one pervasive model[5] utilizes *planning, organizing, directing,* and *controlling* as major functions in its management functions model.

The simple relationship between the two models is shown in the diagram below.

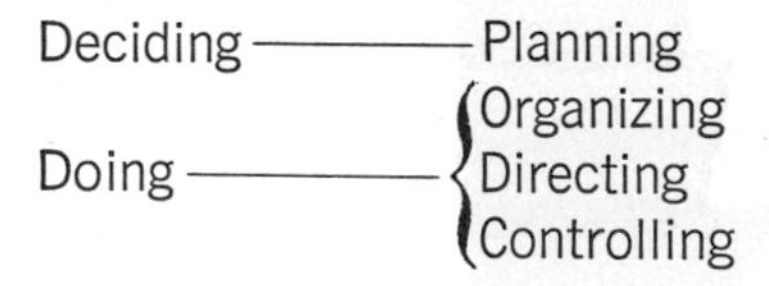

Thus, planning is essentially a decision function and organizing, directing, and controlling are action functions. The same sort of simple correspondence can be made among all of the functional models of management.

The Interrelatedness of Management Functions

The day-to-day world of the manager is actually much more complex than any of these simple models describe. In fact, all of the functions of the manager are interrelated.

Planning and organizing precede control. The plan provides a standard against which to compare results and to correct performance in accordance with a predetermined goal. Thus, management functions may be visualized as a continuous process. Figure 5-2 is one way of illustrating the interdependency of the functions.

Perhaps a "word game" which we shall play offers the best way of illustrating these interrelationships. The manager must:

Plan for organizing
Plan for directing

[5] See reference 3 in Table 5-1.

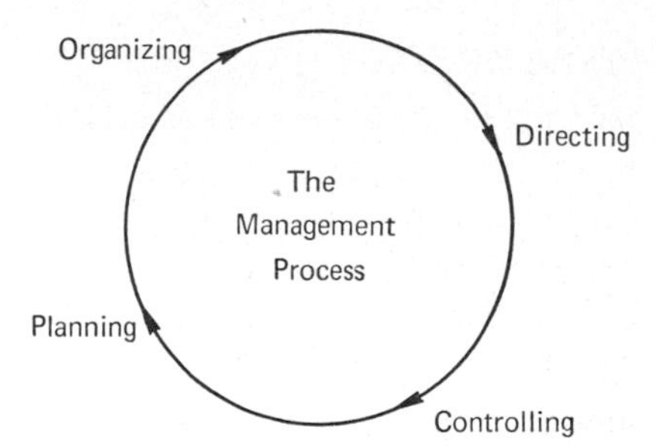

FIGURE 5-2 *The interrelatedness of the management functions.*

Plan for controlling
Organize for planning
Organize for directing
Organize for controlling
Direct planning
Direct organizing
Direct controlling
Control planning
Control organizing
Control directing

However facetious this delineation may seem, it is indeed valid. For instance, even if one views planning in the restricted sense of making strategic decisions which affect the long-range future of the organization, the manager must make decisions regarding the long-range future of his resources and their relationships (organizing), the long-range future of his personnel and their needs (motivating), and the long-range future for the comparison of performance with standards (controlling). Thus he is planning for organizing, motivating, and controlling. A broader view of the planning function leads inexorably to similar interrelationships.

Similarly, the manager must direct (or, in its more humanistic semantics, *motivate* those who will carry on) activities involving planning, organizing, and controlling. And, therefore, a little thought leads one to the conclusion that our "word game" is no game at all. Rather, it is a realistic statement of the existing interrelationships of the management functions. Or, as the systems viewpoint often leads us to conclude, "everything is related to everything else."

Of course, each of these relationships is not equally important. So, it will be the task of much of the remainder of the book to place their relative significances in context.

"TRADITIONAL" MANAGEMENT

The study and practice of management long revolved around a basic model of the organization, called the "bureaucracy," and a set of "principles"—universal truths for "good" management—which are applicable in a wide variety of organizational settings.

Bureaucracy

The bureaucracy is characterized by a number of fixed jurisdictional areas, each with official duties and with individuals who have authority regarding the discharge of these duties. The "system" operates according to fixed rules of superior and subordinate. Individuals are appointed to official positions by superiors, and their status with respect to subordinates is guaranteed by rules of rank.

Bureaucracy is the basic pattern of organization found in the traditional model of management. *Although the bureaucratic form is usually associated with governmental organizations, the structure and processes of bureaucracy are found in many contemporary business and industrial organizations.*

The bureaucracy's primary advantages have been argued by Max Weber:

> *Bureaucratization offers above all the optimum possibility for carrying through the principle of specializing administrative functions according to purely objective considerations. Individual performances are allocated to functionaries who have specialized training and who by constant practice learn more and more. The "objective" discharge of business primarily means a discharge of business according to calculable rules and without regard for persons.*[6]

Weber's phrase "without regard for persons" is of central importance to the bureaucratic concept, for the roots of bureaucracy lie in basic assumptions about people and the way in which they are motivated.

The bureaucratic view is that the "passions" of human beings must be strictly controlled by the organization in order to effectively direct their energies toward the accomplishment of the goals of the organization. The motivation of people under bureaucracy theoretically rested principally on economic matters. Even Frederick Taylor, "the father of scientific

[6]Max Weber, *Essays in Sociology,* ed. and trans. by H. H. Gerth and C. Wright Mills, Oxford University Press, Fair Lawn, N.J., 1946.

management," perpetuated the bureaucratic concept by concentrating his attention on improving the efficiency of the individual in the work situation, to the detriment of the human relations aspect of management. Keith Davis summarizes Taylor's attitude toward the human element by stating: "To Taylor and his contemporaries, human problems stood in the way of production, and so should be removed."[7]

Classical management theory developed and was based on an organizational model of vertical organizational structure. In such a "vertical" organization, problem solving and the resolution of salient issues could be accomplished through the hierarchical structure of superiors and subordinates. In such a system, one simply has to continue up the chain of command until the level is found which possesses the breadth of knowledge and capability to solve the problem. The organization is designed as a pyramid, with a chief executive at the top and working people down the "chain of command."

Perhaps the most incriminating assumption of the bureaucratic model concerns the verticality of the organizational form. When compared with the actual flow of work and organizational deliberations, many of the assumptions surrounding the bureaucratic model seem to be based on nonexistent organizations. Modern organizations take on many different forms, thus severely testing the bureaucratic model of management, which emphasizes the vertical flow of authority and responsibility. Structure in these modern organizations is being subordinated in favor of *processes* or *flows* of resources and the relationships necessary to sustain the organization in its competitive environment.

Principles of Management

"Principles" of management are usually expressed as fundamental truths, universal in application, having value in predicting organizational behavior. A principle is a primary law or doctrine—a settled rule of thought.

For example, the *unity of command* principle holds that an employee should receive orders from only one superior. Consistent with the principle of unity of command is another well-known principle: *parity of authority* (the legal right of a superior to issue orders) *and responsibility* (the obligation to be responsive to orders from a higher superior). The principle holds that the degree of authority and responsibility must be equal for a manager to be effective.

[7] Keith Davis, *Human Relations at Work,* McGraw-Hill Book Company, New York, 1967, p. 9.

A CRITIQUE OF THE "PRINCIPLES" APPROACH A major criticism concerning the traditional "principles" approach to management centers around the lack of data to verify the existence and validity of a principle. Perhaps it is best that we do not consider a principle as a fundamental truth having total and universal application. Rather, it is more useful to consider principles as hypotheses—tentative beliefs or suppositions provisionally selected to explain certain phenomena and to provide guidance for the investigation of others.

The dictionary definition of "hypothesis" might make the point more clear. A hypothesis is "something assumed or conceded merely for the purposes of argument or action."[8] A hypothesis implies insufficiency of current evidence; it is a tentative explanation of facts. It guides in the investigation of circumstances and in predicting what will happen in the future if the future circumstances and facts are reasonably similar to those circumstances and facts which were used to develop the hypothesis.

Thus, a principle should not be thought of as a fundamental truth to be held inviolate or rigid. Rather, it should be used as a *guide* to manager thought and performance. In truth, many of the principles of management are valuable in guiding managerial behavior, and many also have theoretical value in understanding the management discipline. Yet, as the management discipline becomes more mature, certain of our oldest principles become obviously inadequate to explain forces encountered in new environments.

Consider, for example, the principle of *parity of authority and responsibility.* The principle is simply this: "If you are going to give a man responsibility for a job, also give him the necessary authority to discharge that responsibility." This has long been a fundamental basis for organizing people. Yet, many multiproduct organizations have given responsibility for a particular product to "product managers" who must deal with all phases of marketing the product. To do this job, the product manager must deal with individuals in various departments of the business over whom he has no authority. The principle of parity of authority and responsibility appears to be violated in such instances.[9]

If we studied the behavior of many organizations and found this principle followed successfully in practice only 20 percent of the time, it would still have some value in predicting how one might organize. Despite our eagerness to develop new ideas in management, we must not set aside as inappropriate or invalid the concept of management principles. They have both operational merit and predictive value. Moreover, as practical

[8] *Webster's New Collegiate Dictionary,* G. & C. Merriam Co., Publishers, Springfield, Mass., 1961.

[9] For an analysis of the product manager's role, see D. J. Luck and T. Nowak, "Product Management—Vision Unfulfilled," *Harvard Business Review,* May–June, 1967.

men, we must recognize that the application of principles has led to considerable success for many managers and firms.

FAYOL'S PRINCIPLES OF MANAGEMENT Fayol, while pursuing the principles approach, recognized the flexibility of principles by stating, ". . . I shall adopt the term principles whilst dissociating it from any suggestion of rigidity . . . therefore principles are flexible and capable of adaptation to every need."[10]

Fayol listed 14 principles of management which he found most frequently to apply in his work; at the same time he recognized that there was no limit to the number of principles of management. His principles of management can be summarized as follows:

1. *Divison of Work.* This refers to the specialization of labor where the individual is always concerned with the same matters. Division of work is applicable to both technical and managerial work.
2. *Authority and Responsibility.* Authority is the right to give orders and to be obeyed. Responsibility is a corollary of authority—a natural sequence and counterpart—and arises whenever authority is exercised. Authority is official, deriving from office, and personal, "compounded" of intelligence, experience, ability to lead, etc.
3. *Discipline.* Fayol sees discipline as "obedience, application, energy, and respect between employers and employees." Discipline is essential to have a smooth-running operation. So much of the organization's discipline depends on the quality of the leadership, "clear and fair" agreements, and a judicious application of sanctions.
4. *Unity of Command.* This principle means that an individual should receive orders from one superior only. If violated, "authority is undermined, discipline is in jeopardy, order disturbed, and stability threatened."
5. *Unity of Direction.* Each group of activities with the same objective must have one head and one plan. "Unity of direction" referred to organization of the body corporate, whereas "unity of command" referred to the functioning of personnel.
6. *Subordination of Individual Interest to General Interest.* This principle simply states that when individual and organizational interests conflict, the latter must prevail. Fayol recognizes the role of the superior in this principle by setting a good example.
7. *Remuneration of Personnel.* Remuneration for services rendered should be fair and satisfy both employee and employer.

[10] Henri Fayol, *General and Industrial Management,* Sir Isaac Pitman and Sons, Ltd., London, 1949, p. 19.

8. *Centralization.* Centralization is a natural state of affairs in organisms and organizations. Fayol refers to centralization in the context of authority.
9. *Scalar Chain.* A chain of superiors is found in organizations ranging from the top authority down through descending levels to the lowest ranks. Fayol recognizes that departure from the chain is necessary, but the chain should not be disregarded needlessly.
10. *Order.* Fayol conceives of a "place for everything and everything in place." He deals with order in material things as well as social order, whereby employees are in their appointed place.
11. *Equity.* Personnel must be treated with kindness and equity if devotion and loyalty are expected of them. Equity requires good judgment in its application and does not exclude the use of forcefulness and sternness when required. The degree of equity found throughout the organization depends on that equity provided by the head of the business.
12. *Stability of Tenure of Personnel.* Fayol points out some of the problems that occur when people are not given sufficient time to learn their jobs. Frequent turnover will cause the work to be improperly done.
13. *Initiative.* Fayol believes that in thinking out a plan and ensuring its success there is a keen satisfaction for a man to experience. The power of thinking through and executing is what is called "initiative." Initiative on the part of the members of an organization can be a great source of organizational strength.
14. *Esprit de Corps.* In "union there is strength." Harmony among organizational personnel is a source of strength. Union among personnel is accomplished by communication, emphasizing verbal contacts where possible.

In subsequent traditional management literature, much attention has been given to further development, explanation, and expansion of a basic set of principles. For instance, Koontz and O'Donnell give 11 principles related solely to the development and training of managers.[11]

We shall here adopt the posture that *much value can be gained from the study that goes into the development of a principle, whether the principle itself is valuable or not.* Thus, while the modern manager cannot view principles as rigid rules to be followed, part of every manager's philosophy of management should be a willingness to examine and consider putting to use those principles which have been developed, analyzed, and tested.

[11] H. Koontz and C. O'Donnell, *Principles of Management,* McGraw-Hill Book Company, New York, 1964, pp. 446–470.

Line and Staff

One of the concepts of management to which most practitioners and academicians have subscribed is the division of organizational elements into *line* and *staff* groupings. One author, in commenting on the line-staff phenomenon, states: "The concept that all functions or departments of a business enterprise are either 'line' or 'staff' is now so firmly entrenched in management theory that any attempt to dislodge it may well seem doomed to failure."[12]

The traditional view of the line-staff dichotomy comes about because of authority relationships in the organization. *Line* managers make the salient decisions by exercising command authority, whereas *staff* officials advise and counsel, with no authority to command except within their own staff "chain of command." As an adviser, the staff official functions as an extension of the personality and authority of the line official whom he serves.

The *line* official is often described as the individual who stands in the primary chain of command and is directly concerned with accomplishing the organic objectives of the organization. Line elements provide decision authority and a central means for the flow of communications through a scalar chain of authority; staff elements facilitate the decision process by bringing in expert and specialized knowledge.

The concept of line and staff can be found in the history of military and ecclesiastical institutions. The use of staff assistants to assist primary managers came about because of the need to provide special counsel and assistance to the manager who was unable to carry out the demands of a particular position. The use of assistants by a manager provided the means for the manager to extend himself and to enlarge the scope of his influence. A brief examination of the evolution of the staff will add to a better understanding of the concept.

EVOLUTION OF STAFF ORGANIZATION Growth in organizations brings about the requirement for the chief executive to appoint individuals to assist him in his responsibilities. Initially, he will find the administrative responsibilities of his job so demanding that he has to appoint assistant managers to perform primary duties. Eventually the organization will become so complicated that the chief executive's responsibilities will grow beyond his personal relationships to the job. At this time the need will appear for a personal staff official to provide the chief executive with specialized personal assistance, much as the alter ego of the chief execu-

[12] H. Logan Hall, "Line and Staff: An Obsolete Concept," *Personnel,* January–February, 1966, p. 26.

tive. As organizational growth continues there will exist the need for specialized assistance in labor relations, law, finance, marketing, quality control, computer technology, and so forth. Then, specialized staffs are created and developed within the organization. They are designed to give counsel and assistance in each special field of effort. This development of staff is traced in Figure 5-3.

The authority position of the staff official is described by Fayol as "an adjunct, reinforcement and sort of extension of the manager's personality. There are no levels in it and it takes orders only from the general manager."[13] Fayol further envisioned the staff as serving the executive in carrying out his personal duties, yet not exercising any executive part in subordinated organizations.

CLASSICAL CONCEPTS OF LINE AND STAFF The following descriptions serve to clarify the differences between line and staff:

1. "Line" refers to those functions that have a *direct responsibility* for accomplishing the end purpose of the organization.
2. "Line" refers to those individuals who are *directly* concerned with producing the good or service offered by the organization.
3. "Line" refers to those executives who have *operational* responsibilities.
4. Line executives tend to be *generalists.*
5. Line executives *make and execute decisions* related to the attainment of primary organizational objectives.

Thus, "line" refers to those organizational positions whose holders have responsibility and are held accountable for accomplishing primary objectives. Line officials are in the "chain of command" running from the highest executive in the organization to the lowest individual. This is referred to by Fayol as the *scalar chain*[14] and is illustrated by a line of authority running from the board of directors down through subordinate managers to the workers. Each successive manager exercises direct line command over his subordinate. The primary purpose of line is to make things happen—to order things into existence.

Staff, as contrasted with line, exists to advise and counsel, rather than to command. The following characteristics summarize the nature of staff:

1. Staff members perform purely *advisory* functions.
2. Each staff member reports to a boss who is appended to the line or-

[13] Fayol, *op. cit.,* p. 63.
[14] *Ibid.,* p. 34.

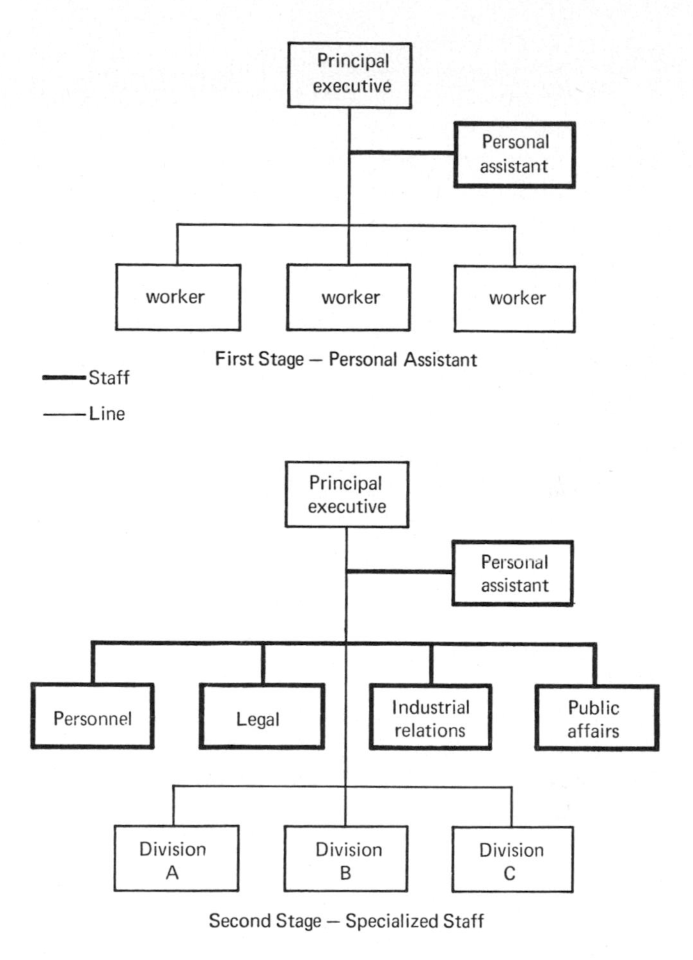

FIGURE 5-3 *Evolution of staff.*

ganization or is a member of the line organization. Therefore, there is *no pure chain of command in staff.*

3. Staff members are typically technical *specialists.*
4. Staff *investigates* and supplies information and recommendations to managers who make decisions.
5. Staff provides *service and functional guidance* and control through the exercise of functional authority.

A CRITIQUE OF LINE AND STAFF Placing organizational groups into categories of line and staff creates opportunity for conflict because of the way the individual perceives his role. Why might this conflict exist?

1. Line executives tend to be suspicious of staff officials, who advise a course of action but do not bear the direct responsibility for seeing it implemented.
2. Staff personnel, who are experts in their field, can easily become overzealous and, seeing the world through parochial glasses, can frequently push their recommendations well beyond their authority to advise and counsel.
3. When a staff official infringes on the authority of a line official, the result can easily be resentment, hostility, and, open or inner reluctance to follow the staff official's recommendation.
4. Line people may not know how to use staff properly. If the line official employs his staff on minutiae in order to keep them busy, there will be resentment. Conversely, if the staff's advice and counsel are not sought, the staff people can feel unneeded and indifferent.[15]
5. Staff may perpetuate itself by broadening its scope of operation until there emerges a proliferation of staff activities at many different levels in the organization. The simple majority of staff people over line executives can create problems of span of control with consequent frustration on the part of line officials dealing with current problems.
6. The elements of demarcation between line and staff authority are rarely clear. When an executive follows the recommendation of a staff official, the legal authority may be his, but the judgment is that of the staff official.[16]
7. Many jobs in line and staff almost defy description except in general terms or in terms related to the results expected of a position rather than of *how* the results are to be accomplished. When job positions are nebulous and care is not taken otherwise to clarify the relationships between line and staff people, the opportunity for overlaps and gaps in authority and responsibility can aggravate personal relationships.[17]
8. There may be overemphasis on the direct relationship of line officials to the primary objectives of the organization, and hence they are

[15] This does not mean that the line officials have to accept the advice of the staff official; but the line official should consult with the staff before taking any action. This mandatory consultation is often described as the principle of compulsory *staff service* and had its genesis in the organization of the Catholic Church.

[16] Adapted from Herbert A. Simon, "Staff and Management Controls," *The Annals of the American Academy of Political and Social Science,* March, 1964, p. 99.

[17] The traditional means of charting organizational structure does not help much to clarify these relationships. David I. Cleland and Wallace Munsey have suggested a new method of charting interpersonal relationships incorporating systems theory. See "Who Works with Whom," *Harvard Business Review,* September–October, 1967, pp. 84–90.

held solely accountable for operating decisions and profit responsibility.

9. People may fail to understand the nature of functional authority, particularly when it is exercised by a staff official. An attitude of "no staff man is going to tell me what to do" on the part of a line official can cause the best of line-staff harmony to deteriorate.
10. Both line and staff may fail to recognize that authority is a function of a de jure organizational position that one holds *as well as* of such things as knowledge, ability to build alliances, reciprocity, and other de facto sources.

Even with relationships between line and staff well defined, conflict will continue to exist. A certain amount of conflict is inevitable and even desirable. If the conflict breaks into open and continuous warfare and degenerates into personal attacks, organizational inefficiency can result. Line and staff should work together as a team to enhance the smooth functioning and productivity of the organization.

Many things depend on what makes good line-staff relations. Education of the people involved is essential. This can be accomplished through training sessions in which the reciprocal authority and responsibility of line and staff officials are discussed; the use of brochures, pamphlets, and other media with messages such as those shown in Figure 5-4 will often prove useful in this function.

MODERN MANAGEMENT

In earlier chapters we introduced the manager's job in terms of two principal functions—deciding and doing. Since the functions are really more complex than is described by those simple terms, we shall refer to them by the more appropriate titles of *planning* and *execution*.

Planning

Planning is the process of thinking about the job to be done, considering what is needed to do it in terms of equipment, people, facilities, and other resources, and coming up with the plans necessary to delineate how the job can best be accomplished. Planning encompasses several major subfunctions:

1. *LONG-RANGE PLANNING* Planning for periods often in excess of one year, which involves all of the functional areas of the business and is effected within the framework of the economic, social, and technological

THIS IS A LINE ORGANIZATION

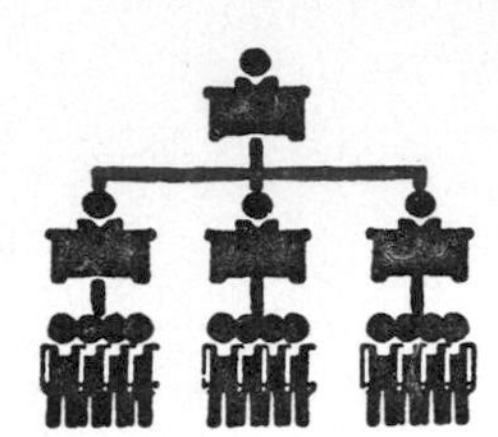

It means that each person has ONE BOSS—Unity of Command

HUMBLE HAS A LINE AND STAFF ORGANIZATION

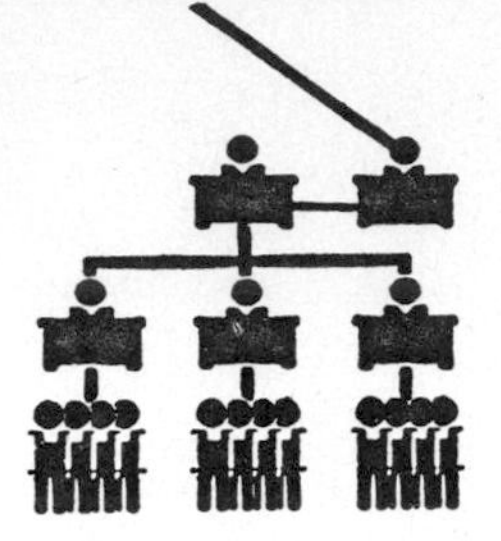

This still means that each person has one boss. The STAFF MAN helps the boss do a more effective job.

THE STAFF MAN
investigates, plans, advises, SERVES

A supervisor directs the work of others. He delegates portions of his authority to others. He needs help on problems involving policy interpretation, company plans, engineering changes, technical information, training, etc. Staff services—THE STAFF MAN—PROVIDES THIS HELP.

THE STAFF MAN
doesn't give orders to line
doesn't perform line duties
HE SOLVES SPECIAL PROBLEMS

He must be familiar with all of the operations and functions of the unit in which he works. Alert observation of conditions enables him to detect potential trouble and do his part in keeping everything operating smoothly.

THE STAFF MAN
DOESN'T TURN IN HALF-BAKED IDEAS

Completed STAFF work is the study of a problem, and presentation of a solution, by a STAFF ASSISTANT in such form that all that remains to be done on the part of supervision is to indicate its approval or disapproval.

THE STAFF MAN
must be tactful and diplomatic, as he contacts many people.
HE CREATES HARMONY BY COOPERATING WITH OTHER AGENCIES IN THE COMPANY

Close cooperation between STAFF MEN with similar assignments, either in the same or in different organizational levels, results in a rapid solution of many problems.

THE MORE COMPLICATED THE BOSS'S JOB IS THE MORE HE NEEDS A STAFF MAN

THE STAFF MAN
IS A SUPPORT FOR THE BOSS

WHEN THE LOAD GETS TOO HEAVY OR THE PROBLEM TOO DIFFICULT, CHECK TO SEE IF THERE IS A STAFF SERVICE AVAILABLE TO ASSIST.

Certain departments such as personnel, training, etc., are a STAFF for everyone. Use their services if you can. The measure of a competent supervisor—is he making full use of all management tools and services?

FIGURE 5-4 *Humble Oil Company—line and staff organization.* SOURCE: *Comparison of Line and Staff Characteristics. Copyright, 1956, by National Industrial Conference Board, Inc. Taken from* Studies in Personnel Policy *no. 153, "Improving Line and Staff Relationships," p. 73. Used by permission by N.I.C.B. and the Humble Oil Company.*

environment is generally referred to as "long-range planning." This concept of planning as "the process of preparing for the commitment of resources in the most economical fashion, and, by preparing, or allowing this commitment to be made less disruptively,"[18] is termed "long-range" since it involves explicit consideration of the (sometimes distant) future.

2. *SHORT-RANGE PLANNING* On the other hand, short-range planning is usually held to involve periods of one year or less.

3. *STRATEGIC DECISION MAKING* In discussing the management function called "planning" in this book, we shall also consider the process of *strategic decision making* to be an element of planning. Strategic decision making involves consideration of the alternative allocations of resources which allow the organization to pursue its goals.

The relationship of the long-range planning and strategic decision-making phases of the planning function can be readily illustrated in the marketing context. Long-range marketing planning encompasses the consideration of the alternative ways of achieving the goals of the firm.[19] Thus, if the future pattern of per-share corporate earnings is the relevant measure of the degree of attainment of the corporation's objectives, the future goals might be achieved by penetrating new markets, introducing new products, expanding sales of existing products, etc. The strategic decisions of the company encompass specific consideration of alternative ways of achieving these desired states. For instance, the firm might wish to consider the relative worth of a new product versus an expanded advertising outlay. Or it might desire to choose a "best" new product. In either case, strategic choices—those related to corporate goals—are being made.[20]

The distinction between long-range and short-range planning may be thought of by relating the planning to the product or service that is involved. Using this criterion, we may assume the following:

Long-range Planning: Deals with future generations of products and/or services.
Short-range Planning: Deals with the current generation of products and/or services.

[18] E. Kirby Warren, *Long-Range Planning: The Executive Viewpoint,* Prentice-Hall, Inc., Englewood Cliffs, N.J., 1966, p. 21.

[19] See Mark Stern, *Marketing Planning,* McGraw-Hill Book Company, New York, 1966, for an extensive discussion of planning in marketing.

[20] See William R. King, *Quantitative Analysis for Marketing Management,* McGraw-Hill Book Company, New York, 1967, for a treatment of strategic decision making in marketing.

Thus, the view of planning that encompasses long-range planning, short-range planning, and strategic decision making defines an element of management which each of us must perform in our daily lives, regardless of whether or not we think of ourselves as managers. Everyone must be concerned with allocating resources. Even one who chooses to lead a life in relative isolation from the rest of the world does not have unlimited resources at his disposal, but must also face the issue of the economical allocation of those which are available to him. Planning simply involves extending these considerations to the long run to think about the future, one's goals, and the various resource allocations which can be used to achieve them.

Execution

The execution phase of management involves several different subfunctions. The most important of these are *organizing, directing,* and *controlling.*

1. *ORGANIZING* Organizing involves a formalization of the thinking and planning required for the organization into some orderly structure of processes and flows so that a stable base for the grouping of activities and equipment is provided. Organizing contributes directly to the planning process, for it is through organizing that the individual plans of action that are developed in the organization are brought to reality. Organizing involves the procurement of the necessary human and nonhuman resources, the grouping and layout of the resources, the delineation of authority and responsibility relationships within the organization structure, and the recognition of the authority and responsibility patterns existing between the participants in the overall system.

2. *DIRECTING* Direction has to do with the face-to-face leadership between superiors, subordinates, peers, and associates. Sometimes called "motivating," directing entails giving sympathy and encouragement, delineating instructions, doing counseling, interpreting policy, and such related activities which set the organization into motion and keep it moving toward the objective. Directing is a function of management which bears special consideration for scholars and practitioners of management, for it is the process of getting people to cooperate effectively regardless of their status or perceived role. The directing functions deal largely with the human subsystem of organizations and become particularly important when one has to do a job which requires the crossing of many different organizational lines wherever located in accomplishing the objective.

3. *CONTROLLING* Controlling is the process of making events conform to plans; it involves such diverse activities as timing and scheduling the work of the organization and relating it to the efforts of individuals and organizational segments. Coordination is often described as an essential part of the control function having to do with the synchronization of activities with respect to time and place in order to make the best use of the resources directed toward the same or related purposes. Controlling also involves checking progress made against plans, setting up standards of individual and organizational performance, checking against these standards, and by other means making sure the objectives are being accomplished as originally anticipated in the organization's plans.

Planning and Execution

The manager performs all of these functions—planning (in any or all of the long-range, short-range, or strategic decision-making contexts), organizing, directing, and controlling—more or less continuously, and regardless of his organizational level, although the emphasis placed on each function is different at the different organization levels. The operational manager who is charged with the responsibility for accomplishing a specific mission is most concerned with the control function in performing the mission itself. On the other hand, a staff official who is charged with the development of the overall plan of a project is more involved with the planning function than with organizing or controlling. Indeed, Drucker holds that "planning and doing are separate parts of the same job; they are not separate jobs. There is no work that can be performed effectively unless it contains elements of both—advocating the divorce of the two is like demanding that swallowing food and digesting it be carried on in separate bodies."[21]

CHANGING MANAGEMENT PATTERNS

Management, like most disciplines, is subject to change. As we become more knowledgeable about how organizational systems function, it is expected that new concepts will be developed and old ideas modified. Relative to other scientific disciplines, management is still in its infancy. Most of the literature that has appeared has come forth in the present century, yet the process of managing organizations has been with us since antiquity.

A rough history of classical management theory has been given

[21] Peter F. Drucker, *The Practice of Management,* Harper & Row, Publishers, Inc., New York, 1954, p. 284.

which describes early concepts of the management process. It is difficult to explicitly identify the period of time which saw the greatest development of these beliefs. One very rough way is to say that these concepts of management appeared during the first half of this century and prevailed as a way of thinking about management until the mid-fifties. Since then, there has been a virtual explosion in management theory. New schools of thought have emerged and new techniques have been tested and proven to facilitate the management of human and nonhuman resources.

We have chosen to call these earlier theories and techniques "traditional," or "classical," management. By "classical" we do not mean "backward" or "outmoded." This term is simply a way of portraying a type and style of management which served admirably in earlier times and which is the basis for today's more advanced techniques. Classical management has its unquestioned place as the core of thought about how organized groups pursue means in accomplishing desired ends. At times the reader may feel that the tenor of the approach taken here represents an unrelenting attack on classical theory. This is not our purpose—we wish to examine both the *suitability* and the *inadequacy* of classical management concepts. Examining traditional theory in this light does not result in throwing out everything that is traditional. Rather, it results in a combination of the best of traditional thought with the best of modern techniques. It is hoped that this combination will prove superior to either view taken alone.

SUMMARY

Management has traditionally been viewed in terms of the manager's *functions.* Various authors use different terms—planning, motivating, controlling, directing, etc.—to describe these functions, but most agree on the job definition of the manager.

Perhaps more important than the names which are used is the interdependence of management functions. Managers must not only plan, organize, motivate, and control, they must plan for organizing, plan for motivating, plan for controlling, etc. Thus, the manager's job is complex, and simple functional models of it do not adequately describe its complexity.

This book utilizes a variety of models of the manager's job from the simple "Decide and Do" one, through "Plan-Organize-Motivate-Control" to more complex models which will be introduced later. As with any model, the key question is not, "which model is right?"; rather it is, "which is most useful?"

Traditional management utilizes the bureaucratic model of the organization—fixed jurisdictional areas, fixed authorities, and the "rule of rank." Such a model is not very descriptive of many modern organizations—even the military and ecclesiastical ones which the model was originally intended to describe.

Another aspect of traditional management thought is the "principles" approach. Principles are sometimes thought of as universal truths, although their adaptation to the modern world might best be accomplished by viewing them as partially-tested hypotheses which are to serve as guides and be further evaluated.

Traditional management also focused on the line-staff dichotomy in which line officials were concerned with the direct seeking of the organization's objectives, while staff officials were advisors to the line officer. Again, this "military-oriented" model has value in modern management, but it is not, and should not be, slavishly followed in structuring an organization.

Modern management is built on the use of systems concepts and formal decision analysis, together with a recognition of the importance of the human element in organizations and the social responsibility of organizations. The modern manager *plans*—in the long-range, short-range, and strategic decision making contexts—and executes by organizing, motivating, and controlling. He does these things more or less continuously regardless of his organizational level.

Thus, traditional management and modern management have merged to a degree through the integration of the best of the classical approach with the newer concepts of systems, formal decision analysis, human behavior, and social responsibility. However, much is left to be done to provide a truly integrated unified approach to management. In later chapters, we shall attempt to develop a model of modern management. This model is one which should not quickly become outdated, as have some of the precepts of traditional management, because it explicitly recognizes the need for change in both management practice and management theory.

EXERCISES

1. What are management principles? Define the role which they can play in modern management theory and practice.
2. What are management functions? What is their role in management theory and practice?
3. Identify several actions that you have seen taken recently by a professor, dean, or manager that you know. Try to classify them in terms of the functions of the manager.
4. Refer to Figure 5-1. Start at the center of the circle.
   a. What meaning does the "elements" circle have? How are these elements related to the management functions? Explain.
   b. Refer to the "tasks" circle and relate the qualifications necessary to perform the various tasks to the functions of the manager. Can one person do all of these?

c. Relate the "activities" circle to the discussion of "Management Subfunctions" on pages 119–120. How are the descriptions the same and how are they different? Why?

d. Give an illustration of an instance in which you have seen someone perform each of these "activities." Note their "definitions" in the outermost circle.

5. Explain the meaning of the "word game" discussed on pages 122–123 under the heading "The Interrelatedness of Management Functions."

6. What is a bureaucracy? Does your university qualify under this definition? Name some other organizations that so qualify and argue why they do.

7. The text argues that the "father of scientific management," F. W. Taylor, furthered the bureaucratic concept. Why? After all, isn't "scientific management" incompatible with bureaucracy?

8. What is meant by a "vertical organizational structure"? In what sense is it unrealistic?

9. What is the "chain of command"? What is the rationale for its existence?

10. Give an illustration of a situation in which it might be desirable to violate the "unity of command" management principle.

11. If you are not given authority to match your responsibility, the principle of "parity of authority and responsibility" has obviously been violated. What would you do in such a case? What special talents would be required on your part to be successful?

12. Discuss Fayol's principle of "subordination of individual interest to general interest" as it may have changed in application from his time to ours.

13. Consider all of Fayol's 14 principles in terms of illustrations from your everyday experience. Which have you seen put to use? Where? Which have you seen violated? Where? Make some assessment of the value of each principle in each situation which you describe.

14. Differentiate between line and staff organizations.

15. In his *The New Industrial State* Galbraith argues that the chief executive's actions are severely constrained because of the extensive study, research, and discussion which has been previously conducted on an issue by lower level "technocrats." How does this relate to line and staff concepts and practice? If this is true, does it mean that the chief executive is a mere figurehead?

16. Relate the planning-execution dichotomy to the model of Figure 5-1. Describe the geometric shape of each of these two major functions in terms of that figure.

17. Is it meaningful to distinguish between "long-range" and "short-range" in terms of fixed durations? What is a better way?

18. The strategic problem of marketing is often thought of in terms of a table such as the one below.

		Markets	
		Existing	*New*
Products	Existing	X	
	New		O

The cross indicates a choice to continue doing what the organization is now doing—selling existing products in existing markets. The circle represents an entirely new venture. If the goal of the company is profit maximization, discuss each of the four opportunities in the table in terms of their relationship to the achievement of the goal.

19. How does the problem in exercise 18 relate to planning?

20. Define "directing," "controlling," and "organizing."

21. One author has differentiated line and staff by defining line functions as ". . . those that contribute directly to accomplishment of organization's primary objectives." How might this statement be related to research and development management within an industrial organization?

REFERENCES

Bassett, Glen A., "Qualifications of a Manager," *California Management Review,* Winter, 1969.

Boynton, Robert E., "Policies of the Successful Manager," *California Management Review,* Fall, 1970.

Cleland, David I., "Completed Staff Action," *Manage,* August, 1966.

Davis, Ralph C., *The Fundamentals of Top Management,* Harper & Brothers, New York, 1951.

Drucker, Peter, "Management's New Role," *Harvard Business Review,* November–December, 1969.

Fayol, Henri, *General and Industrial Management,* Sir Isaac Pitman & Sons, Ltd., London, 1949.

Haimann, Theo, *Professional Management,* Houghton Mifflin Company, New York, 1962.

Kepner, Charles H., and Benjamin B. Tregoe, *The Rational Manager,* McGraw-Hill Book Company, New York, 1965.

Koontz, Harold (ed.), *A Unified Theory of Management,* McGraw-Hill Book Company, New York, 1964.

Levinson, Harry, "Management by Whose Objectives?" *Harvard Business Review,* July–August, 1970.

Logan, Paul H., "Line and Staff: An Obsolete Concept," *Personnel,* January–February, 1966.

Mackenzie, R. Alec, "The Management Process in 3-D," *Harvard Business Review,* November–December, 1969.

Mee, John F., "Pioneers of Management," *Advanced Management - Office Executive,* October, 1962.

Middleton, C. J., "How to Set Up a Project Organization," *Harvard Business Review,* March–April, 1967.

Rockwell, Jr., W. F., "The Eight Hats of the Chief Executive," *Dun's Review,* June, 1967.

6

the systems approach to management

In previous chapters, the systems approach to management of organizations has been briefly introduced. In this chapter, we seek to enlarge on basic systems concepts to demonstrate their direct applicability to the "deciding" and "doing" phases of the manager's job.

THE SYSTEMS VIEWPOINT

The essential element of a systems viewpoint is a perspective of the organization as a conglomerate of *interrelated* and *interdependent* parts. No one of the parts (subsystems) can perform effectively without others, and any action taken on (or by) one will have effects which can be traced throughout the organization and throughout the complex environment in which the organization exists.

The systems view of organizations is a dynamic one which recognizes that the interaction of many different elements is necessary to produce desired effects and that "things (both desired and undesired) are always happening" which have effects and implications transcending their immediate and obvious impact.

Indeed, the expression that "everything depends on everything else" is perhaps the best way of thinking about the systems viewpoint. Although that thought is not a calming one, it does enable us to raise ourselves above the level of provincialism to see the right problem, however complex it may be.

Chains of Effects

The "chain of effects" which is established by an action on or by one part of a system is significant to the systems viewpoint. If we view the development of our modern society in systems terms, we can see how such "chains of effects" have evolved. For example, technological innovations have produced both direct and indirect benefits, and at the same time, they have created new problems. The automobile is a good illustration. The immediate impact of the invention of the automobile was the creation of a nuisance which frightened horses and created confusion. Those with foresight could readily see that the potential benefit of this device for transportation was great, but few could have foreseen the scope of the impact which was to be created by the new industry on the nation's economy and on the day-to-day lives and habits of its people. Moreover, it is apparent that the profound problems which would be created or amplified by the automobile—e.g., traffic congestion and air pollution—were not envisioned by anyone in the early days of its use. Thus, the new invention set up a complex series of events, initially negative, then positive, then again negative, which have directly influenced the life of every living American in rather significant ways.

Other technological advances have triggered similar chains of events. The development of indoor plumbing made high-rise office buildings and apartments feasible. This led to center-city congestion and parking problems which motivated an exodus to the suburbs for both business and personal housing. This, in turn, led to problems in the suburbs similar to those that had been experienced in the center cities.

Within the organization, similar phenomena occur. An executive order issued to managers eventually creates impact on the lowest-level workers. An order affecting one department changes that department's behavior in dealing with other departments. This induces changes in them as well, and, in turn, their behavior affects those with whom they deal. The ultimate effect may be far-reaching.

STATE TRANSITIONS In understanding the subtleties of systems concepts and chains of effects it is useful to think in *state transition* terms. At any point in time, any entity is in a *state.* This state may be described by a number of characteristics. For instance, at the time of this writing, the authors are (*a*) informally dressed, (*b*) at room temperature, (*c*) hungry, and (*d*) in a state of panic caused by the need to complete the manuscript. At a later point in time, after the manuscript is mailed to the publisher and we are on our way to our homes, our state will be (*a*) informally dressed and outerwear, (*b*) at 32° F, (*c*) hungry, and (*d*) less panicky.

Thus, *the state of an entity changes from time to time as a result of actions taken on or by it.* Such a change is referred to as a "state transition."

One can define states to be as comprehensive and as complex as is necessary. For instance, a person's state may involve numerical characterizations of hundreds of measurements of his body state—blood pressure, blood count, heart rate, etc. Such a state description would be useful to a physician, but perhaps not so useful if we wished to determine something about his soap purchasing habits. In other words, like any model, a state transition description can be made to suit a given purpose.

DECISION TREES The state transition model can be used both to illustrate and to predict the potential impact of chains of effects. To do this, consider the simple "decision tree" shown in Figure 6-1. This representation depicts a series of decisions—choices among alternatives. Starting at the left, a circle depicts a state, described in whatever terms are useful and here labeled state #1. In that state, the decision maker is faced with two alternatives, labeled a and b. If he selects alternative a, he proceeds to a new state—#2 represented by the next circle along the topmost main branch of the tree. In that state, he has three alternatives from which to choose—labeled c, d, and e. If, on the other hand, he selects alternative b in the first state, he arrives at a new state—#3—which presents him with three other alternatives—f, g, and h.

This abstract decision tree model illustrates how each state transition involves two aspects:

1. An immediate outcome
2. A new state and a new problem situation

Thus, when the decision maker chooses to make the state transition from #1 using alternative a, he obtains as a result:

1. The immediate consequences of alternative a
2. State #2, with its associated choice among c, d, and e

If he makes the transition from state #1 by using b, he obtains:

FIGURE 6-1 *A simple decision tree.*

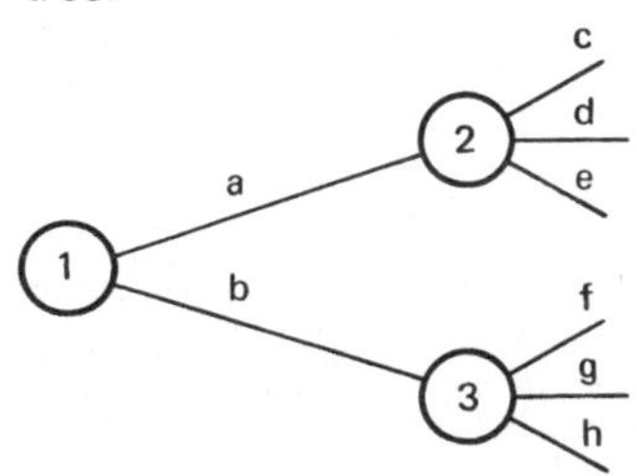

1. The immediate consequences of b
2. State #3, with its associated choice among f, g, and h.

Now, let us relate state transitions to the illustrations used previously. Our society was, in 1895 or so, in a state in which primary personal transportation was by horse-drawn vehicles—state #1. Society "chose" an alternative—"build automobiles"—to lead it to state #2. What it got from this was:

1. A whole new mode of personal transport with multitudinous attendant benefits
2. New decision problems involving air pollution and traffic congestion

Now society has transited to state #2 and it is faced with the need for a solution of that decision problem.

Similarly, if a business firm introduces a new product, it can decide to adopt a "skim the cream" pricing policy, since it is first with the product in the marketplace. This represents a choice of one pricing policy from a set of alternatives which are available. As a result, competitors know that they can sell for less, and they hurry to develop and market competitive products. The effect of the transition from a state in which the innovative firm has no competitors and can charge whatever it wishes has been:

1. High initial profits resulting from the "skim the cream" policy
2. A new state involving many competitors and a new pricing problem which is created by competitive price cutting

Of course, such chains of benefits and new states can be carried out for many stages. Very "tall" trees of the nature of Figure 6-1, can be used to characterize multiple stages.

Moreover, the same sort of phenomenon and the same sort of model can be applied to a wide variety of situations, whether they involve the conscious-choice process of an individual or organization or not. The automobile illustration was clearly of the latter variety, since no individual was faced with the problem of deciding whether or not to build automobiles. However, society did choose between steam-powered and gasoline-powered vehicles, and indeed, it chose between the "newfangled" auto and horse-drawn vehicles. It made its choice over a prolonged period of time through the aggregation of a large number of individual vehicle purchase decisions which eventually forced carriages and steam autos out of existence (or at least drove them to play a minor transport role). But, however long it took and however complex was the decision process, society did choose, and it did obtain, as a direct consequence of that choice, immediate benefits and new problems, many of which have only recently become apparent.

USING SYSTEMS CONCEPTS

The key utility of the systems viewpoint, with its emphasis on a view giving consideration to interactions, interdependencies, and chains of effects, is not its academic value, but rather its applicability to the real world.

SYSTEMS PERCEPTIONS One of the greatest benefits and utilities of systems concepts is the better understanding or perception of complex systems which they bring about. Focusing on chains of effects, interdependencies, and interrelatedness generally leads to either better understanding or a recognition of the areas in which understanding is lacking. This, in turn, can lead to better decisions and, thereby, better management.

An illustration of this sort of benefit lies in the view that the organization takes of the product or service that it offers. Few modern organizations sell simply a physical entity; rather they have come to sell a "package" of interrelated products and services—a *product system.* For example, the manufacturer who color-coordinates towels, bedspreads, sheets, pillowcases, and draperies is utilizing interdependencies to encourage extra purchases. So, too, is the manufacturer of industrial equipment who provides consulting services, economic feasibility studies, time and motion studies, and maintenance and repair services along with machinery.

A good illustration of a firm offering systems-oriented "products" is Ogden Corporation's assumption of responsibility for everything from market analysis to packaging and financing on many projects which may superficially look like simple transportation problems. With regard to Ogden's involvement in Morocco's effort to improve its balance of payments via the export of oranges, a company official is quoted as saying:

> *The first thing a team looks for is an analysis of markets. Then we look for a packaging system that lends itself to materials handling through unitizing or palletizing—or perhaps for a bulk handling system. Then we try to design a total system capability that integrates the products and the markets. Finally, we try to work out financing so the system will pay for itself.*[1]

[1]"Transportation: Ogden Engineers a Total System," *Business Week,* Mar. 15, 1968, pp. 58–62.

In the consumer area, the same systems approach is apparent. Nobody sells toothpaste anymore. Toothpaste manufacturers sell a system which will clean teeth, sweeten breath, improve your social life, and whatever else you want and are likely to believe. Diverse products are displayed in supermarkets in "diet centers" to encourage impulse and "tie-in" and "extra-item" purchasing. Before improved systems understanding, the dieter would have had to seek out diet vegetables in one section, diet fruits in another, etc. Now he is served conveniently, and he undoubtedly buys more because it is all presented to him without effort on his part.

The idea of a system of services has been adopted in the organizational context in terms of "projects" for the study of major decisions. Decisions regarding such things as potential corporate mergers or acquisitions, the location of new plants, etc., are so complex and require input information from such a wide variety of sources that it is often impractical to analyze them without the full-time assignment of personnel who can devote themselves to performing the analysis on which a decision will be based. Such a project organization has the ability to concentrate on the decision at hand and to bring to bear the many different specialists which may be required. The existence of such an organization illustrates an understanding of the many interdependencies associated with such complex decisions. Hence, it is a direct application of the systems approach.

SYSTEMS PREDICTION The basic benefit of systems concepts and models is better understanding. One who recognizes the systems concept is usually better able to understand today's problems in the light of yesterday's happenings and yesterday's choices.

However much understanding may be valued for itself, it is not terribly useful to the manager unless he is able to use it to make *predictions* of the future. The manager's job can be thought of as one of predicting the future consequences (the "entire" chain of effects) of choices made in current states. If he can accurately predict the consequences of a pricing policy in terms of the benefits (profits) and the new problems which will be created by it, he is able to make a rational choice and to intelligently guide his organization into the future. He can be expected to do little more.

Of course, no manager can even be expected to foresee all of the interdependencies and the "entire" chain of effects, since there are an infinity of interrelationships and effects in any complex system or decision situation. He can do so, however, within the context of a *systems model.* For instance, he can use a decision tree model such as that in Figure 6-1. The tree model is developed to extend into the future as far as can be reasonably foreseen, and it incorporates into the state descriptions all of those factors which are deemed to be salient to the purpose at hand.

In this way the manager can make systems predictions without being clairvoyant or mystical. He then integrates the various predictions into the model and uses it to make overall predictions and evaluations of alternative actions.

Some illustrations will serve to show that these predictions can be useful without necessarily being precisely accurate or even quantitative in nature.

Jay Forrester at MIT has developed a computerized systems simulation model which he refers to as "industrial dynamics." While his modeling technique is much more advanced than the approach used in Figure 6-1, it is based on the same concepts. He applied his approach in an advertising situation and the model revealed the degree to which the effect of advertising on consumers is displaced in time from the day of appearance of the ad. Coupled with this, it revealed time lags in the impact of consumer purchases on factory shipments and on the scheduling of new advertising. All of these produced cumulative effects which were not at all apparent before analysis using a systems model.[2]

The importance of such analyses is clear. If the product is one which is largely consumed on hot summer days, should the seller advertise only on hot summer days or in the spring before the heat arrives? Or could the advertising-effect time lags be so long that he should advertise in the winter? Such questions can be answered intelligently only in the light of an understanding of the myriad factors which combine to produce the total lagged effect of advertising. Moreover, such understandings can best come through a systems model, since such a model incorporates the many interdependent factors which combine to make up the time lag in advertising effect.

To illustrate that the manager may need only to be able to make gross qualitative—perhaps directional—predictions of effects, consider the illustration of time lags again. However, this time let us deal with the time lags involved in a manufacturer's perception of the current sales rate of his product. Suppose that a manufacturer's orders tell him that sales of an item are down, so that he discontinues production of that item. After all, he might reason, many of the items are already in inventory in retail stores, in wholesale warehouses, and in his own warehouse. If sales are down, he might as well turn his production facilities to other more salable items.

However, he may fail to perceive that the time lags between purchase

[2] J. W. Forrester, "Advertising: A Problem in Industrial Dynamics," *Harvard Business Review,* March–April, 1959, pp. 100–110.

at the retail store and orders to the manufacturer are amplified by the various stages in the chain of effects created by a consumer purchase. This is so precisely because of the various inventories. When a customer purchases the item, it comes from the retailer's inventory. The retailer orders from the wholesaler only when his inventory is low. So, too, the wholesaler orders from the factory only when his inventory is low. The impact of a consumer purchase may therefore not be felt at the factory until many weeks after it occurs.

Indeed, a virtual sales explosion at the retail level may be so masked by the time lags induced because orders are filled from inventories at several levels that factory orders are a pale reflection of what is happening at the retail level. The effect of the time lag between a customer purchase and a retail store order is amplified so that the end result of the chain of effects may lead the manufacturer to exactly the wrong conclusion. He may decide that sales are down because of declining wholesaler orders, when they are actually up at the retail level. If he acts to stop production on the basis of this information, he may be deluged with unfillable orders when the effects of increased retail sales begin to reach him. Thus, without an understanding of the chain of effects, he is led to do the wrong thing at the right time.

If, however, he had a systems model which would enable him to accurately predict the *direction* of changes in the consumer sales rate at a given time, either in terms of seasonal patterns or through the use of sample retail sales data, he would have a much better basis for decision making. After all, if he knew that a rush was taking place, at least he would be producing the demanded items, even if not in precisely the proper quantity. In fact, perhaps he has little need to know anything else, since if consumer demand is sufficiently strong, he need only produce as much as possible.

THE DYNAMICS OF SYSTEMS

Since just about anything can be thought of as a system, the claim could be made that systems ideas are universally applicable. In fact, however, the utility of systems ideas is greatest in a dynamic context—i.e., in terms of *systems which are evolving over time.*

Most complex systems are dynamic in the sense that they move from state to state as time progresses. An operating automobile engine is a physical system which is dynamic in this respect since its *state*—which might be characterized by a number of quantities, such as temperature, quantity of fuel available, pressure in each of the cylinders, etc.—is constantly changing. So, too, is the organization a dynamic system,

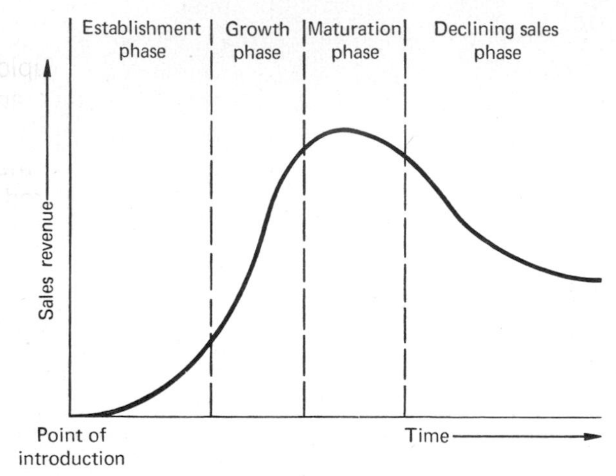

FIGURE 6-2 *Product sales life cycle.*

since each of the many descriptors which characterize its state at any given time—resources available, problems to be faced, decisions to be made, inventory level, financial position, etc.—is also constantly evolving.

The dynamic nature of a system is one of the most significant characteristics which must be accounted for when a system is to be designed and utilized. *This is the case whether the system is a product to be marketed, a system to be used as a management aid, or an organizational system.*

Sales Life Cycle

Every dynamic system has natural phases of development. Recognition of these phases permits the manager to properly control what is happening and to use characteristics of the various phases to advantage. A product, for example, moves through various phases of a *sales life cycle* after it has been placed on the market. One of the authors[3] has referred to these life cycle phases as *establishment, growth, maturation,* and *declining sales* phases. Figure 6-2 shows these phases in terms of the sales revenue generated by the product during its period of slow establishment in the marketplace, then a period of rapid sales increase, followed by a peaking and a long, gradual decline. Virtually every product displays these dynamic characteristics, although some may have a sales life cycle which is so long or so short that the various phases are not readily distinguishable.

[3]William R. King, *Quantitative Analysis for Marketing Management,* McGraw-Hill Book Company, New York, 1967, p. 113.

For example, a faddish product such as "super balls," "hula hoops," or "flying saucers" will have a very high-peaked sales curve with a rapid decline. Many such products will have a long, slow decline after an initial rapid decline from the peak. With other products, the maturation phase is very long and the declining sales phase very gradual. But the general life cycle concept is virtually unavoidable for a successful product, for without product improvements its competition will eventually lure away customers, because consumers' attitudes, habits, and needs will change as time passes.

Of course, the sales portion of the life cycle of a product is really only one aspect of its entire "life." Indeed, only products which are marketing successes ever get to experience the sales life cycle of Figure 6-2. Most new consumer products have from the beginning of their sales period ". . . an infinitely descending curve. The product not only doesn't get off the ground; it goes quickly under ground—six feet under."[4]

Systems Development Life Cycle[5]

All products—sales successes or otherwise—begin as a gleam in the eye of someone and undergo many different phases of development *before* being marketed and subjected to the sales life cycle considerations of Figure 6-2. For instance, the U.S. Department of Defense (DOD) and the National Aeronautics and Space Administration (NASA) have extensively defined and detailed phases which should be encountered with hardware systems development. Their *system development life cycle* concept recognizes a natural order of thought and action which is pervasive in the development of many kinds of systems—be they commercial products, space exploration systems, or management systems.

New products, services, or roles for the organization have their genesis in ideas evolving within the organization. Typically such "systems" ideas go through a distinct life cycle, i.e., a natural and pervasive order of thought and action. In each phase of this cycle different levels and varieties of specific thought and action are required within the organization to assess the efficacy of the system. The "phases" of this cycle serve to illustrate the systems development life cycle concept and its importance.

[4]Theodore Levitt, "Exploit the Product Life Cycle," *Harvard Business Review,* November–December, 1965, p. 82. For instance it has been estimated that 80 to 90 percent of newly introduced packaged grocery products are marketing failures. See Peter J. Hilton, "New Product Introduction for Small Business Owners," Small Business Management Series no. 17, U.S. Government Printing Office, Washington.

[5]The material in this section is based on concepts developed by the United States Air Force under the title "systems management."

THE CONCEPTUAL PHASE The germ of the idea for a system may evolve from other research, from current organizational problems, or from the observation of organizational interfaces. The conceptual phase is one in which the idea is conceived and given preliminary evaluation.

During the conceptual phase, the environment is examined, forecasts are prepared, objectives and alternatives are evaluated, and the first examination of the *performance, cost,* and *time* aspects of the system's development is performed. It is also during this phase that basic strategy, organization, and resource requirements are conceived. The fundamental purpose of the conceptual phase is to conduct a "white paper" study of the requirements in order to provide a basis for further detailed evaluation.

There will typically be a high mortality rate of potential systems during the conceptual phase of the life cycle. Rightly so, since the study processes conducted during this phase should identify projects that have high risk and are technically, environmentally, or economically infeasible or impractical.

THE DEFINITION PHASE The fundamental purpose of the definition phase is to determine as soon as possible, and as accurately as possible, cost, schedule, performance, and resource requirements and whether all elements, projects, and subsystems will fit together economically and technically.

The definition phase simply tells in more detail what it is we want to do, when we want to do it, how we will accomplish it, and what it will cost. The definition phase allows the organization to fully conceive and define the system before it starts to physically put the system into its environment. *Simply stated, the definition phase dictates that you stop and take time to look around to see if this is what you really want* before the resources are committed to put the system into operation and production. If the idea has survived the end of the conceptual phase, a conditional approval for further study and development is given. The definition phase provides the opportunity to review and confirm the decision to continue development, create a prototype system, and make a production or installation decision.

Decisions that are made during and at the end of the definition phase might very well be decisions to cancel further work on the system and redirect organizational resources elsewhere.

PRODUCTION OR ACQUISITION PHASE The purpose of the production or acquisition phase is to acquire and test the system elements and the total system itself using the standards developed during the preceding phases. The acquisition process involves such things as the actual setting up of the system, the fabrication of hardware, the allocation of authority and responsibility, the construction of facilities, and the finalization of supporting documentation.

THE OPERATIONAL PHASE The fundamental role of the manager of a system during the operational phase is to provide the resource support required to accomplish system objectives. This phase indicates the system has been proven economical, feasible, and practicable and will be used to accomplish the desired ends of the system. In this phase the manager's functions change somewhat. He is less concerned with planning and organizing and more concerned with controlling the system's operation along the predetermined lines of performance. His responsibilities for planning and organization are not entirely neglected—there are always elements of these functions remaining—but he places more emphasis on motivating the human element of the system and controlling the utilization of resources of the total system. It is during this phase that the system is placed in its proper place in the greater system. Eventually the system may lose its identity per se and be assimilated in the "institutional" framework of the organization.

If the system in question is a product to be marketed, the operational stage begins the sales life cycle portion of the overall life cycle, for it is in this phase that marketing of the product is conducted.

THE DIVESTMENT PHASE The divestment phase is the one in which the organization "gets out of the business" which it began with the conceptual phase. Every system—be it a product system, a weapons system, a management system, or whatever—has a finite lifetime. Too often this goes unrecognized, with the result that outdated and unprofitable products are retained, inefficient management systems are used, or inadequate equipment and facilities are "put up with." Only by the specific and continuous consideration of the divestment possibilities can the organization realistically hope to avoid these contingencies.

A detailed outline of the elements of the various phases of the systems development life cycle is shown in Tables 6-1 through 6-5. Of course, the terminology used in these tables is not applicable to every system which might be under development, since the terminology generally applied to the development of consumer product systems is often different from that applied to weapons systems. Both in turn are different from that used in the development of a financial system for a business firm. However, whatever the terminology used, the concepts are applicable to all such systems.

TABLE 6-1 Conceptual phase

1. Determine existing needs or potential deficiencies of existing systems.
2. Establish system concepts which provide initial strategic guidance to overcome existing or potential deficiencies.
3. Determine initial technical, environmental, and economic feasibility and practicability of the system.
4. Examine alternative ways of accomplishing the system objectives.
5. Provide initial answers to the questions:
   a. What will the system cost?
   b. When will the system be available?
   c. What will the system do?
   d. How will the system be integrated into existing systems?
6. Identify the human and nonhuman resources required to support the system.
7. Select initial system designs which will satisfy the system objectives.
8. Determine initial system interfaces.
9. Establish a system organization.

TABLE 6-2 Definition phase

1. Firm identification of the human and nonhuman resources required.
2. Preparation of final system performance requirements.
3. Preparation of detailed plans required to support the system.
4. Determination of realistic cost, schedule, and performance requirements.
5. Identification of those areas of the system where high risk and uncertainty exist, and delineation of plans for further exploration of these areas.
6. Definition of inter- and intrasystem interfaces.
7. Determination of necessary support subsystems.
8. Identification and initial preparation of the documentation required to support the system, such as policies, procedures, job descriptions, budget and funding papers, letters, memoranda, etc.

TABLE 6-3 Production phase

1. Updating of detailed plans conceived and defined during the preceding phases.
2. Identification and management of the resources required to facilitate the production processes such as inventory, supplies, labor, funds, etc.
3. Verification of system production specifications.
4. Beginning of production, construction, and installation.
5. Final preparation and dissemination of policy and procedural documents.
6. Performance of final testing to determine adequacy of the system to do the things it is intended to do.
7. Development of technical manuals and affiliated documentation describing how the system is intended to operate.
8. Development of plans to support the system during its operational phase.

TABLE 6-4 Operational phase

1. Use of the system results by the intended user or customer.
2. Actual integration of the project's product or service into existing organizational systems.
3. Evaluation of the technical, social, and economic sufficiency of the project to meet actual operating conditions.
4. Provide feedback to organizational planners concerned with developing new project and systems.
5. Evaluation of the adequacy of supporting systems.

TABLE 6-5 Divestment phase

1. System phasedown.
2. Development of plans transferring responsibility to supporting organizations.
3. Divestment or transfer of resources to other systems.
4. Development of "Lessons learned from system" for inclusion in qualitative-quantitative data base to include:
   a. Assessment of image by the customer.
   b. Major problems encountered and their solution.
   c. Technological advances.
   d. Advancements in knowledge relative to department strategic objectives.
   e. New or improved management techniques.
   f. Recommendations for future research and development.
   g. Recommendations for the management of future programs, including interfaces with associate contractors.
   h. Other major lessons learned during the course of the system.

Assessing the System Life Cycle

Every system evolves according to a natural order of activity and thought from conceptualization to implementation. However, in order for the overall system life cycle concept to be operationally useful to managers, it must be feasible to assess the status of a system as it progresses through its cycle.

VARIOUS "KINDS" OF LIFE CYCLES The measures which may be applied to the evaluation of the status and progress of a system during the various stages of its life cycle vary widely. For example, in developing a new product, one might characterize the various phases of the system's development life cycle in terms of the proportional composition of the work force assigned to the activity. In the beginning, research personnel predominate; subsequently their role diminishes and engineers come to the forefront; finally, marketing and sales personnel become most important. Alternately, the level of expenditures on the development of the product may well be an appropriate way to characterize various phases of development.

Basic life cycle concepts hold for all dynamic systems. Thus, an organizational system develops and matures according to a cycle which is much like that of a product. The measures used to define various phases of an organization's life cycle might focus on its product orientation—e.g., defense versus nondefense—its personnel composition—scientists versus nonscientists—its per-share earnings, etc. For a management information system, the life cycle might be characterized by the expenditures level during the developmental phase together with the performance characteristics of the system after it becomes operational.

A hardware system displays no sales performance after it is in use, but it does display definite phases of operation. For example, Figure 6-3 shows a typical failure rate curve for the components making up a complex system. As the system is first put into operation, the failure rate is rather high because of "burn in" failures of weak components. After this period is passed, a relatively constant failure rate is experienced for a long duration; then, as wearouts begin to occur, the component failure rate rises dramatically.

Perhaps a comparison of Figures 6-2 and 6-3 best illustrates the pervasiveness of life cycle concepts and the importance of assessing the life cycle properly. Figure 6-2 represents a sales life cycle for a product. The most appropriate measure to be applied to this product's sales life cycle is "sales rate." Figure 6-3 shows the operating life cycle of a hardware system—say, a military weapons system. The concept is the same as that of Figure 6-2, but the appropriate measurement is different. In Figure 6-3, the "failure rate" is deemed to be the most important assessable aspect of the life cycle for the purpose for which the measurements will be used.

FIGURE 6-3 *Component failure rate in a system as a function of age.*

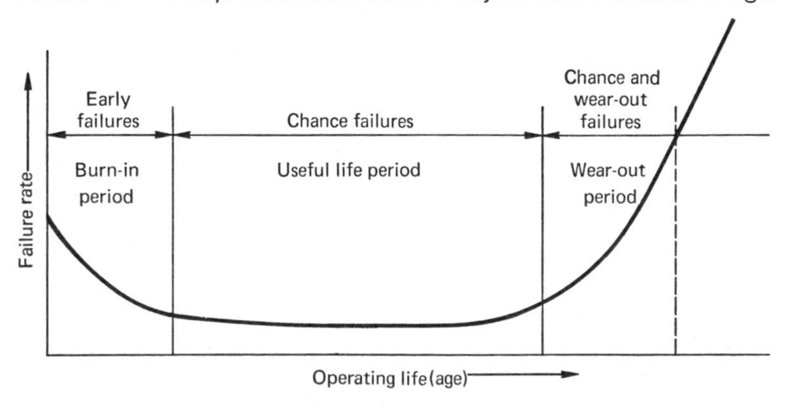

DIMENSIONS OF LIFE CYCLE ASSESSMENT Three critical general dimensions which can be used for assessing the progress of most projects are cost, time, and performance. *Cost* refers to the resources being expended. One would want to assess cost sometimes in terms of an expenditure rate (e.g., dollars per month) and sometimes in terms of total cumulated expenditures (or both). *Time* refers to the timeliness of progress in terms of a schedule which has been set up. Answers to such questions as: Is the project on schedule?, How many days must be made up?, etc., reflect this dimension of progress.

The third dimension of project progress is *performance*—i.e., how is the project meeting its objectives or specifications? For example, in a product development project, performance would be assessed by the degree to which the product meets the specifications or goal set for it. Typically, products are developed by a series of improvements which successively approach a desired goal—e.g., soap powder which does an adequate cleaning job is too sudsy, so it is refined into powder with the same cleaning properties but less sudsiness. In the case of an airplane, certain requirements as to speed, range, altitude capability, etc., are set and the degree to which a particular design in a series of successive refinements meets these requirements is an assessment of the *performance* dimension of the aircraft design project.

We shall have more to say about the critical dimensions of cost, time, and performance later. Now, let us consider the entire range of projects which face an organization at a given time.

A Stream of Projects

Every organization can be characterized at an instant in time by a "stream of projects" which place demands on its resources. The combined effect of all of the "projects" facing an organization at any given time determines the overall status of the organization at that time.

The projects facing a given organization at a given time typically are diverse in nature—some products are in various stages of their sales life cycles, other products are in various stages of development, management subsystems are undergoing development, organizational subsystems are in transition, major decision problems such as merger and plant location decisions have been "projectized" for study and solution, etc.

Moreover, at any given time each of these projects will typically be in a different phase of its life cycle. For instance, one product may be in the conceptual phase undergoing feasibility study, another may be in the definition phase, some are being produced, and some are being phased out in favor of oncoming models.

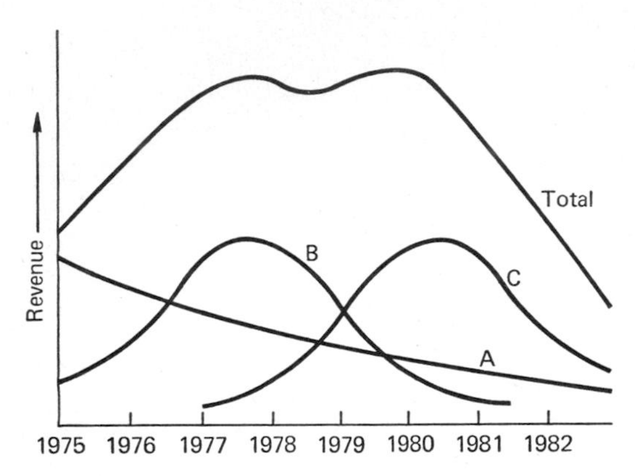

FIGURE 6-4 *Life cycles for several products.*

The typical situation with products which are in the sales portion of their overall life cycle is shown in Figure 6-4 as projected through 1982 for the sales levels of three products, A, B, and C. Product B is expected to begin sales in 1975 and to be entering the declining sales phase of its cycle after 1978. Product A is already in the midst of a long declining sales phase. Product C is in development and will not be marketed until 1977. At any moment in time, each is in a different *state.* In 1978, for example, A is in a continuing decline, B is beginning a rather rapid decline, and C is just expanding rapidly.

Whatever measure is chosen to represent the activity level or state of completion of each of the projects in the stream facing an organization—be they products, product-oriented projects, management system development projects, or decision-oriented projects—the aggregate of all of the projects facing the organization represents a stream of projects which it must pursue. Although the same measures (e.g., revenues, resources employed, percent completed, etc.) will not normally be applicable to all projects, the idea of a stream of projects—each at a different phase of its life cycle—is applicable to assessing the *state* of any dynamic organization.

ORGANIZATIONAL SYSTEMS

The manager can use systems ideas in a variety of ways in accomplishing his objectives. Among the most useful ways are:

1. The use of systems models to enhance his understanding of the operations of systems
2. The use of systems models as predictive devices to enable him to evaluate alternative courses of action
3. The conceptualization of the organization as a subsystem of a larger

environmental system which permits him to assess environmental considerations (see Chapter 4)

4. The development of operating systems which aid in the management of the organization

For instance, the manager can use his understanding of life cycle concepts and the stream of projects to aid in decisions concerning the level and nature of managerial and technical resources to be assigned to various projects at any given time. He recognizes that project activities begin with an idea or concept, progress through definition and production, and end when the product, service, or result that is created is no longer used by the "customer."[6] During each of the phases of the life cycle, different actions are required and different levels of involvement of various people and other resources are necessary. It is part of the manager's job to assess the progress of each project in terms of cost, time, performance, and whatever auxiliary measures are relevant and to determine the level of attention required by each. By doing this, he can allocate appropriate resources to each project so that the overall goals of the organization are best fulfilled.

In previous sections of this chapter, we have focused on various ways that the manager can use his perceptions and assessments of system characteristics to practice better management. Now we turn to the latter two of the aforementioned uses—the view of the organization as a system which is composed of many operating subsystems and, at the same time, is a subsystem of a larger environmental system.

The Systems View of the Organization

In viewing itself in systems terms, the organization thinks of itself as a system which is composed of many subsystems. At the same time, the organization is a subsystem of an overall environmental system. Often, the elements of the environment which are in competition with the organization—be they competition for federal budget allocations, scarce resources, or the consumer's dollar—are themselves best thought of as a separate environmental subsystem.

Thus, a model which views the world as a "complex of systems" is often appropriate as a guide for systems-oriented management thinking. A delineation of such a complex of systems might be:[7]

[6] The "customer" is the individual organization or agency which has a need for the outcome of the project activities. The outcome will be in the form of a product or service demanded by the customer.

[7] This complex of systems is described in somewhat different terms by Richard A. Johnson et al., *The Theory and Management of Systems,* McGraw-Hill Book Company, New York, 1967, pp. 34–36.

1. *The competitive system*
2. *The environmental system*
3. *The internal system*

These terms most readily apply to the business firm as a system, but they also are a useful conceptualization for other organizations.

The Competitive System

The competitive system for an organization is the complex of other organzations which compete for the distribution of scarce resources in a given environment. Since there is never enough of everything to go around, every organization must contend with a competitive system. Even Robinson Crusoe, alone with Friday on an island, had to compete with animals (and potentially with Friday, heaven forbid) for food. If food had been scarce, the competition might have become severe.

So, too, it is with other organizations in modern society. In business, an organization's competitor is thought to be one who is buying or selling in the same market. However, the idea is equally applicable to nonbusiness organizations.

At first, one might think that, say, a public school system has no competitive system. However, if one accepts the idea that a school system exists to facilitate the generation and transmission of knowledge and to abet the learning process, then there are, indeed, other organizations and/or institutions that contribute to the learning process, perhaps in a competitive manner. For example, the ghetto child might well learn more in his street environment than he does in his school. Consider also the "learning" that takes place in environments such as:

1. The church and related organizations
2. Recreational organizations (movies, bowling alleys, restaurants, etc.)
3. The home
4. Sports arenas
5. Television
6. Private companies

Indeed, such entities as private companies are increasingly making it apparent that they are in direct competition with the public school system. Under a concept called "performance contracting," for instance, private companies have contracted to take over certain teaching functions with "guaranteed" learning results. The guarantee is in the form of payment terms which guarantee the company a profit only if the learning goals are achieved. Thus, the company is a potential direct competitor with established public school systems for federal and state funds.

The Environmental System

An organization's environmental system is the economic, political, and social milieu in which the organization operates. We have already discussed the environmental system in Chapter 4. Here we need only refer the reader to that chapter and to point out that the environmental system is disjoint from the competitive system. Both are external to the organization, and we define them to be disjoint for convenience. This is shown in the simple model of Figure 6-5.

In a subsequent chapter on the analysis of management decisions, we shall demonstrate why this simple model is so important and useful in decision making.

The Internal System

The internal system is what is generally thought of as "the" organization. We have emphasized that this internalized view of the organization is too restrictive in and of itself to be useful to the modern manager. The latter must rather think in terms of the "extensive organization," which includes the organization's various clientele.

FUNCTIONAL SUBSYSTEMS The internal system is itself obviously of great importance. We can characterize it in terms of various subsystems. For instance, the internal system can be thought of as being made up of:

FIGURE 6-5 *A simple organizational systems model.*

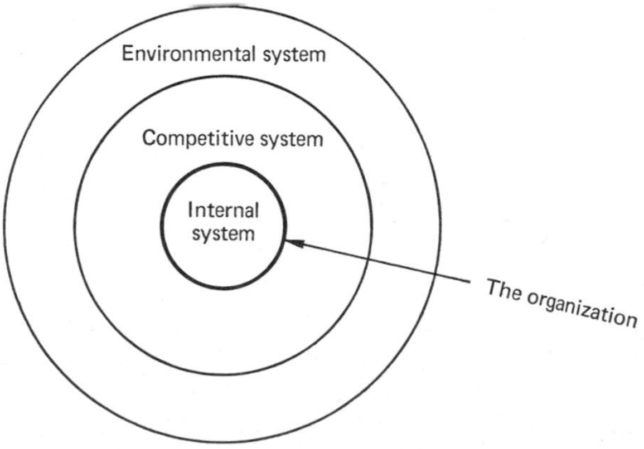

1. A production subsystem
2. A marketing subsystem
3. A financial subsystem
4. A research and development subsystem

Raw materials in the form of matter, information, and energy flow into the *production subsystem* of a business firm. There they are processed and transformed as they flow between a variety of subsystems (e.g., machine centers). Finally, finished products are output from the system.

The *marketing subsystem* has basic responsibility for the flow of finished products from the organization to its environment. It has associated responsibility for the flow of market-based information, obtained through such media as sales records and survey research, back to the internal subsystem of the organization.

Funds enter the *financial subsystem* through a pool of funds drawn from debt and equity sources and are used to purchase material, labor, facilities, and supplies. Then, when the product or service is sold, the funds that are received are used to pay for the goods and services necessary to sustain the business; the residual is returned for disposition to the pool of funds, to creditors (as interest and debt principal), or to stockholders (as dividends).

Ideas enter the *research and development (R&D) subsystem* to be evaluated and developed into potential new products or services. They may come from outside the organization, as with the "mad inventor" who offers his invention to the organization, or from a formalized program of research within the subsystem. In any case, it is the responsibility of the R&D subsystem to develop the ideas to the point where they can be rationally evaluated as potential products of the organization.

MANAGEMENT SUBSYSTEMS The internal organizational system can also be thought of in terms of *management subsystems.* Such subsystems are designed to facilitate effective and efficient operation of the functional subsystems.

For example, a *production control subsystem* is a set of interrelated policies, procedures, reporting requirements, decision rules, etc., for ensuring that proper control is maintained over the production process. It takes operating information from the production subsystem and compares it with "standards" for the operation. Answers are obtained to operational questions such as: Are we producing to acceptable quality? Will we be able to meet our order commitments?, etc.

A *management information subsystem* is a set of data collection elements, data processing devices, reports, etc., which provides managers with the information necessary to make decisions. It collects summary data from all functional subsystems so that management can guide and control the flow of monetary, personnel, and material resources as they address the stream of projects.

Innumerable other management subsystems can be delineated. For instance a *personnel management subsystem* focuses on planning and controlling the development of personnel resources while a *marketing information subsystem* addresses itself to the collection and dissemination of market-based information.

Of course, any of these subsystems may be imbedded in other management subsystems. Both a marketing information subsystem and a personnel management subsystem may be parts of an overall management information system, for instance. The important thing is not which particular categorization is used, but that critical functions, flows, and processes be considered in the overall systems model.

We shall subsequently make use of both functional and management subsystem models as frameworks for discussion. Here our purpose is to illustrate the fact that the organization can apply the systems view and systems concepts at many different levels and to many different kinds of systems. Subsequently, we shall detail many of these systems and study the analysis, design, and management of each.

An Internal Systems Hierarchy

Internal organizational subsystems may also be identified in terms of their level. It is apparent that various internal subsystems have different degrees of interaction with competitive and environmental systems. Of course, the chief executive of an organization is more concerned with identifying and measuring the forces in the competitive and environmental systems than first-line supervisors would be. However, both the chief executive and the first-line supervisor each have their relevant competitive and environmental subsystems. These systems and relationships are different and will be used differently by each incumbent according to how he perceives the overall system.

It is difficult, if not impossible, to categorize organizational systems in a manner representative of all organizations. One way of attempting this is to describe an organization as consisting of the following.[8]

1. *An underlying system* where the goods or services being created are produced and distributed. *Decisions and managerial actions at this*

[8]This description of organizational systems is adapted from Herbert A. Simon, *The New Science of Management Decision,* Harper & Row, Publishers, Inc., New York, 1960, pp. 49–50.

level are programmed; guidance is provided by standard, routine, repetitive behavior. Decisions that are required are guided by predetermined policy or procedural responses. The operation of this system depends upon existing information systems. Extraordinary actions not in harmony with existing policy and procedures are referred upward to the next higher system level.

2. *A layer of a controlled information system* which compares programmed actions with expected accomplishments. This level works in both programmed and nonprogrammed modes. Decisions necessary to plan and direct control of the lower systems are interpreted for the next higher level of effort.

3. *The highest system* consists of the realm of the chief executive, where nonprogrammed decisions involving organizational strategy are made. This is the system where overall organization strategy is developed and supporting plans are created. Such a level designs the structure and processes of the supporting subsystems and modifies this structure and process as necessary to meet changing internal, competitive, and environmental conditions. The work in this area tends to be abstract, projected into the future, and based on the forecasted behavior of economic, social, and political systems. Decisions made in this system by the executive have a high degree of futurity; they are decisions made today which will be executed in the long-range future of the organization. Hence, these decisions carry large elements of risk and uncertainty.

Management processes at the first two levels of the system are more suitable to automation in minimizing the use of human labor in the production and distribution of the product or service. In the case of recurring operations (such as in fabrication and information processing), machines can do the job better, more quickly, and with less error than people. However, the human element will always be required to plan for and guide the use of the machines. *Control,* as a managerial function, takes on added significance in these levels, whereas *planning* is paramount at the top level of the system.

Planning at the top level involves the forecasting of expected future conditions, the establishment of overall strategic objectives, and the design of the systems essential to survival. This is a task of extraordinary proportions since it requires explicit, clear statements of long-range goals and strategy. Organizational managers at this level must learn to live and cope with changing environmental influences and constraints as they design the management system for their organization. Perhaps the most important function of the highest level of the system is to determine how the organization can best adjust to the changing pressures in its environment.

We shall subsequently utilize the concept of various levels of organizational systems in discussing the specifics of the particular management subsystems of an organization. For example, in determining the appropriate information system for an organization, it is necessary to focus attention on the *decisions* which need to be made by its managers. From this viewpoint, it is clear that different levels of informational requirements exist at the various levels of the organizational system.

SUMMARY

The systems view conceives of any system—be it an organization, a product being sold, a management system, or whatever—as a complex of interacting and interdependent parts.

One important system concept has to do with the *chain of effects* produced by an action by, or on, some element of the system. The effects reverberate throughout the system and its environment, sometimes becoming amplified and sometimes damping until they are no longer detectable.

The systems view dictates that insofar as possible, these chains of effects should be taken into account in systems design, planning, and decision making. Thus, the manager who takes the systems view cannot be satisfied with superficial consideration of the immediate and obvious consequences of his actions; he must trace out the chain of effects and make his decisions in the light of the entire sequence of consequences.

One approach which aids the manager in taking this viewpoint is a *state transition model.* A state describes the condition of an entity at a point in time. Transitions from one state to another take place over time.

A *decision tree* describes such state transitions in graphical terms. It illustrates that a single transition involves *(a)* an immediate outcome, and *(b)* a new state and new problem situation. Thus, the decision tree approach imbeds one decision problem within another and provides the manager with a model which can be used in operationalizing the systems approach.

The systems approach requires that the manager should be aware of interactions and interdependencies. He can use this awareness to create new systems, e.g., product systems and management systems, and to make predictions of the future behavior of complex systems.

Every dynamic system can be thought of as transitioning through a number of states in its *life cycle.* A *systems development life cycle* which is widely applicable to product systems, management systems, etc., consists of five well-defined phases—conceptual, definitional, production, operational, and divestment. If the system is a product, it also experiences a *sales life cycle* which involves establishment, growth, maturation, and declining sales phases.

In each of these phases of the overall life cycle, different specific measures are appropriate for assessing the progress of the system under development and different managerial skills and talents are required. The general *cost, time,* and *performance* dimensions, however, are always applicable.

The typical organization faces a *stream of projects* at any given moment. Typically, the various projects are in different stages of their life cycles, so that it becomes the top manager's task to integrate them and to plan for the future in terms of the predictions which he makes for the future of each.

The manager can use systems ideas to enhance his understanding of the operation of systems, to predict developments, to aid in the evaluation of alternative courses of action, to develop managerial systems, and to include considerations from the greater environmental system into all of these.

Two models of the organization which are useful to the manager are the competitive-environmental-internal subsystems model and the internal organizational model which views the organization at three levels. The former model sees the organization as a subsystem of an environmental system. The internal model focuses on the kind of actions required of managers at various levels in the organization and on the nature of their contact with the environment. For instance, at the highest systems level in an organization, strategy is established and the manager must have a good contact with and information concerning the environment. At the underlying system level, actions are largely programmed and the manager has less environmental interaction.

EXERCISES

1. The definition of a system uses the terms "interrelated" and "interdependent." Define these terms and distinguish between them.
2. What is a "chain of effects"? Illustrate the applicability of the concept to your own actions.
3. Describe your "state" at the moment in formal terms. If you were to make a state transition by deciding to take a shower, how would your state change? How would it be the same?
4. A state transition can occur either because of an action taken by an entity or because of an action by something else which affects the entity. Illustrate both in personal terms as in exercise 3.
5. A state transition thought of in decision-making terms produces two kinds of consequences. What are they?

6. Construct a decision tree for the following situation. You must choose to purchase either a Ford, Chevrolet, or Plymouth. Then you must decide whether or not to purchase air conditioning, except that because of autos currently available, if you choose a Ford, you cannot get air conditioning. However, Ford offers an option called an "automatic semi-inverted audio receiver" which is not available with other brands.
7. Construct a decision tree for a new product introduction involving price, advertising expenditure, and distribution policy. Is there a "natural" or "logical" time sequence to these various aspects of the overall decision? Construct alternative trees which treat this time sequence differently.
8. What is a "product system"? Describe several which are familiar to you.
9. In what sense does a project set up to analyze merger possibilities represent a systems view?
10. Describe the manager's predictive problem—i.e., his problem of predicting the future—in systems terms.
11. What is the value of the decision tree model in prediction?
12. What is a dynamic system? How does this concept relate to that of a "state transition"?
13. What is a "sales life cycle"? How does this relate to the "state transition" concept?
14. Describe the general nature of the sales life cycle for *(a)* a new clothing fashion, *(b)* a staple such as salt, *(c)* a successful high-quality beverage, *(d)* a breakfast cereal.
15. What is a "system development life cycle"? How does it differ from a "sales life cycle"?
16. Relate the two life cycle concepts in exercise 15 by characterizing how they relate to the development and marketing of a new product such as a three-dimensional camera.
17. Give examples of how a new idea may evolve from *(a)* research, *(b)* current organizational problems, *(c)* the observation of organizational interfaces.
18. Describe the various phases of the typical *(a)* sales life cycle, *(b)* systems development life cycle.
19. One of the authors[9] has described the process of finding new product ideas as one which involves first a gross evaluation of feasibility and then a stringent economic evaluation in terms of potential costs and profit. How does this relate to the phases of the systems development life cycle?
20. The conceptual and definition phases of the systems development life cycle concentrates on the time, cost, and performance aspects of the system being developed. Apply these time, cost, and performance

[9]William R. King, *Quantitative Analysis for Marketing Management,* McGraw-Hill Book Company, New York, 1967, chap. 5.

parameters by identifying specifically the kinds of measurements which could be used if the system under development were: *(a)* an airplane, *(b)* a new drug, *(c)* a management information system.

21. Describe the activities which might take place in each phase of the systems development life cycle for each of the systems in exercise 20.

22. What different management skills do the various phases of the systems development life cycle emphasize?

23. Characterize the "stream of projects" with which you are faced at the moment. Try to assess each in terms of the life cycle concepts. Which phase is each in?

24. Define the *state* of a project in terms of its cost, time, and performance parameters.

25. Define the state of an organization in terms of the total stream of projects with which it is involved.

26. What are the various general ways in which the manager can use systems ideas in accomplishing his objectives?

27. Characterize the internal, external, and competitive systems for *(a)* a police department, *(b)* a hospital, *(c)* a country club, *(d)* a fraternity.

28. How are the functional subsystems of an organization involved in the various stages of the life cycle of a product being developed and marketed by that organization?

29. Identify the key questions to be asked concerning a system during the various phases of its life cycle.

REFERENCES

Adler, Lee, "Systems Approach to Marketing," *Harvard Business Review,* May–June, 1967.

Anshen, M., "Manager and the Black Box," *Harvard Business Review,* November–December, 1960.

Barkin, Solomon, "A Systems Approach to Adjustments of Technical Change," *Labor Law Journal,* January, 1967.

Churchman, C. West, *The Systems Approach,* Dell Publishing Co., Inc., New York, 1968.

Cleland, David I., and William R. King, *Systems Analysis and Project Management,* McGraw-Hill Book Company, New York, 1968.

Davis, Keith, "The Role of Project Management in Scientific Manufacturing," *I.R.E. Transactions on Engineering Management,* vol. 9, 1962.

DeBono, Edward, "Lateral Thinking—The Searching Mind," *Today's Education,* November, 1969.

DeGreene, Kenyon B., *Systems Psychology,* McGraw-Hill Book Company, New York, 1970.

Drucker, Peter F., "Technological Trends in the Twentieth Century," C. W. Pursell, Jr., and M. Kranzberg (eds.), *Technology in Western Civilization,* vol. 2, chap. 2, Oxford University Press, New York, 1967.

Eberhard, John P., "Technology for the City," *International Science and Technology,* September, 1966.

Fricks, R. E., *Structural Problems in Organization Theory,* Systems Research Center, Case Western Reserve University, Cleveland, 1967.

Gale, Morton, and Paul Alelyunas, "The Systems Man," *Space/Aeronautics,* December, 1966.

Hitch, Charles, "On the Choice of Objectives in Systems Studies," Reports Office, RAND Corporation, Santa Monica, Calif.

Holtz, J. N., "An Analysis of Major Scheduling Techniques in the Defense Systems Environment," RAND Corporation, RM-4697-PR, October, 1966, Santa Monica, Calif.

Johnson, R. A., et al., *The Theory and Management of Systems,* McGraw-Hill Book Company, New York, 1967.

Kast, Freemont E., and James E. Rosenzweig (eds.), *Science, Technology, and Management,* McGraw-Hill Book Company, New York, 1963.

King, William R., *Quantitative Analysis for Marketing Management,* McGraw-Hill Book Company, New York, 1967.

Levine, Robert A., "Systems Analysis and the War on Poverty," paper presented before the 29th National Meeting of the Operations Research Society of America, Santa Monica, Calif., May 18–20, 1966.

Levitt, Theodore, "Exploit the Product Life Cycle," *Harvard Business Review,* November–December, 1965.

Martin, Jr., E. W., "The Systems Concept," *Business Horizons,* Spring, 1966.

Miller, E. J., and A. K. Rice, *Systems of Organization,* Tavistock Publications, Ltd., London, 1967.

Mockler, Robert J., "The Systems Approach to Business Organization and Decision Making," *California Management Review,* Winter, 1968.

Morrison, Edward J., "Defense Systems Management: The 375 Series," *California Management Review,* Summer, 1967.

Rivett, Patrick, *An Introduction to Operations Research,* Basic Books, Inc., Publishers, New York, 1968.

"Systems Analysis? What's That?" *Changing Times,* August, 1969.

Whitehead, Clay Thomas, "Uses and Limitations of Systems Analysis," Reports Office, RAND Corporation, Santa Monica, Calif.

Williams, Edgar G., "A Systems Approach to Manpower Management," *Business Horizons,* Summer, 1964.

7

implementing the systems approach to management

In the previous two chapters we have dealt with the essential elements of management together with the basic concepts of the systems approach. In this chapter, our attention turns to implementing the systems approach to management.

We begin with a conceptual model of management using the systems approach which relates to the functional model of management presented in Chapter 5. Then we concentrate on a number of specific areas of application of this conceptual model in order to demonstrate the wide range of applicability of the systems approach to management.

A SYSTEMS MODEL OF MANAGEMENT

The systems approach to management and the life cycle concept can be integrated into a conceptual model of management which is often more useful than the functional one described previously. This model focuses on the process of management rather than the functions of planning, organizing, motivating, and controlling.

The systems model of management has two essential elements which form a sequential process in the analysis, development, and management of any system—be it a hardware system, a management system, or whatever. The elements are *systems analysis* and *systems management.*

Systems Analysis

Systems analysis is an analytic process designed to help a decision maker select a preferred choice among possible alternatives. It involves a thorough investigation of his objectives, relevant measures of the degree of attainment of the objectives, and a comparison of the cost, effectiveness, risk, and timing associated with alternative ways of achieving the objectives. Systems analysis also includes the design of better alternatives and the selection of other goals if those that have been examined are found wanting.[1]

Thus, systems analysis is a way of reaching decisions which contrasts with intuition-based and unsystematic approaches; the techniques of systems analysis help to make a complex problem understandable and manageable in the sense of offering possible strategies and solutions and establishing, to the maximum extent practical, criteria for selecting the best solution.

The value of systems analysis to the management of an enterprise can be seen in terms of two elements of the manager's job. First, he desires to *achieve the overall effectiveness* of his organization—not to have the parochial interests of one organizational element distort the overall performance. Second, he must do this in an organizational environment which invariably involves *conflicting organizational objectives.* The systems analysis approach gives him a hope of actually being able to perform this difficult "balancing act."

Systems analysis relies on *models.* Because these models are usually symbolic, it is possible to reduce complex relationships to a form that can be put down on paper and, using techniques of logic and mathematics, to consider interrelationships and combinations of circumstances that would otherwise be beyond the scope of any human being. Models permit experimentation of a kind which is unavailable in many environments; thus, the analyst may experiment on the model describing a system rather than on the system itself.

The end product of the systems analysis process as applied to the development of a system is the specification of all aspects of the system—hardware, software, procedures, personnel requirements, training requirements, etc.

Systems Management

Systems management[2] deals principally with executing decisions within the constraints of the systems specifications and of redefining decisions for application of systems analysis. Systems management is specifically concerned with directing, motivating, and controlling; it includes the

[1] This description of systems analysis is based on one given by E. S. Quade in *Analysis for Military Decisions,* RAND Corporation, R-387-PR, November, 1964, chap. 1.

[2] The reader should note that the term "management" is used in two different senses at this point. "Management" involves both planning and execution. The term "systems management" as used here refers primarily to the execution phase. This should cause no difficulty, since the text is about all of management and the term "systems management" will be used to refer to the narrower interpretation.

coordination and integration necessary to pull together various subsystems to ensure overall goal achievement.

The general relationships of the two elements of this systems model to the four elements of the functional model are shown in Figure 7-1. Note that this figure shows only a very general and vague relationship—not a direct correspondence. In other words, systems analysis and systems management are not just new words to describe planning, organizing, directing, and controlling. Rather, the elements of the systems model are process-oriented; they are directly related to the natural order of thought and action exemplified by the life cycle of a system. Conversely, the elements of the functional model represent managerial functions. The two models are related because certain functions tend to predominate during certain phases of the life cycle.

For example, as a manager works with the people who are performing the systems analysis, there are obvious elements of *directing, motivating,* and *controlling* these people. But the principal activity during this systems analysis phase is the performance of studies leading to the collection and evaluation of sufficient data on which to base a decision. In a similar fashion, the manager who is mainly concerned with managing an ongoing system will find ample opportunity to do analysis; yet his main focus will be directed toward the execution function of management, i.e., the management of an existing system. As the system proceeds through its life cycle and grows in maturity, the manager will become more knowledgeable with respect to its capabilities and limitations. The need for redesign of the original system may be realized and the process of systems analysis will be brought into play again. In this redesign, both elements of systems analysis and systems management will be involved.

We should reemphasize that, as with all conceptual models, the systems analysis–systems management dichotomy is meant to be neither definitive nor fixed. It is not possible to put an ongoing dynamic process into neatly defined compartments, since the elements overlap, are inter-

FIGURE 7-1 *General relationships between functional and systems models of management.*

Functional Model	Systems Model
Planning Organizing	Systems Analysis
Motivating-Directing Controlling	Systems Management

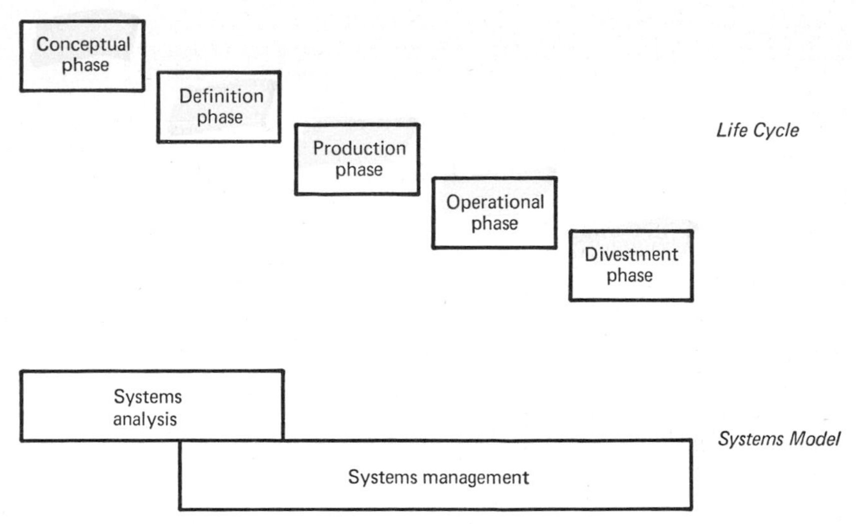

FIGURE 7-2 *Relationship of systems analysis and systems management to system development life cycle.*

dependent, and reinforce one another. All that the systems model does is to relate the systems concept and the managerial process.

Figure 7-2 relates the systems model to the system development life cycle concept which was discussed in the previous chapter. It shows that systems analysis predominates in the conceptual and definition phases while systems management comes to the forefront in the other phases.

APPLICATIONS OF THE SYSTEMS MODEL TO THE MANAGEMENT OF SYSTEMS

The organization can be viewed as a system which exists in the context of an environmental system and which, in turn, is composed of many subsystems. In these terms, the key role of the manager is to develop and prescribe the means ". . . for interrelating and coordinating these various subsystems."[3] In other words, the systems view of the manager's job is tied to the interactions and interdependencies of the various subsystems and of the organization with its environment. The manager must first *recognize and consider* these interdependencies. But also, *he must plan and act to take advantage of them.*

Today, more than ever before, the recognition and use of system interactions and interdependencies are virtually essential. The reasons for

[3] Richard A. Johnson, et al., *The Theory and Management of Systems,* McGraw-Hill Book Company, New York, 1963, p. 56.

this are that the world is evolving so rapidly and that decisions must be made which have such momentous implications that "simpler" approaches to management, such as a reliance on intuition or the application of "principles," will no longer produce satisfactory results with the relative frequency that they may have in the past. Moreover, the measurement of results is becoming easier and more stringent. Thus, bad management actions are more readily apparent than they have been in the past because of the advent of the computer and because of better data collection and processing techniques.

Thus, while management tasks have become more complex, the cost of managerial error (the cost associated with bad decisions and actions) has also become much greater—both to the individual and to the organization. In such a managerial environment, a number of essential requisites for "good management" become clear:

1. The recognition that the whole consists of *both* the sum of its parts *and* the interactions of those parts
2. The recognition that the extent and nature of these system interactions are just as important as are the characteristics of the basic elements which make up the system
3. The recognition that the outside world—the environmental and competitive systems—must be dealt with and treated as of equal importance with the internal system
4. The recognition that the organization and its environment are dynamic—that "things will always be changing," and changing rapidly
5. An explicit recognition of the risk and uncertainty which are inherent in any high-level organizational undertaking

Drucker was quoted in Chapter 3 as summarizing the essence of these requisites in saying: "The systems approach, which sees a host of formerly unrelated activities and processes as all parts of a larger integrated whole, is not something technological in itself. It is, rather, a way of looking at the world and ourselves."[4]

Generally, the systems approach involves an identification of a problem or need, studying the environment, formulating objectives to satisfy the need or solve the problem, and then performing the tasks that will accomplish the objectives. Alternative means of accomplishing the objectives can be evaluated and the most acceptable alternative (considering time, cost, and performance) can be elaborated into a design for the system.

[4] P. F. Drucker, "Technological Trends in the Twentieth Century," in C. W. Pursell, Jr., and M. Kranzberg (eds.), *Technology in Western Civilization*, vol. 2, chap. 2, Oxford University Press, New York, 1967, p. 20.

The systems approach can be applied to policy- and management-oriented systems such as organizational design, the development of information and accounting systems, new-product development, and research project management, as well as to hardware-oriented systems such as military weapons systems, transportation systems, and construction projects. The systems approach has received its greatest acceptance in the management of large-scale systems. Yet, it can be applied to most small problems, or to a small aspect of a problem within a larger system. For example, one could use the concept in developing a strategic plan for a corporation; he might also just as well use a scaled-down version of the systems approach in going about reorganizing an office or in planning a trip with the family.

To begin a sequence of illustrations of the application of the systems model in real-world situations, let us consider just such a simple example —the planning of a family camping trip. Such an illustration serves to provide insights through the use of a context which all of us understand, and it provides a light start for a discussion of a weighty topic.

Systems Analysis of a Camping Trip

Abramson and Kennedy, of the TRW Systems Group, have provided a detailed plan for the accomplishment of a simple project—a family camping trip. Figures 7-3 and 7-4, respectively, show the development of a "work breakdown structure" and a "flow chart" for the trip.[5] The specifics of these analytic techniques are not so important at this point as the general ideas of breaking the total "project" into packages of work which can be assigned and managed—through the work breakdown structure—and of establishing the relationships between the various activities which must be performed and decisions which must be made—through the flow chart.

Weapons Systems Analysis

Systems analysis has been put to extensive use by the U.S. Department of Defense (DOD) and the National Aeronautics and Space Administration (NASA) in the planning and development of weapons and space systems. The salient element of the approach is the *cost effectiveness* or *cost benefit concept.* The systems analysis approach to military strategy, put simply, says that in judging any proposed program in the light of the alternative ways of using the resources which it consumes and the alternative

[5] Bertram N. Abramson and Robert D. Kennedy, *Managing Small Projects,* TRW Systems Group, 1969. Used by permission.

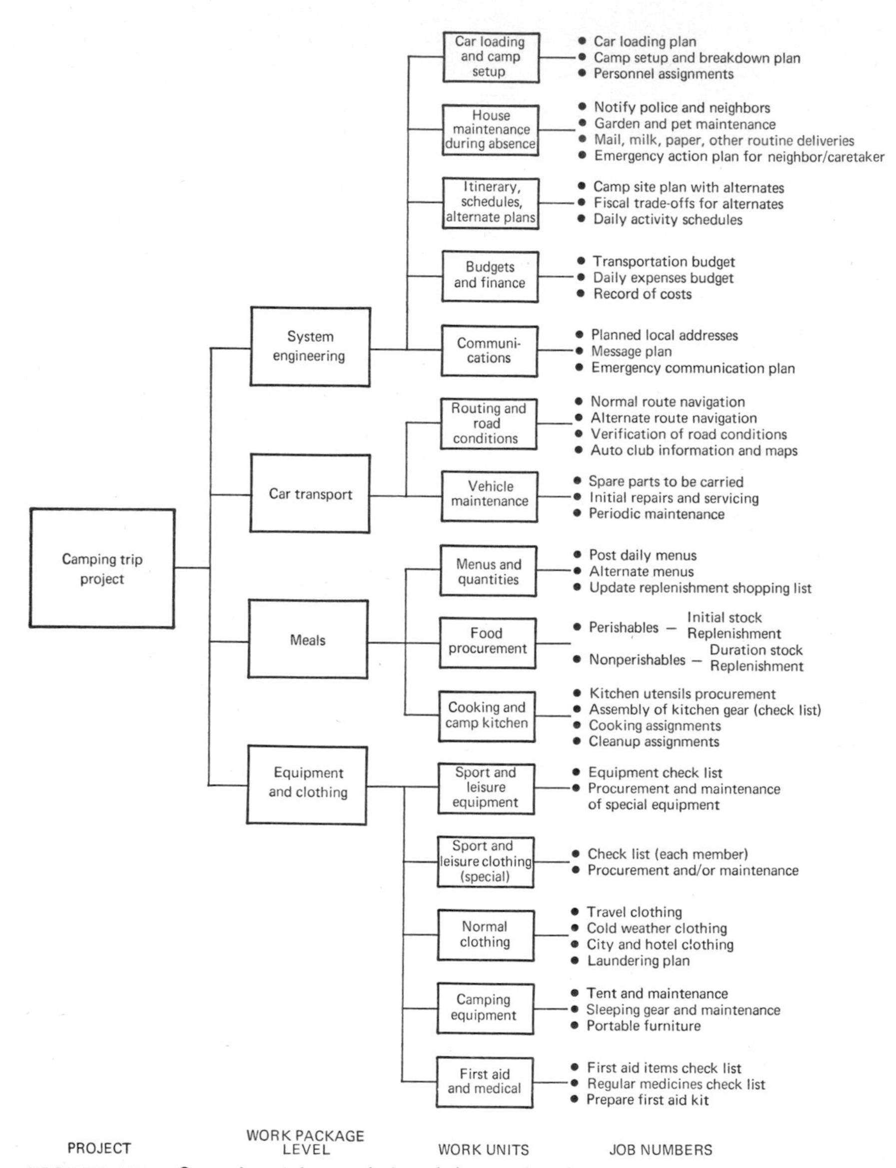

FIGURE 7-3 *Camping trip work breakdown structure.*

ways of achieving the objectives which it attains, the decision maker should consider both the resource cost and the effectiveness (benefit) obtained.

Of course, the cost benefit idea is not new or novel. Indeed, all of us consider cost effectiveness ideas in our everyday lives. If we did not, we would all drive Rolls-Royces instead of Fords and Chevrolets, and we

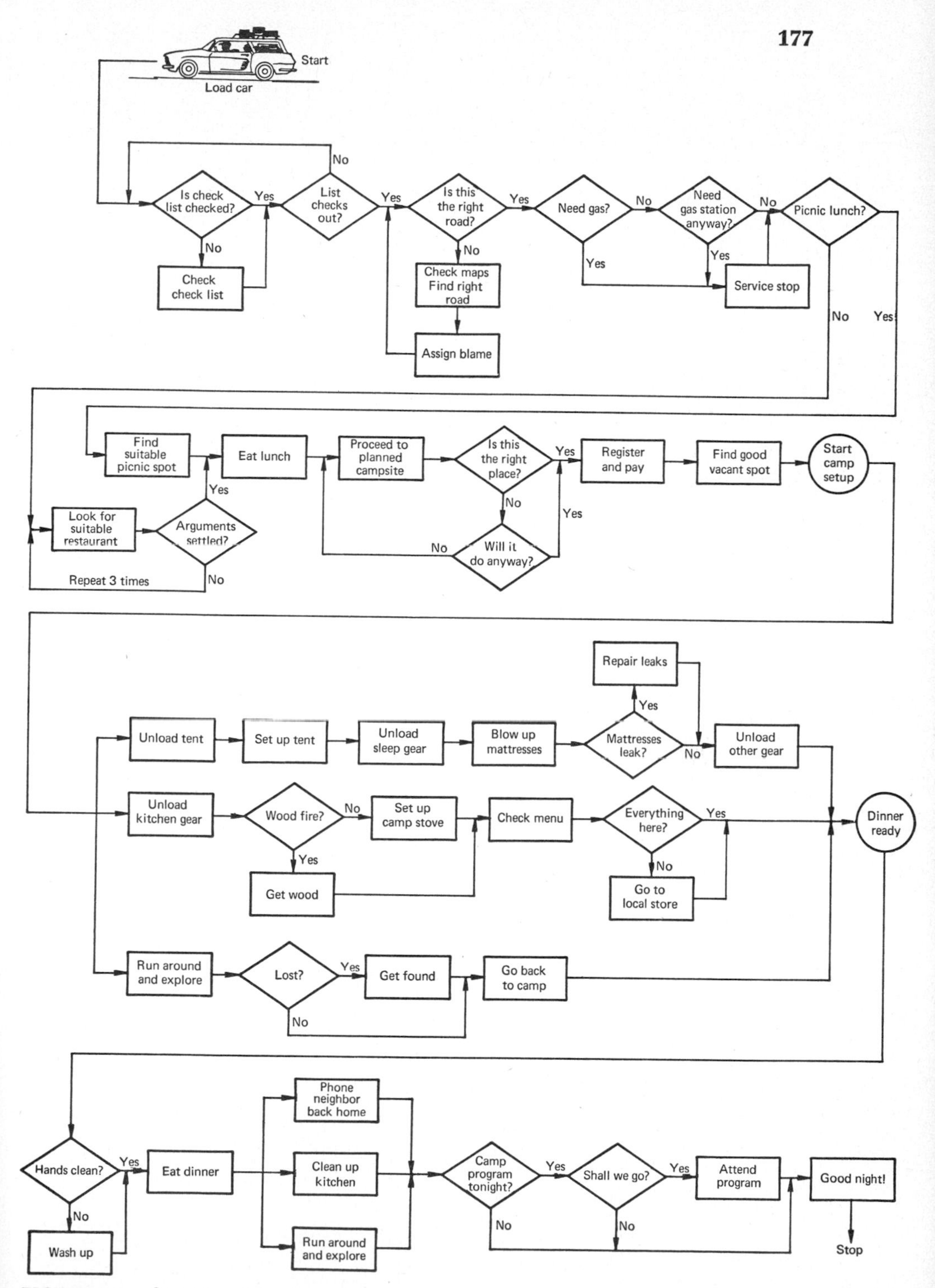

FIGURE 7-4 *Camping trip flow chart.*

would all insist that our children be treated for their everyday lacerations by world-renowned medical specialists.

Prior to 1961, the Department of Defense was not operated in a fashion which facilitated the simultaneous consideration of the cost and the effectiveness associated with military programs. Military requirements were then determined by the services with little reference to the total budget available. The financial management system prepared budgets with little cognizance of military requirements or effectiveness. Thus, costs and effectiveness were treated separately. When they inevitably came into conflict, compromise was necessary—and the compromise did not always necessarily produce good results.

Alain Enthoven, former Assistant Secretary of Defense for Systems Analysis, has illustrated this procedure and its results with a "homey example."

> *Suppose that I want to buy a house and, instead of using the cost effectiveness approach, I do it in the more traditional way. First, I determine my housing requirements without any consideration of costs. I count up the rooms I require: I need a bedroom for myself, one for each of my children, and one for my parents or other guests who come to visit us occasionally. I need a study because I occasionally bring some of my work home with me and need a quiet place to work. My wife needs a sewing room. I need a pool in the basement because my doctor has told me that I must swim every day if I don't want to have another operation on my back. Now, you might laugh when I say that I have a pool in my basement, but I can validate that requirement. I can argue for it very convincingly. I can produce a Doctor's certificate, and you can't prove to me that I don't need that pool. Moreover, I work at the Pentagon and I work long hours. Therefore, I need to live within five minutes' drive of the Pentagon. When I put this all together, I find that I have established a requirement for a house that costs a hundred thousand dollars. Having done that, I review my financial situation and find that I am only able to spend about $30,000. So what do I do? If I am operating under the old concept, I take the $100,000 design and I slice off 70 percent of it and what's left is my house.*
>
> *Now, clearly that's not a very sensible way to design a house. I might find that I left off the bathroom, or included the bathroom but left off the plumbing that is required to make it work. Yet that's a pretty fair description of the way that the Department of Defense did its business. We found in 1961 that we had Army Divisions without adequate airlift or other means of mobility and with far from adequate supplies or equipment. We had tactical air wings without*

supplies of non-nuclear ordnance, and numerous other similar problems. In effect, we had bought a lot of houses without the bathrooms or the plumbing.[6]

Enthoven goes on to say that the rational way of buying either a house or a defense program is "to consider alternative balanced programs each of which correspond approximately to the availability of our resources."[7] Then the extra advantages presumably associated with more expensive alternatives should be weighed against the financial sacrifices required to buy them.

Of course, basic cost effectiveness concepts had been individually applied by the various military services prior to 1961. However, their use on a strategic planning level and their incorporation into a long-range planning, programming, and budgeting process which focused on the *outputs* of the defense establishment *in relation to our national security policy objectives* had not previously been done.

The institution of such a procedure facilitated systems analyses in the Department of Defense. By thinking in terms of goal-related outputs rather than resource inputs (personnel, transportation, etc.), planners could take the systems approach of the national policy level. For example, they could contrast submarine-launched missiles and manned bombers in terms of the degree to which each satisfies our need for strategic retaliatory power and in terms of the various mixes which might be feasible.[8]

Systems Analysis in Government Planning[9]

The widespread application of systems analyses in defense planning was concurrent with the institution of a planning, programming, and budgeting system (PPBS) in the Department of Defense. This system's purpose was to provide necessary information in the proper form for defense decision making. The success of PPBS and systems analyses in defense led President Johnson in 1965 to direct that other government departments ". . . immediately begin to introduce a very new and a very revolutionary

[6]Alain Enthoven, address before the Aviation and Space Writers Convention, Miami, Fla., May 25, 1964.

[7]*Ibid.*

[8]For more detailed discussions of these ideas, see David I. Cleland and William R. King, *Systems Analysis and Project Management,* McGraw-Hill Book Company, New York, 1968, and David I. Cleland and William R. King (eds.), *Systems, Organizations, Analysis, Management: A Book of Readings,* McGraw-Hill Book Company, New York, 1969.

[9]The material in this section is adapted from William R. King, "Systems Analysis at the Public-Private Marketing Frontier," *Journal of Marketing,* January, 1969, pp. 84–89.

system of planning and programming and budgeting through the vast Federal Government."[10] This directive has now been implemented in most federal agencies with results that have indeed been revolutionary.

Of course, systems analysis or cost benefit analysis is not new to some federal agencies. Cost benefit techniques have long been applied in assessing public works projects.[11] However, often cost benefit analyses have been confined to considering the cost and benefit aspects of individual projects without comparing and evaluating alternative projects.

For example, the Army Corps of Engineers' practice has been to consider each of its proposed major harbor improvement projects as an entity and to recommend project approval if the benefit/cost *ratio* is 1 or more. If a project could be undertaken at various levels of intensity, the project was chosen which maximized the present value of the benefit-minus-cost difference. In each case, no "competition" among various proposed projects was conducted as a part of the formal analysis.

Such an approach does not fulfill the systems concept in that it does not consider *overall* effectiveness in relation to national objectives and it does not consider interdependencies between various projects. For example, improvements in the Boston harbor may well have an effect on the demand placed on the New York harbor. If these interdependencies are not considered, benefits are normally estimated poorly because the negative benefit at unconsidered locations is overlooked.

Generally, the greatest difficulties in systems analyses in nondefense governmental areas lie on the benefits side of the ledger.[12] First, there is the problem of *incommensurate benefit measures* which make it impractical to compare programs of widely differing varieties. Ideally, the systems approach to planning in the federal government would entail the comparison of a wide variety of different government programs—e.g., poverty programs, harbor improvement programs, transportation systems, etc., in terms of the degree to which each program and combination of programs achieves the objectives recognized by a free society for its government. In practice, the benefit measures associated with such diverse

[10] As reported in *Time*, Sept. 3, 1965, p. 20.

[11] For example, The River and Harbor Act of 1902 required a board of engineers to report on the desirability of Army Corps of Engineers river and harbor projects, taking into account the amount of commerce benefited and the cost. The Flood Control Act of 1936 authorized federal participation "if the benefits to whomsoever they may accrue are in excess of estimated costs."

[12] Of course, the same basic problem exists in military analyses. For instance, the natural benefit measure associated with artillery and aircraft weapons may be quite different. However, the problem is generally neither so pronounced nor apparently insoluble in the defense context. See, for example, R. N. Grosse, "An Introduction to Cost-Effectiveness Analysis," Research Analysis Corporation, McLean, Va., RAC-P-5, July, 1965, for a discussion of this point in military analyses.

programs are incommensurable (not measurable in the same terms).[13] How can one compare the lives saved through the institution of federal requirements for auto safety features and promotional campaigns to increase the public's use of those devices with the benefits from an educational program, for example?

Partial answers to such questions are indeed available, but they are not simple to implement. One way in which this might be done is by translating all benefits into a common measure such as dollars. A life saved can be partially valued in terms of the present value of the person's future earnings, and an educational program can be similarly valued in terms of the increased earnings of the participants. However, many unquantifiable social benefits are omitted from such calculations—e.g., the avoidance of grief in the case of a life saved, as well as the benefit associated with not having children raised in a fatherless family. Other questionable social values are implicit in such simple dollar measures. For example, there is a potentially undesirable higher value which is implicitly placed on male lives versus female lives and young lives versus old lives in such an earnings-related measure.

Such *secondary benefits* as these social implications are often of significant importance to a systems view of governmental decisions, for often the summation of the indirect benefits can be of greater import than the direct ones. One illustration of the importance of detailed analysis and the difficulties in assessing benefits was much publicized in the studies conducted of the supersonic transport aircraft (SST). In many ways, the experience gained in systems analysis in the Department of Defense proved useful in the SST analysis because of the obvious physical similarity of the systems.

One easily quantifiable aspect of both military and nonmilitary hardware systems is the returns from sales to other nations. Our weapons hardware is frequently sold to friendly powers, thus providing a way for recouping some of our investment. The same is true of commercial aircraft. In the early stages of discussion of the proper role which the federal government might play in SST financing, the effect of overseas sales of American-made SSTs on our gross national product was put forth by proponents of vigorous government participation as itself nearly providing a justification for the project. Detailed analysis of this benefit aspect revealed that the interdependencies which it ignored relegated this aspect of benefits to an

[13]General discussions of the difficulties involved in applying systems analyses in various contexts are treated by A. R. Prest and R. Turvey, "Cost-Benefit Analysis: A Survey," *The Economic Journal,* December, 1965, and by E. S. Quade, "The Limitations of a Cost-Effectiveness Approach to Military Decision-Making," The RAND Corporation, Santa Monica, Calif., P-2798, September, 1963.

insignificant, or possibly negative, role.[14] The ignored interdependencies involved substitution effects, such as the proportion of foreign plane purchases which have traditionally been financed within the country, the purchase of SSTs in lieu of other United States-built aircraft, the loss of passengers by United States' airlines to foreign SST-equipped lines, etc.

Equally difficult problems of analysis arise in most other areas of strategic governmental decision making. Transportation planning, for example, is another area in which demonstrable success has been achieved through systems analysis. There, a major benefit aspect is the time saved by passengers. The question is, What is time worth? Indeed, more complex interactions arise: for example, is a one-hour time saving for one person equal to 60 single minutes saved by 60 different people?

Major interdependencies which must be considered in transportation planning include the effect on other transport media, the effect on the insurance industry (which is heavily involved in insuring people, autos, etc.), the productivity lost through transportation—incurred disability and death, etc. None of these are easily quantifiable or estimable, and, of course, the "natural" measures which one would choose are usually incommensurable.

Systems Analysis Using Technological Forecasting

The interdependence and interrelatedness of technological advances have been put to use in a systems context through *technological forecasting.* The methods of technological forecasting are varied, but the essential element is an attempt at setting timetables for the meshing of the need for products and services and the technological capability for producing them. Some companies[15] use networks which show the interrelated technological developments which are necessary to achieve various products together with the by-products which will result at the various stages. The company can use the network as a basis for deciding whether it should attempt to develop the product itself or whether it should focus attention on the development of components or subassemblies.

Figure 7-5 illustrates a network used by TRW, Inc.,[16] to analyze the development of holographic three-dimensional color movies. The key to the network shows the distinction between events directly related to the

[14] See Stephen Enke, "Government-Industry Development of a Commercial Supersonic Transport," *American Economic Review,* vol. LVII, May, 1967, pp. 71–90.

[15] TRW, Inc., is reported to have extensively applied this approach in "New Products—Setting a Timetable," *Business Week,* May 27, 1967, pp. 52–56.

[16] Taken with permission from Robert V. Ayres, *Technological Forecasting and Long-range Planning,* McGraw-Hill Book Company, New York, 1969, pp. 154–155.

program and corollary developments. In this case, many by-product techniques and devices would result along the way to the primary goal. Thus, any company which chose to undertake such a program would find that it would have a sequence of by-products to market long before the primary product. Added advantages might well be enough to warrant a firm's undertaking a project even though the end product, because of the long time period required to develop it, might not justify it.

Systems Analysis in Industry

The major difference between systems analyses in government and those conducted in industry is the ease with which benefits are measured. In industry, the natural-benefit measures of sales revenue and profit are commensurable with the natural dollar cost measures. However, as businesses become more complex and businessmen become more sophisticated, it is less clear that benefit measurement is indeed as simple as it appears.[17] Today, short-term versus long-term objectives have become intertwined and social objectives take on greater significance than they have had previously. Thus, the problems of performing systems analyses in industry are not so different from those encountered in government.

The marketing process itself may profitably be viewed as a system composed of interdependent activities—production, pricing, promotion, distribution, etc. In analyzing the market system, the suboptimization approach is usually adopted—i.e., the "best" price is determined, given some assumptions concerning the other marketing controllables, then the best level of promotional effort is determined, then the pricing decision is reevaluated in the light of the knowledge gained in promotional analysis, etc.

One of the most significant business areas in which systems ideas have been applied is in product decisions. Product systems, consisting of interacting individual products, are being developed, for example. When a manufacturer color-coordinates linens, bedspreads, towels, and bathmats, he is creating a product system. Such an application of systems ideas is based on the recognition that customers do not really buy products; rather, they buy satisfactions. Thus, the housewife buys nutrition for her family rather than milk, beauty rather than cosmetics, and esthetically pleasing and utilitarian surroundings rather than linens.

A new-product idea—be it a product system or an individual component—must be evaluated on the basis of the satisfactions which the product produces and its interdependencies with other products, both existing

[17] See W. R. King, "Performance Evaluation in Marketing Systems," *Management Science*, vol. 10, no. 4, July, 1964.

FIGURE 7-5 *Network description of program for the development of 3-D color holographic movies.*

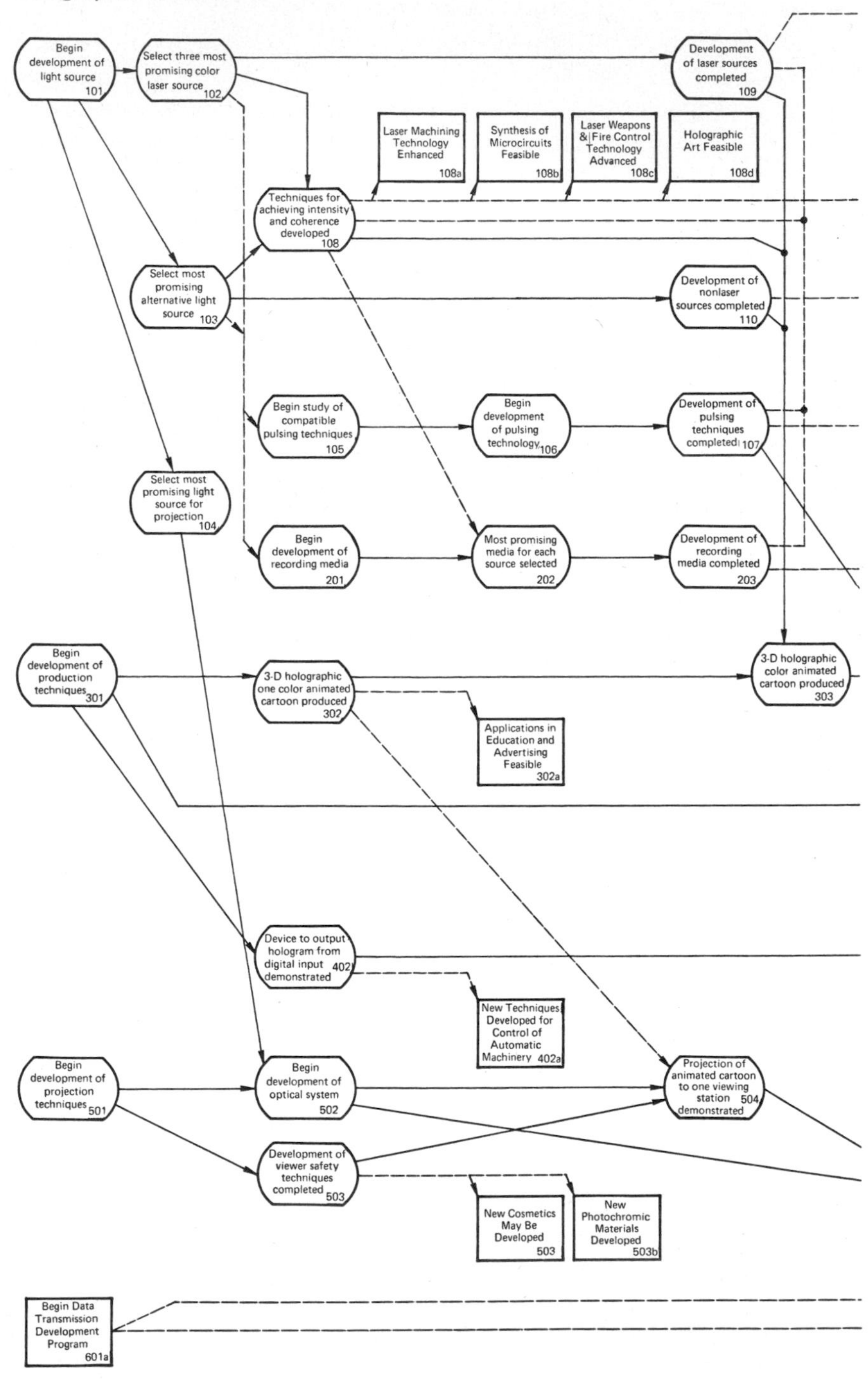

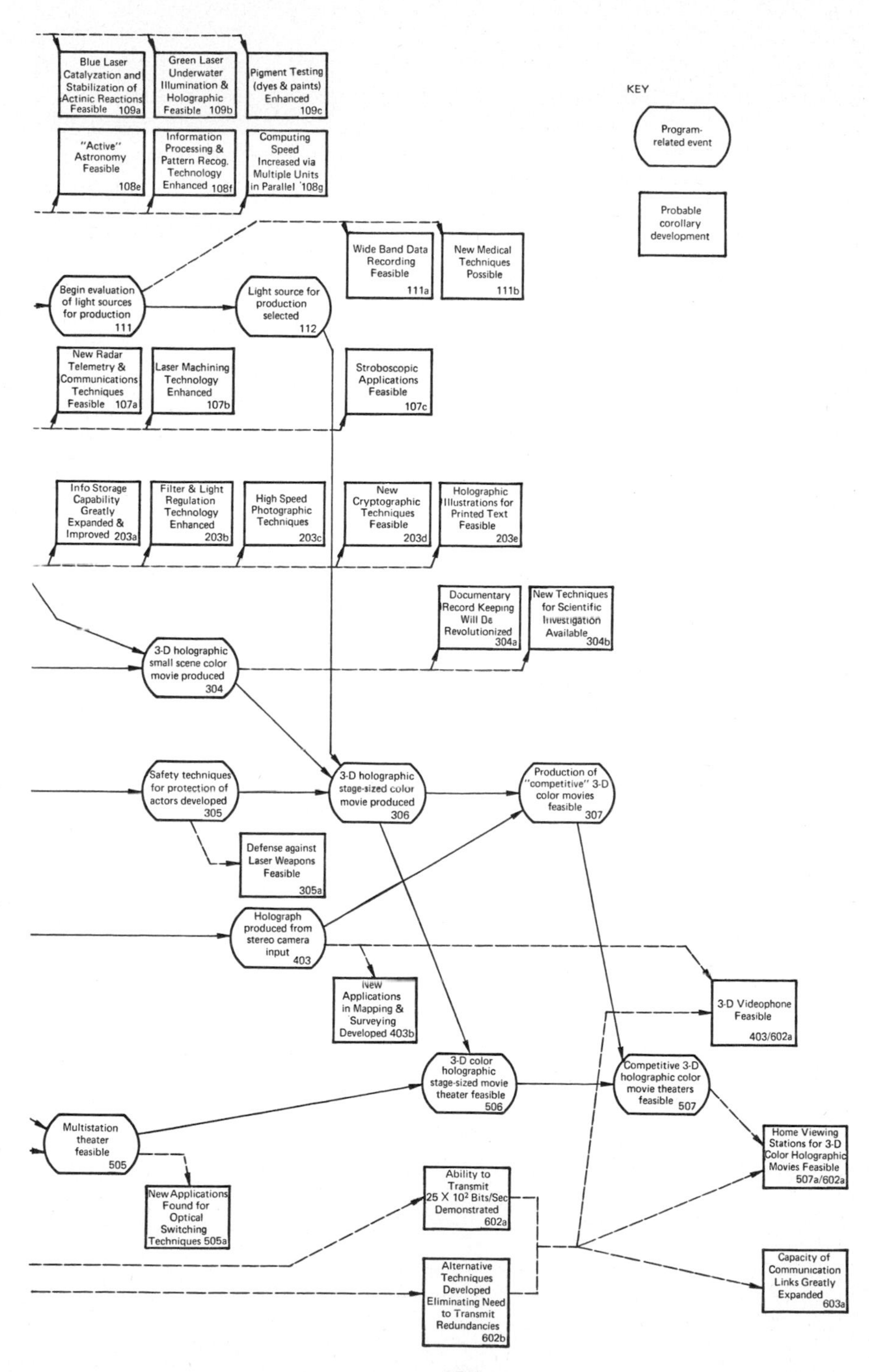

Blue Laser Catalyzation and Stabilization of Actinic Reactions Feasible 109a
Green Laser Underwater Illumination & Holographic Feasible 109b
Pigment Testing (dyes & paints) Enhanced 109c
KEY
Program-related event
Probable corollary development
"Active" Astronomy Feasible 108e
Information Processing & Pattern Recog. Technology Enhanced 108f
Computing Speed Increased via Multiple Units in Parallel 108g
Wide Band Data Recording Feasible 111a
New Medical Techniques Possible 111b
Begin evaluation of light sources for production 111
Light source for production selected 112
New Radar Telemetry & Communications Techniques Feasible 107a
Laser Machining Technology Enhanced 107b
Stroboscopic Applications Feasible 107c
Info Storage Capability Greatly Expanded & Improved 203a
Filter & Light Regulation Technology Enhanced 203b
High Speed Photographic Techniques 203c
New Cryptographic Techniques Feasible 203d
Holographic Illustrations for Printed Text Feasible 203e
Documentary Record Keeping Will Be Revolutionized 304a
New Techniques for Scientific Investigation Available 304b
3-D holographic small scene color movie produced 304
Safety techniques for protection of actors developed 305
3-D holographic stage-sized color movie produced 306
Production of "competitive" 3-D color movies feasible 307
Defense against Laser Weapons Feasible 305a
Holograph produced from stereo camera input 403
New Applications in Mapping & Surveying Developed 403b
3-D Videophone Feasible 403/602a
3-D color holographic stage-sized movie theater feasible 506
Competitive 3-D holographic color movie theaters feasible 507
Multistation theater feasible 505
Home Viewing Stations for 3-D Color Holographic Movies Feasible 507a/602a
New Applications Found for Optical Switching Techniques 505a
Ability to Transmit 25 X 10² Bits/Sec Demonstrated 602a
Capacity of Communication Links Greatly Expanded 603a
Alternative Techniques Developed Eliminating Need to Transmit Redundancies 602b

and planned. Systems analyses related to decisions involving the selection of new products must also consider questions about other interactions. For example: How would the new product fit into the existing production facilities? To what degree would it be compatible with the skills of the people who make up the organization? Would it require new sources of raw materials, new distributors, new consumer groups? How would it affect the sales of existing products?

The myriad of costs and benefits affecting each of these interactions are complex and difficult to measure. As a result, techniques have been developed to make use of the judgments of experts in making gross overall evaluations of product ideas on a total systems basis.[18] These techniques generally employ scoring schemes which are really subjective utility estimates made for each of the interactions associated with a new-product idea. These estimates can then be combined and an overall gross product idea evaluation made. Those products which appear to have high overall value can then be subjected to stringent economic evaluation.

Other aspects of the marketing system have been approached by systems analysts using simulation techniques. Simulations offer greater flexibility and an opportunity for a more complex (and, hence, more accurate) description of the existing system than do analytic models. Thus, a complex model of an entire marketing system can be built into a computer and the operations of the system simulated. In this way, a wealth of simulated history can be accumulated under various market conditions and economic environments and using various combinations of marketing controllables—e.g., price, product, promotional campaigns, etc.

The most comprehensive marketing system simulations are *microbehavioral* in nature—i.e., they entail a set of models of individual customers who are representative of the overall market and sets of retailers, wholesalers, and competitors. The model through which these entities interact within the computer embodies the firm's best understanding of the structure of the marketing system as developed from experience, surveys, and statistical sampling studies. A number of such simulations are now in use and more are currently being developed.[19]

The work of Jay Forrester at MIT probably represents the most comprehensive systems simulations which have been usefully performed. Forrester's *industrial dynamics* approach[20] involves modeling the entire business enterprise and its environment in terms of flows of information,

[18] See W. R. King, *Quantitative Analysis for Marketing Management,* McGraw-Hill Book Company, New York, 1967, chap. 5.

[19] See "Pillsbury Finds a New Mix That Pays," *Business Week,* June 25, 1966, p. 178, or A. A. Kuehn and D. L. Weiss, "Marketing Analysis Training Exercise," *Behavioral Science,* January, 1965.

[20] Jay W. Forrester, *Industrial Dynamics,* The MIT Press, Cambridge, Mass., 1961.

resources, products, etc. Various levels of the company's controllables can be tested through the simulation model.

Product Life Cycle Planning[21]

The concept of product life cycle has long been accepted by marketing managers, but it has been little used to advantage. The systems approach recognizes the dynamic nature of the product's existence and naturally leads to questions of *how the life cycle can be used to enhance product sales.*

Most elementally, a knowledge and understanding of the nature of the life cycle of a product can be used to advantage in the making of decisions associated with the sale of a product. For example, it is important that a new product make a favorable first impression on consumers. If the impression is bad, those who are dissatisfied will tell others, and the product will probably never be successful. To maximize the probability of an initially successful consumer experience, a manager may well decide to make use of his knowledge of a life cycle by distributing the product first through small specialty stores where it can be demonstrated and where the potential purchaser can have personal instructions in the product's use. Then when the product becomes established, a switch from distribution solely through small specialty shops to volume outlets may be undertaken so that the growth phase of the product's life cycle may be used to advantage.

Also, the manager may act to influence the life cycle. For example, he may set the price of the product low in order to discourage competitors from entering the market with competing products. If he is successful, he may well extend the life cycle of his brand.

The concept of life cycle planning involves not only this sort of life cycle-influence decision but also the attempts to extend the life of the product which are undertaken at the outset of its sales life. A life cycle plan prescribes a sequence of actions to be taken during the life of the product in an attempt to sustain sales and profits. For example, after new original users of the product are virtually exhausted, promotions would begin to promote more frequent use among current users, to develop more varied uses among current users, and to create new users, and attempts would be made to find new uses for the product. Illustrations of products which have successfully accomplished this are rather easy to find. For example, Jello began by increasing the number of flavors which were available, then introducing the concept of Jello as a basis for salad, then they

[21] Many of the ideas in this section are adapted from Theodore Levitt, "Exploit the Product Life Cycle," *Harvard Business Review,* November–December, 1965.

pinpointed potential users who required another appeal (e.g., by a fashion-oriented weight control promotion when weight consciousness was prevalent, and by a flavorless Jello for strengthening of fingernails and as a bone-building agent).

Whether or not these actions were taken on a preplanned basis in this or any other particular case is not important. The important point is that timing and preplanning are indeed essential to take maximum advantage of a product's life cycle. Timing is essential in terms of assessing the point of consumer readiness for a product or for a new version of an existing product. Sequencing of such actions is also important so that the interdependency between various products may be accounted for. For example, a hair coloring boom could not but have been preceded by consumer acceptance of hair sprays, which created a consciousness of hair fashion.

A plan which utilizes life cycle concepts becomes the basis for assigning priority to the various activities which might be undertaken to enhance a product's sales. Under such a plan these priorities are determined rationally rather than accidentally. Most important, such a plan prevents the organization from trying to do too many things at once.

Life cycle planning also forces the organization into a systems view of its business. For example, Jello expanded into puddings, pie fillings, and Whip and Chill—in effect, into the dessert technology business—as a result of its various versions of the basic product. Moreover, life cycle planning forces consideration of competitive action. In a sense, the life cycle is in part determined by competition. Thus, in accounting for the interdependencies between products, the dynamic nature of the life cycle, consideration of competition, and a broadened view of the scope of the organization's activities, life cycle planning represents a direct application of systems ideas.

Project and Product Management

Systems ideas have also been implemented on a large scale in organizing for the execution of projects. This approach—called "project management" or "systems management" in hardware-oriented industries and "product management" in many consumer products industries—involves the creation of an organization to handle a well-defined project.

Usually the project to be accomplished by project management techniques is of significant magnitude and involves major efforts of diverse personnel on highly interrelated activities for its successful accomplishment. Also, such projects usually are of rather unique character; for example, repetitive production operations are probably not adaptable to project format.

The utility of project organizations to diverse applications became apparent after they performed successfully in weapons systems development activities. The military services, NASA, and major aerospace contractors have developed project organizations to the degree that they represent major management philosophies.[22]

In consumer product industries, the same ideas have been applied to the marketing of products under the title "product management." The creation of product manager positions resulted from the recognition that top management could not be expected to know all of the details about each of the organization's products. Similarly, functional managers properly show more concern for their function than they do for products. Thus, the need for a manager who can cut across traditional functional lines to bring together the resources required to achieve product goals is clear.

The product manager is able to organize the diverse objectives and motivations of the various functional units of an organization so that the total effort is directed toward the accomplishment of overall product goals. Thus, controllable aspects of the enterprise can be viewed by one person (or group) whose concern is with the product and its contribution to the organization rather than with one of the specific methods for achieving the goals. In this way effective implementation and control of the product's sales program may be enhanced.[23]

In a later chapter, we shall more extensively discuss project management as it illustrates the application of systems ideas to the execution function of management.

SUMMARY

Systems analysis and systems management constitute the most significant elements of a systems model of management. Systems analysis is an analytic process designed to help managers make decisions; hence, it is closely related to the planning aspect of their job. Systems management deals with the execution of decisions within the framework of the choices made in the planning process.

The systems model requires the manager to recognize system interactions and interdependencies and to act to take advantage of them. The approach is applicable at all levels of organizational activity, although its

[22] *Ibid.*

[23] For discussion and evaluation of product management, see William R. King, *Quantitative Analysis for Marketing Management,* McGraw-Hill Book Company, New York, 1967, or Robert M. Fulmer, "Product Management: Panacea or Pandora's Box," *California Management Review,* Summer, 1965, pp. 63–74.

primary impact has been in high-level policy-oriented decisions and actions.

Weapons systems analysis is the first area in which systems concepts became widely visible at the national level. The application of cost-effectiveness (benefit) ideas in defense planning has resulted in the spread of systems analysis activity throughout the federal government and into many state and local governments as well. The planning, programming, and budgeting systems (PPBS's) which have been instituted at various governmental levels are a direct consequence of weapons systems analysis in the federal government.

Other areas of systems activity in pervasive use today are technological forecasting, product life cycle planning, and project management. In all of these, systems concepts play a predominant role, and the success achieved by these systems applications foretells of future successes to be achieved by further applications of systems ideas.

EXERCISES

1. What is systems management?; systems analysis? How are they related to the functions of the manager?
2. Discuss the role of models in systems analysis. Why are they necessary?
3. Relate the systems analysis-systems management model to *(a)* the functional management model of Chapter 5, *(b)* the system development life cycle.
4. Why is it that "bad," or ineffective, management is more apparent (and therefore less allowable) today than it was previously?
5. The Flood Control Act of 1936 authorized federal participation in projects, "if the benefits to whomsoever they may accrue are in excess of estimated costs." What does this wording indicate concerning the government's view of its organizational claimants?
6. How does systems analysis relate to traditional intuition-based techniques for decision making?
7. Why are systems analysis and systems management best thought of as *not* directly corresponding to the planning and execution phases of the manager's job?
8. Interactions and interdependencies are important to the systems concept. But it is not enough for the manager simply to recognize these aspects of the system in question. He must do more. Explain.
9. What is ". . . the risk and uncertainty which are inherent in any high-level organizational undertaking"? Give an illustration.
10. Develop a simple work breakdown structure, in the fashion of Figure 7-3, for a family picnic.
11. Develop a flow chart, as in Figure 7-4, for a family picnic.
12. Sophisticated application of the cost effectiveness concept requires consideration of both *(a)* alternative ways of employing resources, and

(b) alternative ways of achieving the desired objectives. Explain by example.

13. If we did not use cost benefit ideas in our everyday lives, we would all be driving Cadillacs (or Rolls Royces, Mercedes, etc., depending on your tastes). Explain.

14. When costs and benefits are considered separately in decision making, what can happen? Relate to both DOD military decisions and to Enthoven's "homey example."

15. Relating again to Enthoven's "homey example," describe in detail how such a decision *should* be made. For instance, how should you decide which home models to consider? How should you decide whether or not to include a pool?

16. The Army Corps of Engineers has evaluated projects in terms of a cost-benefit *ratio* or *difference.* What does the use of these criteria imply about the sort of measures being used to assess costs and benefits?

17. Why is a "competition" among various proposed projects, each of which has been evaluated in cost-benefit terms, a vital part of management decision making?

18. What are "incommensurate benefit measures"? Why do they create a problem for the manager?

19. Give an illustration of how it may sometimes be possible to resolve the problem of incommensurate benefit measures. Why is this approach not always good?

20. Why do we not speak of "incommensurate cost measures" as well as "incommensurate benefit measures"? Can they exist?

21. What are secondary benefits? How does this concept relate to the chain of effects produced by an action?

22. What is technological forecasting? Give examples of various ways in which it might be used in planning for the future for *(a)* a business firm, *(b)* a police department, *(c)* a hospital.

23. Business firms have always been faced with relatively simple cost-benefit assessment problems because the cost and benefit measures are commensurate. What factors are causing this to change?

24. What is suboptimization?

25. Product systems are discussed in this chapter in terms of customer satisfactions. Why is it important to create product systems with this in mind? Give an illustration of a product system which could be created in association with bed sheets if it were believed that the customer wanted *(a)* beauty, *(b)* convenience, *(c)* sanitation, *(d)* all three.

26. New product systems require consideration of both customer-oriented interactions and producer-oriented interactions. Give examples.

27. Life cycle concepts are reintroduced in this chapter under the heading "Product Life Cycle Planning." What is this concept and how does it differ from the basic sales life cycle concept introduced earlier?

28. It has been said that certain concepts and products which were accepted by our society and by consumers were "ideas whose time had come." Relate this to life cycle planning.

29. How does project management illustrate an application of systems concepts?
30. The systems approach to the resolution of problems helps to reduce the parochial view that a manager might take. Why does this parochial view exist? How does the systems approach reduce this parochial factor?
31. In explaining systems analysis, the authors make the statement that every problem should be viewed in as large a context as possible. How would you explain this statement?
32. Select the following business organization functions and demonstrate how the systems approach could be applied in each of these functions: *(a)* production, *(b)* finance, *(c)* marketing, *(d)* research and development, *(e)* advertising, *(f)* manpower.
33. One of the important concepts of systems has to do with flows and processes as opposed to functions. For instance, *Business Week* reports[24] on the concept of an "office landscape" in which offices are laid out and arranged to follow natural lines of work flow and communication. This results in a deemphasis on paneled offices, private offices, carpets, etc. How complete do you think this application of systems concepts is? Can you take a broader systems view?

REFERENCES

Ayres, R. V., *Technological Forecasting and Long-Range Planning,* McGraw-Hill Book Company, New York, 1969.

Cleland, D. I., and King, W. R., *Systems Analysis and Project Management,* McGraw-Hill Book Company, New York, 1968.

——— (eds.), *Systems, Organizations, Analysis, Management: A Book of Readings,* McGraw-Hill Book Company, New York, 1969.

Fisher, G. H., *Cost Considerations in Systems Analysis,* American Elsevier Publishing Company, Inc., New York, 1971.

Grosse, R. N., "An Introduction to Cost-Effectiveness Analysis," Research Analysis Corporation, McLean, Va., RAC-P-5, July, 1965.

Johnson, R. A., et al., *The Theory and Management of Systems,* McGraw-Hill Book Company, New York, 1963.

Kellehner, G. J. (ed.), *The Challenge to Systems Analysis,* John Wiley & Sons, Inc., New York, 1970.

King, W. R., "Systems Analysis at the Public-Private Marketing Frontier," *Journal of Marketing,* January, 1969.

Levitt, T., "Exploit the Product Life Cycle," *Harvard Business Review,* November-December, 1965.

[24] "Management: Making Office Walls Come Tumbling Down," *Business Week,* May 11, 1968, pp. 56–58.

Prest, A. R., and Turvey, R., "Cost-Benefit Analysis: A Survey," *The Economics Journal,* December, 1965.

Quade, E. S., "The Limitations of a Cost-Effectiveness Approach to Military Decision-Making," The RAND Corporation, Santa Monica, Cal., P-2798, September, 1963.

——— and Boucher, W. I. (eds.), *Systems Analysis and Policy Planning,* American Elsevier Publishing Company, Inc., New York, 1968.

section four

PLANNING

8

the planning function of management

In the simple models which we have used in this book to characterize the manager's job, planning is one of the two major functions of management. Planning has been associated with the "deciding" aspect of the manager's job, while the other major function—execution—involves the implementation of decisions.

In this chapter, we seek to provide a conceptual framework or "way of thinking" about planning and decision making. Then in the subsequent chapters in this section, we will adopt the "how to do it" approach.

BASIC PLANNING CONCEPTS

Dr. Fred Polak has developed a theory which argues that the future of a civilization, a country, or a people is determined in large measure by its "images of the future." He contends that it is possible to measure these images of the future and that it may be possible to alter or adjust them, and thus to guide a nation's or a people's future. According to Polak, if a society has optimistic ideas, dynamic aspirations, and cohesive ambitions, the civilization will grow and prosper. If it exhibits negative trends, uncertain ideals, and hesitant faith, it is in danger of disintegrating. The basic concept is that *by thinking about the future, one may create that future according to his image.*[1]

Planning in organizations involves exactly this—thinking about the organization's future in relationship to its present in such a way that the future can be affected in ways which benefit the organization.

More explicitly, planning is the process of thinking through and making explicit the strategy, actions, and relationships necessary to

[1] Weldon B. Gibson, as cited in "Guideposts for Forward Planning," in David W. Ewing (ed.), *Long-range Planning for Management,* Harper & Brothers, New York, 1958, pp. 488–489.

accomplish an overall objective or purpose. Planning involves the use of judgment and discretion in determining what one wants to be or what he wants his organization to be, and then determining what actions will be most effective in accomplishing his end objective. Planning means looking ahead, predicting what a future environment holds, estimating future needs and situations, and drawing up a program of action for the future needs and situations.

Planning is an important part of everyone's life and every organization's life because it has to do with these fundamentals:

1. The purpose, goal, objective, or mission of an individual or an organization
2. The process of reasoning (thinking through) what resources and what action must be accomplished to reach those goals
3. The determination of the most effective allocation of resources and actions
4. The establishment of methods for appraising how effectively predetermined strategy and goals are being pursued

Because planning deals with these fundamentals, it furnishes the necessary guidance as to *what to do and how to do it.*

Development of the Planning Concept

When the word "plan" entered the English language, it had a highly refined meaning. It was derived from the Latin *planus,* connoting "flat," and it is etymologically the same word as "plane." The *Oxford English Dictionary,* which attempts to present words with all relevant facts concerning their form, sense history, pronunciation, and etymology in order to provide an idea of how the word developed and a good indication of the development of the concept which the word connotes, defines and dates "plan" as:

> *To devise, contrive, design (something to be done, or some action or proceeding to be carried out); to scheme, project, arrange beforehand. . . . 1737 Popel Hor. Epist. II. e 374. We needs will write Epistles to the king; . . . . Be called to court to plan some work divine. 1782 Miss Curney Ceciliar V. XI., Cecilia the whole time was planning how to take her leave . . . . 1868 Freeman, Norman Conquest. II.X. 470. Never was a campaign more ably planned.*[2]

[2] *The Oxford English Dictionary,* vol. 7, N-Poy, The Clarendon Press, Oxford, 1933, pp. 941–942.

Earlier use of the word in English (1678) describes a drawing or sketch made by projecting an object on a plane surface; apparently it was something similar to a "blueprint."

Contemporary meaning of the word "plan" and its gerund has evolved into a broader meaning. Nevertheless, today's concept of planning still retains much of its seventeenth-century meaning of a drawing or sketch. A fundamental difference is that while the early usages of the word and its derivatives dealt with things and people, current connotation of the word contemplates these together with ideas, concepts, abstractions, ideologies, and philosophies; and whereas early usage of the term did not reflect time dimensions, present employment of the word, especially in business and industry, considers definite temporal connotations—e.g., in the use of the "long-range" and "short-range" modifiers.

Today, planning is concerned with the development of a plan; it involves mental activity directed toward what should be done, how it should be done, where action is to be effected, who is responsible, and why such action is necessary. In its modern organizational context, planning involves the element of futurity, with the degree of futurity becoming greater as higher organizational levels are involved.

Historical Development of the Planning Process

Military leaders visualized "plans" and did "planning" many centuries ago. In the late 1870s and 1880s many executives in Eastern manufacturing institutions concerned with metal working and the construction of machinery began to evaluate production planning problems, especially in relation to standards and efficiency. These factors were generally identified as matters which could be investigated in a scientific manner and planned so as to optimize the resource employment available to the enterprise.

Frederick W. Taylor focused attention on planning when he described his "planning room" in his system of management. Some twenty-five years after Taylor used the term, the word was applied to the layout of municipalities or other areas. "City planning" and "regional planning" are now definite and understandable uses of the term. Industrial planning, military planning, and municipal planning remained dominant fields of usage of the word until the period of the establishment of the Union of Soviet Socialist Republics and the economic depression of 1929. Then the terms "national planning" and "social-economic planning" came into general use.

Fayol placed considerable emphasis on the managerial function of *prevoyance,* or looking ahead.[3] He considered the process of thinking out a plan and ensuring its success as one of the greatest satisfactions that an intelligent man could experience, as well as a powerful stimulant to human endeavor. In describing the planning process Fayol said:

> *The maxim, "Managing means looking ahead," gives some idea of the importance attached to planning in the business world, and it is true that if foresight is not the whole of management, at least it is an essential part of it. To foresee, in this context, means both to assess the future and make provisions for it; that is, foreseeing is itself action already. Planning is manifested on a variety of occasions and in a variety of ways, its chief manifestation, apparent sign and most effective instrument being the plan of action. The plan of action is, at one and the same time, the result envisaged, the line of action to be followed, the stages to go through, and methods to use. It is a kind of future picture wherein proximate events appear progressively less distinct, and it entails the running of the business as foreseen and provided against over a definite period.*[4]

The difficulties of creating a good plan of action were recognized by Fayol, who considered the management function as playing a significant part in planning but also noted that the planning process calls into play every important department and function of the business. His general features of a good plan of action, namely, *unity, continuity, flexibility,* and *precision,* are as appropriate today as they were when he outlined them.

Fayol developed a concept of long-range planning in his writing relative to the drawing up of a plan of action in a large mining and metallurgical firm. The entire plan was contemplated as being made up of a series of forecasts or separate plans with yearly, ten-yearly, monthly, weekly, daily, long-term, and special forecasts. Provision was provided for the correlation of long-range and annual forecasts through the process of annual review of the forecasts. This ensured the maintenance of a unity of planning for each year. The comprehensiveness of his early forecasting and planning can be exemplified by a brief examination of the contents of such forecasts as are reflected in Figure 8-1.

The special forecasts envisioned by Fayol were for activities which

[3] Henri Fayol, *General and Industrial Management,* Sir Isaac Pitman & Sons, Ltd., London, 1949. (Henri Fayol's observations on the principles of general management first appeared in French in 1916 and did not appear in an English version until about 1929.)

[4] *Ibid.,* p. 43.

YEARLY AND TEN-YEARLY FORECASTS

Technical Section

Mining rights, premises, plant
Extraction, manufacture output
New workings, improvements
Maintenance of plant and buildings
Production costs

Commercial Section

Sales outlets
Marketable goods
Agencies, contracts
Customers, importance, credit standing
Selling price

Financial Section

Capital, loans, deposits
Circulating assets
Available assets

Supplies in hand
Finished goods
Debtors
Liquid assets

Reserves and Sundry Appropriation

Creditors
Sinking funds
Dividends
Bankers

Wages
Suppliers
Sundry

Accounting

Balance sheet
Profit and loss account
Statistics

Security

Accident precautions
Works police, claims, health service
Insurance

Management

Plan of action
Command
Control

Organization of personnel selection
Coordination, conferences

FIGURE 8-1 *Contents of forecasts as envisioned by Henri Fayol, as reflected in his* General and Industrial Management, *Sir Isaac Pitman & Sons, Ltd., London, 1949, p. 47.*

exceeded one or several of the ten-year forecasts. The integration of all the forecasts—yearly, ten-yearly, and special—constituted the firm's general plan; such a plan, in turn, after approval by the board of directors, served as the guide, directive, and law for the entire corporate staff.

Without question Fayol had an extraordinary insight into the problems which beset business management in any era. While others were primarily concerned with shop management, Fayol directed his writings to an examination of the functions of the general administrative manager.

The attention given to planning, both in his writings and in his business practice, seem to qualify him as the first real pioneer in the development and evolution of long-range planning thought.

PLANNING AND THE MANAGEMENT OF CHANGE

The planning which takes place in modern organizations is directed toward the *effective management of change.* If change were not to take place, or if the organization did not wish to introduce it, there would be no necessity for planning.

Day-to-day changes in an organization are typically so imperceptible that people miss them. Then, when the changes cumulate sufficiently to be noticed, they appear as a drastic and sudden change, often reflected in the organization's operating reports. These reports, such as the balance sheet and the profit and loss statement, actually describe the *result* of changes which have already occurred. The reports *reflect a change* but they do not identify *why* the change occurred.

For example, if a competitor develops and produces a new product, the company that has failed to assess what is happening in its competitive environment may find that the first identifiable symptom of the change is a drastic reduction in the organization's profitability. By then, the situation may be irretrievable and management may readily be panicked into a "crisis" reaction. One of the more regrettable outcomes of such reaction is the tendency to permit short-term expediencies to take precedence over long-term goals and the alternative strategies which should be considered when the organization contemplates its future. One of the manifestations of this is the familiar situation in which the manager finds himself so busy "running about fighting fires" that he has no time to think about the future.

Guidelines for "Change Management"

But how can we manage change? There is no simple answer to this question, but there are several basic guidelines for managers to use in guiding their thinking about the management of change:

1. Accept the idea of change and the need to anticipate change. To do this means to develop the necessary forecasts or predictions about when, where, and why change is likely to happen.
2. Develop the necessary plans and controls whereby management will be able to organize its resources so that unanticipated change can be "reacted to" faster.

3. Develop a strategy whereby the organization will be able to take advantage of the changes that do occur and to *influence the changes that will occur* in the future.

STRATEGY AND PLANNING

The term "strategy," as used in the foregoing guidelines, warrants some additional attention. The term has been used by various authors in different fields in a variety of ways. As used in these guidelines, "strategy" is the *complex of plans for bringing the organization from a given posture to a desired position in a future period of time.* Learned et al. describe strategy as ". . . the pattern of objectives, purposes, or goals and major policies and plans for achieving these goals, stated in such a way as to define what business the company is in or is to be in and the kind of company it is or is to be."[5]

A useful analogy might be drawn between an organization which must develop a strategy in order to survive in its environment and an individual who must survive as a productive member of society. Each person has the problems of:

1. Establishing objectives in life—What do I want to become?
2. Conceiving and accepting a policy on how he will spend resources, usually time and money, to accomplish his objectives.
3. Developing a series of personal plans or standards that will bring him to desirable positions in life.
4. Identifying a flow of information in his life which gives him some insight into how well he is doing in the things he set out to do.

Unfortunately, most of us do not approach life with any real strategy. Our lives evolve as a series of *conditioned* responses rather than as *contrived* responses. Contrived responses must be the result of deliberate strategy. Strategy is the process of thinking through what one wants to be, the explicit evaluation of alternatives, the assessment of risk and uncertainty, and the integration of personal value systems into a plan of action.

For instance, the many alternatives that are open to the young person of today offer a wide range of career choices. For many careers, particularly those of a professional nature, long and careful preparation is required. Yet, many young people approach a career objective with little planning as to how that objective will be obtained. Many, for example,

[5] Edmund P. Learned et al., *Business Policy, Text and Cases,* Richard D. Irwin, Inc., Homewood, Ill., 1965, p. 17.

do not select courses of study which will provide them the opportunity to study toward specific professional careers. This is particularly pervasive at the high school level, where many intelligent and qualified students opt for "easy" courses and thereby forgo the opportunity for college admission. They must then either completely sacrifice the opportunity to go on to college or waste valuable time taking remedial work. A little effort directed toward long-range planning at the high school entrance level would go a long way for most in assuring that their time is not so wasted.

Such personal illustrations vividly show that *strategy and planning are important even if one's objectives are ill defined or transient.* Even if an individual does not know what he wishes to become, the idea of developing a strategy does not imply that a fixed, known, and irrevocable goal has been established. Often, the goal is vague, and it is certainly not irrevocable. In such a situation, one may plan by choosing actions which provide progress while still leaving options open. For example, a college major can be chosen which will allow the individual to proceed on to graduate study in any of a number of fields in which he may be interested. Thus, even though the immediate course of action—the college major—may not fulfill the individual's (unknown) goals, he has made progress toward a degree and *kept open his options for the future.*

The Importance of Planning

The concept that an entity has a much better chance of becoming what it would like to be through a deliberate planned strategy than through conditioned responses is as applicable to an organization as to an individual. This aversion on the part of planners to conditioned responses is made clear by George Dively, board chairman of Harris-Intertype Corporation, who has been quoted as saying, "We plan ahead instead of letting daily decisions shape the direction of our company."[6] (Since Mr. Divvely's company has more than doubled its sales each five years for a quarter-century, his view of the importance of planning to an organization warrants one's attention.)

Planning has been described as the process of determining what the future organizational environment is going to be and then drawing up a set of documents outlining the desired position of the organization in the future. The basic tenet of planning is that the organization must not tacitly accept its fate in the environment of the future, but rather that it must attempt to influence the future by studying it and thinking about it.

[6] Quoted in *Investor's Reader,* publication of Merrill Lynch, Pierce, Fenner, & Smith, Apr. 23, 1969. p. 17.

Even with the planning ideas which have been an important part of management thinking for hundreds of years, it is interesting to note that the planning view of management was not always generally accepted. Successful entrepreneurs of the past relied primarily on their ability to *adapt* to a changing environment, rather than attempting to predict the future and then acting to take advantage of it. Modern managers must contend on a day-to-day basis with unprecedented advancements in technology, changes in consumer attitudes and needs, and changes in the other aspects of their environment. Indeed, we are living in a world where change is a way of life, and if the organization is able to do no more than react to a changing environment, it has submitted its future to the vagaries of chance.

The systems view of the organization serves to put the importance of planning into perspective. If an action taken by a manager initiates a "chain of effects," the manager should consider the entire chain and its associated sequence of consequences *before he takes the initial action.*

For instance, a decision to develop a new product today will affect the entire organization for many years in the future, since production, marketing, and other supporting functions must be developed and arranged. The importance of such chains of effects has been well recognized in many organizations, including the federal government, where actions taken have sometimes required unforeseen expenditures extending for many years into the future. For example, in the past, the U.S. Congress sometimes authorized the undertaking of projects without recognizing the future implications of those projects and the expenditures which they would subsequently require. With effective long-term planning, such chains of effects can be laid out in advance and, subject to the ability of the planner to predict the future, the future implications of current actions can be considered before the actions are undertaken.

The Manager and Planning

Planning is inseparable from managing. Of course, many organizations have established "long-range planning" functions or departments. This leads to the mistaken conclusion on the part of many nonmanagers that planning can somehow be separated from the job of the organization's managers—i.e., that planning can be handled by staff personnel. In truth, planning is a responsibility that can be *shared* with many groups in an organization; many of the technical aspects of planning, such as the collection of data and the interpretation of information, can indeed be carried out by staff specialists. Yet, since planning involves the making of assumptions, the forecasting of the future, and the strategic decision making involved in the allocation of basic resources, *it is not a function of the manager which can be delegated.*

THE PRESENT VERSUS THE FUTURE Few would argue the point that the effective management of change requires planning. However, the manager is constantly engaged in a battle between the emergencies of today and the need for thinking of the future. Both contend for his attention and both are important to him.

Indeed, the authors have heard unbelieving managers respond to suggestions concerning planning for the future by saying, "Why worry about that? We'll all be dead in the long run anyway!" or, "If we don't solve today's problems, we may not have a future." Both of these comments obviously have an element of truth, but they beg the question. *The organization will go on in the long run—if managers spend time thinking about the long run.* Moreover, while a failure to solve some of today's problems may preclude tomorrow from even happening, a modicum of planning today can prevent tomorrow's problems from occurring.

A PLANNING "SYSTEM" In order to properly prepare for the future, one must plan according to a "system," i.e., he must plan in a systematic way. Clairvoyant visions of the future would certainly be desirable to have available, but if they appear in random and unordered ways, they are difficult to use to full advantage. For us who are not blessed with clairvoyance, the need for systematization in planning is even more acute.

The planning process must be systematized so that the various kinds of planning which are necessary for the management of change are integrated into a single effort directed toward the most basic management objective—the enhancement of the organization's future. Planning builds on the basic objectives of an organization and attempts to devise systems whereby these objectives can be realized with economy, effectiveness, and due regard to the supporting role that business organizations play in the greater environmental systems.

THE ADAPTIVE NATURE OF PLANNING While no planning system can completely avoid the necessity of reacting to unforeseen events, *proper planning requires that each "fire" which is extinguished be analyzed so that similar conflagrations can be avoided in the future.*

However, a planning system can also itself adapt to "unplanned" contingencies. To successfully deal with change, the organization must develop a strategy which is *dynamic and adaptive* rather than fixed and static in nature. Thus, the organizational instruments of adaptation—the plans—must not only involve forecasts and predictions of the future, but they should entail *delineations of appropriate responses to contingencies which may occur.*

For example, the organization may have available a prediction of future demand for its product. Around this prediction it establishes a set of planning premises, develops its goals, and establishes its strategy. However, it should clearly not "place all of its eggs in a single basket" and assume the very considerable risk that the predictions are incorrect. Rather, it should establish mechanisms which can adapt to changing conditions, so that if predictions do not materialize, the organization may *act to take advantage of the developing conditions.* Such an adaptive planning process is depicted in Figure 8-2.

The adaptive planning process of Figure 8-2 illustrates how planning differs from simple forecasting. A nonadaptive plan might be so tied to a particular forecast of the environment that the failure of the forecast would imply the nonachievement of the plan's goals. However, an adaptive mechanism will permit the forecast to be changed and actions taken which take advantage of the change. Thus, forecasts and the actions based on them are not rigidly "believed" in adaptive planning. Rather, they are thought of as dynamic quantities which are subject to reevaluation and revision.

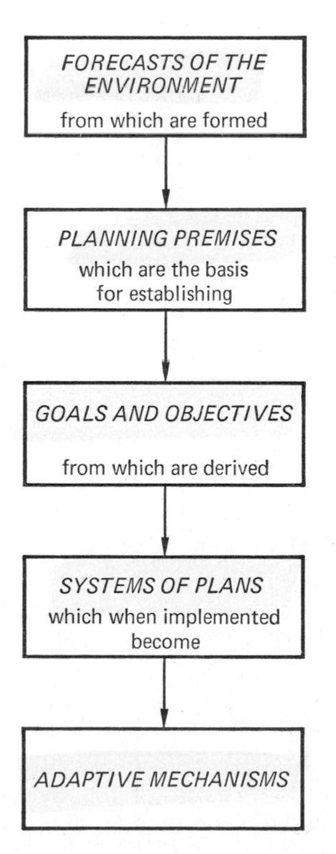

FIGURE 8-2 *The adaptive planning process. Adapted from Francis D. McCarthy and James W. Starr, "A Systems Concept of Interaction and Adaptation in Business Organizations," unpublished thesis at Air Force Institute of Technology, Wright-Patterson AFB, 1968.*

The Time Dimension of Planning

A planning system must involve strategies and plans with different degrees of futurity. Some aspects of the organization have requirements that entail planning for many years into the future, while others require planning over only a short horizon. For instance, financial planning usually must be done with a rather extended *planning horizon,* since commitments are made which do not terminate for many years. Product planning may be of the long- or short-run variety, depending on the industry. If the product is a machine tool, product planning requires that a "long look" be taken into the future, because such tools are long-lived. In the fashion industry, new-product planning may involve only a matter of a few months, since the "fashions" of the day are short-lived and dynamically changeable.

The time dimension of planning is not so obvious as it might first appear, however. Many complex factors interact to determine an organizational planning period—industry peculiarities, the market demand, the availability of resources, the lead time involved in the product life cycle, and the strategic objectives of the organization.

In speaking of "how far ahead organizations plan," we should also be careful to specify the particular planning period which we are discussing, since there are usually a number of different periods utilized by an organization—e.g., the overall organizational long-range planning period, the planning period for a specific functional area of effort, or the period for a particular element of long-range planning, such as that directed toward the availability of resources.

An organization should plan as far ahead as is useful, but only as far as can be done with some reasonable degree of accuracy. It is axiomatic that the farther one extends his planning horizon into the future, the less certain his predictions become. Also, his perception of the impact of his predictions becomes less sharp as his planning period is extended.

Three to five years is a common long-range planning duration. However, when one considers the problem of determining an overall organizational long-range planning period in the context of the "other" plans which are produced by an organization, he recognizes that the span of an overall planning period may be determined in any of a variety of ways. Illustrative of various planning periods are:

1. The average planning period for the functional areas of effort
2. The longest single period of functional-area long-range planning
3. The time period required to provide for the amassing of necessary resources
4. An arbitrary period which in the judgment of the executive group best fits the long-range objectives of the firm

5. A period which encompasses the diverse requirements of most of the critical areas of importance within the organization
6. A period which provides for the best market advantage in terms of economic cycles and long-term growth

The above list suggests the feasibility of *multiple planning horizons.* The precise period of time is less important than the determination of *the ability of the organization to realize a return on the resources that have been committed in the planning process.* According to Koontz and O'Donnell:

> *Since planning and forecasting that underlie it are costly, a company should probably not plan for a longer period than is economically justifiable; yet it is risky to plan for a shorter period. The logical answer as to the right planning period seems to lie in the "commitment principle," that planning should encompass the period of time necessary to foresee (through a series of actions) the fulfillment of commitments involved in a decision.*[7]

For example, pulp, paper, and lumber companies plan in terms of a 40-year horizon because of the nature of their "production process," whereas a cosmetics manufacturer would have no need for such a long time frame.

Long-range and Short-range Planning

A distinction between "short-range" and "long-range" planning is often made on the basis of the period of time involved. Although there is indeed a clear correlation between these kinds of planning and the length of the planning horizon, the more important distinction is on the basis of *the nature of the planning elements.* One can make statements as to the "typical" lengths of various planning horizons, but he must do so with the realization that what is long-range in one industry or organizational area may well be short-range in another.

TACTICAL OR OPERATIONAL (SHORT-RANGE) PLANNING Tactical planning determines what efforts will be taken to sustain the organization in its production and distribution of current products or services to existing markets. Typically, tactical plans cover a time period of less than one year

[7] Harold Koontz and Cyril J. O'Donnell, *Principles of Management,* 3d ed., McGraw-Hill Book Company, New York, 1964, p. 87.

and directly affect the functional groups—production, marketing, finance, administration, etc. The objectives of tactical planning are:

To subdivide work into logical and assignable work units in the organizational structure
To set immediate-term schedules, budgets, goals, and *modus operandi*
To establish necessary authority and responsibility patterns in the organization
To provide a suitable layout of the human and nonhuman resources

Within its time dimension, tactical planning answers pertinent questions about a particular function as follows:

1. Why is the action required?
2. What action is to be taken?
3. What will the action accomplish?
4. When are the results of the action required?
5. What objectives and conditions must be met?

STRATEGIC OR LONG-RANGE PLANNING Strategic planning is usually thought of as involving periods of time in excess of one year. These plans encompass all the functional areas of the business and are effected within the existing and long-term future framework of economic, social, and technological factors. Strategic planning is more than forecasting; it is an outline of how the organization will compete in the future. The fundamental idea behind strategic planning is simply that: ". . . by thinking about the future, man creates that future to his image."[8]

Successful strategic planning involves more than just establishing the organizational expectations. It also involves a deliberate analysis of the changing environmental conditions, particularly with respect to how the organization relates to its competition and environment. For example, a strategic corporate plan may specify financing through a substantial issue of securities, in which case the acceptability of the plan to investment bankers and other potential investors must be considered.

An interdependence exists, therefore, between the business system and the greater environmental system of which it is a part. Strategic planning is required continuously in order to forecast future environments and to draw up the necessary strategies for remaining competitive in those environments.

When strategic planning is conducted by an organization, plans are

[8]Weldon B. Gibson, as cited in "Guideposts for Forward Planning," in David W. Ewing (ed.), *Long-range Planning for Management,* Harper & Brothers, New York, 1958, pp. 488–489.

developed. Plans are simply the documents which evidence that planning activity has been accomplished. The creation and dissemination of these plans are an important process, but over a period of time the documents are often the least significant outputs of strategic planning. *Basic changes in the organization's structure and activities are the real output of the planning system.* For example, as a result of strategic planning, we would expect to see such changes as:

The establishment of new product lines
Research and development directed toward new product lines
Product diversification
Mergers and acquisitions
Reorganizations and realignments of authority and responsibility patterns within the organization
Divestment and liquidations
The designing and building of new facilities
Establishment of executive development and other training programs
Development of new markets

The Dynamics of Planning

Since management itself is a continuous activity, so, too, is its planning component. A plan which is soundly conceived and evaluated for feasibility, desirability, and practicality provides the basis for countless day-to-day decisions and actions. Hence, the plan is a document which must be kept up to date. The plan of action should contain a chronological list of specific decisions and actions to be required of the organization's executives. For example, one company's plan of action contained an "action item list" of considerations that had to receive executive attention at stipulated periods of time. These considerations entailed such actions as:

> *Decide whether the company should continue efforts on the design of manned, reusable space vehicles, If so, establish desired level of effort.*
>
> *Decide whether the company should construct a worldwide support vessel capability to maintain Deep Quest at sea.*
>
> *Review the marketing organization to assure that it has all the functions necessary for the attainment of its objectives.*
>
> *Determine if additional facilities are needed for the overhaul and repair of future deep-submergence vehicles.*[9]

[9]Speech by L. Eugene Root, President, Lockheed Missiles and Space Company, keynote address, Western Regional Conference, American Institute of Industrial Engineers, San Francisco, Cal., Nov. 3, 1967.

While we cannot expect the planning process to anticipate all of the problems of the future, we can expect that it will alert management to the trends in technology and in economic and social affairs. By recognizing some of the problems and opportunities in advance, the organization will be in a better position to cope with these issues when they become current and critical to the organization's survival.

PLANNING, PROGRAMMING, AND BUDGETING

The federal government has instituted a planning system which is referred to as a *planning, programming, and budgeting system* (PPBS). This PPBS promises to revolutionize government administration at all levels. Although the federal PPBS has received the greatest amount of attention, similar systems are operating in business and industry and in state and local governments. In fact, many of the basic ideas of PPBS were developed in the business world and borrowed by government planners.[10]

In this section, we shall discuss PPBS in the federal context while recalling its widespread applicability in other organizations. In later chapters, we shall demonstrate this applicability in a variety of contexts.

Programs

Essential to PPBS is the concept of a program. However, this concept is not precisely defined, because it must be given a somewhat different interpretation in different organizations. *A program is closely related to the objectives of the organization.* In fact, it is in part because the objectives of large organizations are often so difficult to define that no single definition of a program is satisfactory. In practice, a number of criteria may be applied in the definition of a program.

One essential feature of a program is clear: it is *output-oriented.* In other words, programs are defined first in terms of what the organization is trying to achieve, rather than in terms of the resources which the organization can bring to bear (inputs). The Bureau of the Budget used the following principles in their initial efforts at guiding the development of appropriate output categories:

1. *Program categories are groupings of agency programs (or activities or operations) which serve the same broad objective (or mission) or which have generally similar objectives. Succinct*

[10] For example, see George Steiner, "Program Budgeting: Business Contribution to Government Management," *Business Horizons*, Spring, 1965, pp. 43–52.

captions or headings describing the objective should be applied to each such grouping. Obviously each program category will contain programs which are complementary or are close substitutes in relation to the objectives to be attained. For example, a broad program objective is improvement of higher education. This could be a program category, and as such would contain Federal programs aiding undergraduate, graduate, and vocational education, including construction of facilities, as well as such auxiliary Federal activities as library support and relevant research programs. . . .

2. *Program subcategories are subdivisions which should be established within each program category, combining agency programs (or activities or operations) on the basis of narrower objectives contributing directly to the broad objectives for the program category as a whole. Thus, in the example given above, improvement of engineering and science and of language training could be two program subcategories within the program category of improvement of higher education.*
3. *Program elements are usually subdivisions of program subcategories and comprise the specific products (i.e., the goods and services) that contribute to the agency's objectives. Each program element is an integrated activity which combines personnel, other services, equipment and facilities. An example of a program element expressed in terms of the objectives would be the number of teachers to be trained in using new mathematics.*[11]

For instance, a program structure for the U.S. Forest Service might be made up of such categories as timber production, outdoor recreation, and natural beauty. The Coast Guard's structure might include search and rescue, aids to navigation, law enforcement, etc. A university might use science and humanities categories, involving teaching and research subcategories. In business, natural programs are products or product lines. The obvious differences in these program structures imply that the designation of appropriate program structures is not subject to any inviolable rules.

Developing Program Structures

Since there are no hard-and-fast rules involved, the determination of a program structure is a pragmatic undertaking, involving the consideration

[11] *U.S. Bureau of the Budget Bulletin* no. 66-3, Oct. 12, 1965, p. 4.

of alternative structures and a recognition that the structure chosen can itself have an important impact on the decisions which are to be made.

An important criterion to be used in this determination is that *the program structure should permit comparison of alternative methods of achieving objectives,* however vaguely defined these objectives may be. However, programs may also consist of a number of interactive components—the effectiveness of each of which depends on the others. Thus, in the Defense Department, Strategic Retaliatory Forces is a program because it relates to a national defense objective and can be broken down into elements—manned bombers, ICBM's, submarine-launched missiles, etc.—which are to some extent substitutes for one another and which involve questions of "resource mix."

On the other hand, the elements of a program may be complementary to one another, as in the case of the research and teaching elements of a university program.

Programs should also emphasize extended planning horizons, say, five or ten years. This enhances the value of programs as bases for long-range planning.

Other practicalities and peculiarities of particular organizations enter into the definition of programs. Frequently, for instance, the time period over which the goal will be pursued is a natural criterion. Pure research is the best illustration of an activity which is not easily related to organizational outputs. Thus, its long-range nature naturally defines it as a distinct program of many organizations. However, applied research and development is most often related with specific organizational objectives, and thus it should be included in the same program category as the other activities related to the same objective.

It is sometimes necessary to have programs defined in terms of intermediate outputs rather than final outputs. Thus, while "Strategic Retaliatory Forces," "Continental Defense," and "General Purpose Forces" are Defense Department programs related to final outputs (national objectives), "Airlift and Sealift" is a program which is clearly related only to an intermediate output. However, since it lends itself nicely to the comparison of alternatives (mixes of airlift and sealift, for example), it is considered a program.

There is also a "General Support" program in the Department of Defense's program structure. This includes items which cannot be identified with another specific program but for which an accounting must obviously be made. In most organizations, it will be necessary to have some sort of catch-all program as this. It is axiomatic to say that such catch-alls are potential devices for hiding activities which are in fact closely identifiable with objectives and for which alternatives are conceivably available for consideration. Care should be taken at all levels to avoid having this happen either purposefully or accidentally, since such perversion of the program structure can easily negate the greatest value of the concept—that of comparing alternatives.

The program structure of an organization should not be allowed to become either (1) a reflection of the organization's administrative structure or (2) a way of putting new labels on old budget activities. *The program structure need not reflect the organization's structure.* It is often desirable to have basic program categories which cut across organizational lines to facilitate the comparison of alternative elements which are potential substitutes for one another. So, too, should a relabeling of old budget activities be avoided. As we shall see, there is nothing incompatible about a traditional budgeting process such as that which has been used by the federal government and the program budget. The former can be developed from the latter. However, to do this meaningfully requires that planning and programming be done in the light of objectives and not in terms of inflexible budget categories.

Program Budgeting

A *program budget* is a financial expression of a program plan, just as any budget is a plan for future expenditures. Usually the time period involved in such a plan is a year or two. However, every organization must orient its thinking toward a planning horizon which is much beyond one year. In industry, for example, the development and testing of a new product often require several years, and consumer product life cycles—which often extend over many years, but seldom over many decades—require that business organizations look toward the distant as well as the short-term future. In government, the same need for advanced planning arises; e.g., a weapons system requires years for development, and a welfare program must be instituted, changed, and evaluated over a long period of time.

Traditionally the budget of the federal government has had two important characteristics which its critics regard as limiting the effectiveness of government expenditures. First, it is an annual budget, and thus it has limited usefulness as a basis for comprehensive long-range planning. Second, it is broken down into functional or object-class categories such as pay and allowances, construction, etc., rather than into categories which are related to governmental objectives.

In the context of defense planning, it became apparent in the 1960s that a budget of this variety was inadequate. The short-range nature of the annual budget precluded its use as a long-range planning document. Using the budget information, it was extremely difficult to gather together all the costs which would eventually result from a decision to undertake a program or to purchase a weapons system. Since the costs involved were scattered among a variety of appropriations and the budget projected requirements for only a year, the United States government found itself in the position of having purchased and operated many weapons systems and programs with little regard to, or knowledge of, the *total* cost. For example, operating costs, which may often be more significant than development and procurement costs, often tend to be bigger in the initial stages of the system's development.

Clearly, this confusion concerning the amount and timing of costs did not discourage various governmental units from proposing projects that could not be funded to completion. The visible evidence of this was the larger number of weapons system development projects which were canceled, at least partly because adequate advance planning was not required by the budgeting process. Some important weapons systems had indeed been specially "costed" under the traditional federal government budgeting process for purposes of decision making, but such a procedure was not a natural product of the budget. The institution of a program budget, as a complement to the traditional budget, was provided as a basis for costing weapons systems and output-oriented programs of the government. Such costings are essential to the comparison and evaluation of alternatives, which are, in turn, a vital part of the rational analysis of decision problems.

Programs and the Organizational Structure

The interaction of the programs, operating units, and staff functions of an organization in program budgeting is illustrated in Figure 8-3, which gives a hypothetical description of the U.S. Department of Defense. Each of the Armed Forces and each staff function conceivable cuts across each major program. Thus, by budgeting on a program basis, duplications can be eliminated and valid requirements for the accomplishment of objectives can be determined. For example, both the Navy and the Air Force contribute toward the strategic retaliatory mission via submarine missiles and long-range bombers. Also, all staff functions must exert efforts toward accomplishing these objectives.

In the civilian sector, we may construct a hypothetical corporation to illustrate a similar interaction. Figure 8-4 depicts the interaction of operating units, staff functions, and programs for a corporation in the same fashion that Figure 8-3 does for the U.S. Department of Defense. One major program of the fictitious organization is plant nutrition. The corporation's chemical-products division is obviously involved, as is the agricultural-marketing division (which may sell bulk fertilizers to

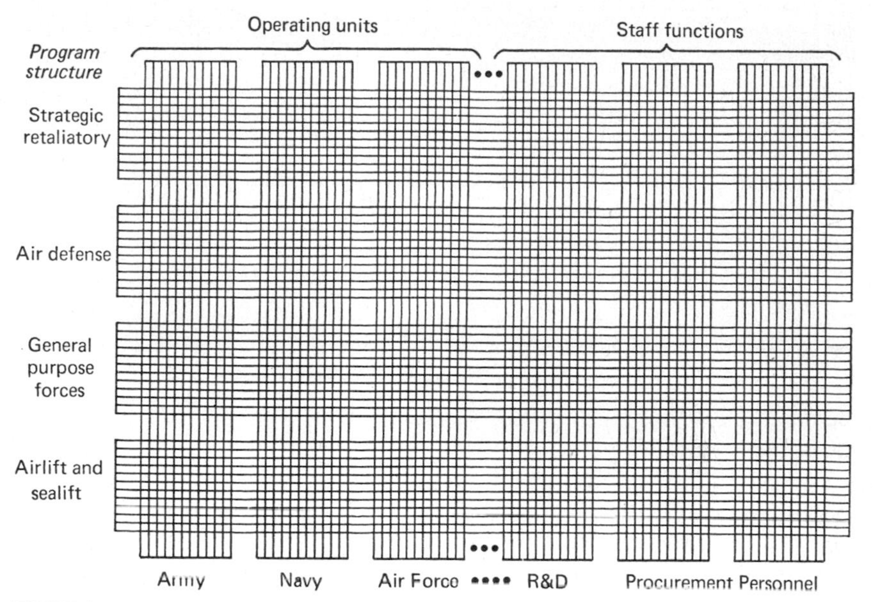

FIGURE 8-3 *Programs and functions in the Department of Defense.*

farmers) and the consumer-products division (which may sell both lawn-care chemicals and equipment to individual consumers). Similarly in the animal nutrition program, the chemical-products and agricultural-marketing divisions have interests in farm animal foods, just as the chemical-

FIGURE 8-4 *Program and functions in a corporation.*

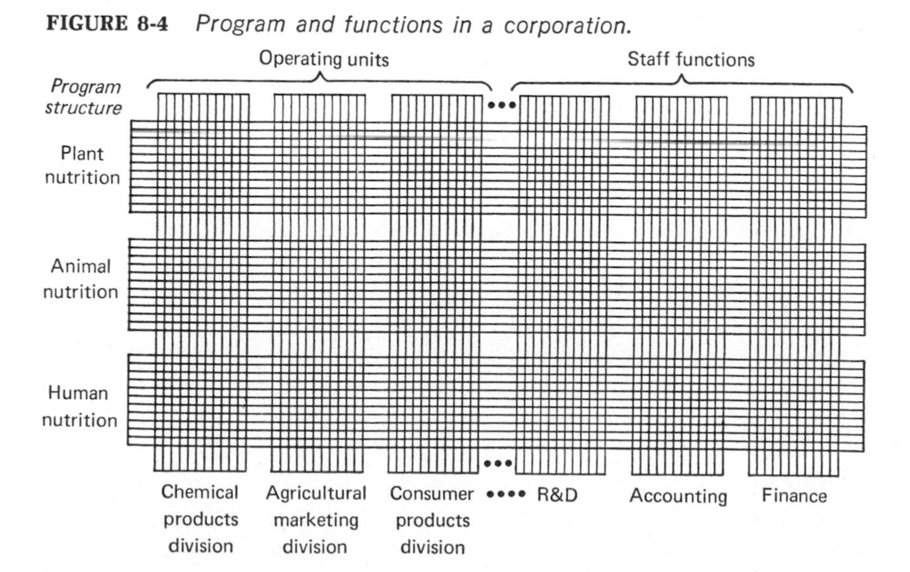

and consumer-products divisions might be concerned with pet-food products.

SUMMARY

Planning is the process whereby the manager thinks about the future and thereby creates the future in his own image. Thus an organization which does planning thinks about a desired future position for the organization in terms of its present posture and the means available for achieving the desired position.

Planning is an essential part of the management of change. Change is occurring around us each day, and those who possess a plan are able to adapt to change and to influence the changes that occur. Those without plans are reduced to reacting to change and to having little influence on it.

The development of strategy through planning is important if day-to-day decisions are not to shape the future of the organization. Tactical decisions should emanate from an overall strategy so that they will be consistent and contribute to the achievement of desired objectives. If tactical decisions are made outside the framework of strategy, they are likely to be uncoordinated. Taken together, they will be likely to lead the organization into its future in a haphazard and inefficient way.

Planning is an essential function of the manager; not a function to be left to staff planners. If it is done systematically, it requires no clairvoyance. Rather, it requires logical skills in foreseeing the chains of effects resulting from the organization's current position and its contemplated actions.

However, planning is neither static nor inflexible. Plans may be made in terms of contingencies which may occur. Plans may (indeed, should) be revised as conditions change. Thus planning is a continuous process which involves all levels of the organization; it is the framework around which organizational management takes place.

Planning, programming, and budgeting systems (PPBS's) are new approaches to planning which have been instituted in various governmental units. The concept of a program—an output-oriented major activity of the organization—is basic to PPBS. These programs are integrated into *program structures,* which define the organization's activities, and into *program budgets,* which represent the financial plans for achieving its objectives.

EXERCISES

1. How does planning involve "thinking about the future"? Relate to Polak's ideas as discussed at the beginning of Chapter 8.

2. Explain how planning involves *(a)* objective setting, *(b)* resource allocation, *(c)* control.
3. How does prevoyance differ from clairvoyance? Which is essential to planning?
4. How is the necessity for planning related to change? What is meant by "change management"?
5. How is the concept of "strategy" related to contrived responses and conditioned responses?
6. "Strategy and planning require specific objectives. If one's objectives are unclear, he has no use for planning." Do you agree?
7. "Plans are too restrictive. If I have a plan, I must follow it, even if I may not later wish to." Do you agree with this statement?
8. George Dively is quoted in the chapter as saying that he does not want daily decisions to shape the direction of his company. What does he mean by this? Relate this to the decision tree model of decision making.
9. Relate the concept of planning to the systems concept of a "chain of effects."
10. Suppose that Congress is presented with a proposal to allocate $300 million for this fiscal year on a new advanced weapons system. What should it know about the future of the system's development in order to make a rational decision? Why?
11. "The manager should leave planning to the professional planners." Do you agree?
12. What is a planning system? Why is it needed? Relate to exercise 3.
13. What is meant by an "adaptive planning system"? Illustrate.
14. Relate adaptive planning to the utilization of forecasts. What kind of forecasts are most useful for adaptive planning?
15. Why does the duration of one's *planning horizon* depend on the context in which the planning is being done? Illustrate.
16. Can one organization have multiple planning horizons? Why might this be desirable?
17. What is the "commitment principle" for determining a planning horizon?
18. Long-range and short-range planning differ in more important respects than the duration of the time period which is involved. Explain.
19. Describe the important objectives of short-range planning. How is it related to the "organizing" function of management?
20. Why can it be said with some truth that plans are the least important part of planning?
21. What is a program? How does it relate to an organization's objectives?
22. What is a program structure? What guides should be used in evaluating alternative program structures?
23. What is a program budget? Relate to traditional budgets such as the federal government's object-class budget.
24. Why is program budgeting important to the *total cost* concept of evaluating alternatives? Why is the total cost concept itself important?
25. Relate program budgeting to the organization's structure.
26. What is the difference between planning and forecasting?

27. Projects have their genesis as ideas evolving out of the mainstream of organizational activity or as a perception of potential "customer" need. What relationship do projects have to the overall planning system of an organization?
28. Decision making has been broadly categorized by the authors as consisting of two types: *strategic* and *tactical.* What are the differences and similarities of these two types of decisions?
29. Identify the decisions that you have made concerning your personal life in the last six months. Which of these decisions have been strategic? Which have been tactical?
30. What is the relationship between tactical decisions and organizational strategy?
31. Forecasts of the future should be made in terms of the ranges of the environmental variable which have reasonable likelihood rather than in terms of a single numerical quantity. Why?
32. What are the assumptions that might be made about the changing roles of universities in the next 10 years?
33. Are there any competitors to universities and colleges as we know them today?

REFERENCES

Bell, E. C., "Practical Long-Range Planning," *Business Horizons,* December, 1968.

Bowles, Samuel, *Planning Educational Systems for Economic Growth,* Harvard University Press, Cambridge, Mass., 1969.

Brown, Harold, "Planning Our Military Forces," *Foreign Affairs,* January, 1967.

Dale, Ernest, *Planning and Developing the Company Organization Structure,* American Management Association, New York, 1952.

Hekimian, James S., and Henry Mintzberg, "The Planning Dilemma: There Is a Way Out," *Management Review,* May, 1968.

Henrie, Jr., Samuel N., and Higgins D. Bailey, "Planning Carefully or Muddling Through: An Educator's Choice," *Journal of Secondary Education,* December, 1968.

Kast, Freemont E., "A Dynamic Planning Model," *Business Horizons,* June, 1968.

Kopkind, Andrew, "The Future Planners," *New Republic,* Feb. 25, 1967.

Warren, E. Kirby, *Long-range Planning: The Executive Viewpoint,* Prentice-Hall, Inc., Englewood Cliffs, N.J., 1966.

Williams, Harry, *Planning for Effective Resource Allocation in Universities,* American Council on Education, Washington, D.C., 1968.

Wright, Lyle O., "Comprehensive Planning: A Progress Report," *Audio Visual Instruction,* January, 1970.

9

decisions and organizational objectives

E. Kirby Warren has delineated four "realistic expectations" for long-range planning:

1. *Clear understanding of likely future impact of present decisions*
2. *Anticipating areas requiring future decisions*
3. *Increasing the speed of the flow of relevant information*
4. *Providing for faster and less disruptive implementation of future decisions*[1]

Thus, the values of long-range planning are intrinsically interwoven with *decision making*—in terms of both better current decisions and better and more effective future decisions.

Of course, decisions pervade any organization and the day-to-day existence of managers. Indeed, the process of management can itself be viewed as simply the sequential solution of various decision problems, both large and small, both important and unimportant.

All rational approaches to formal decision analysis presume the existence of explicit objectives. Since the establishment of the objectives of the organization is one of the primary steps in a comprehensive planning process, this chapter deals with the general characteristics of decision problems and with the question of establishing objectives on which plans can be based and which, in turn, serve as a basis for the analysis of decision problems. We then turn, in the next chapter, to the analysis of management decisions.

The relationship between planning, decision making, and objective setting is a direct and intrinsic one. Perhaps this is best exemplified by Koontz and O'Donnell, who say that planning "is the function of selecting the enterprise's objectives and the policies, programs, and procedures

[1] E. Kirby Warren, *Long-range Planning: The Executive Viewpoint,* Prentice-Hall, Inc., Englewood Cliffs, N.J., 1966, pp. 29–31.

for achieving them. Planning is, of course, decision making, since it involves choosing among alternatives."[2]

STRATEGIC AND TACTICAL DECISIONS

The determination of appropriate plans for an organization intrinsically involves the making of decisions. However, decision making pervades the organization far beyond the long-range strategic planning activity. Two varieties of decisions are usefully distinguished for purposes of discussion —strategic and tactical.

A *strategic* decision is one which is made during a current time period but whose primary effect will be felt during some future time periods. Strategic decisions affect organizational structure, objectives, facilities, and finances. Thus, a company's decision to issue stock or bonds is of the strategic variety, since it clearly affects long-term finances. So, too, are the decisions to embark on the development of a new product and to negotiate for the acquisition of a firm which possesses products or skills which are complementary to one's own.

Other decisions—*tactical* in nature—*are made within the framework of the strategic decisions and are designed to implement strategy.* Tactical decisions are found in the day-to-day life of the organization and determine how resources are to be employed in support of the current generation of products and services. Tactical decisions support and complement organizational strategy.

Thus, there are a variety of levels of decision making which must constantly be in progress in any organization. Although decisions are important to the planning function, all decisions made in an organizational context are obviously not of the strategic-planning-related variety. Perhaps this relationship of planning and strategic decision making is best understood by a business illustration. Long-range marketing planning encompasses consideration of alternative ways of achieving the goals of the firm.[3] Thus, if the future pattern of per-share corporate earnings is the relevant measure of the degree of attainment of the corporation's objectives, future goals might be achieved in various ways—e.g., penetrating new markets, introducing new products, expanding sales of existing products, etc. The strategic decisions of a company encompass specific consideration of alternative ways of achieving these desired states. For instance, the firm might wish to consider the relative worth of a new prod-

[2] Harold Koontz and Cyril J. O'Donnell, *Principles of Management,* 2d ed., McGraw-Hill Book Company, New York, 1959, p. 35.

[3] See Mark Stern, *Marketing Planning,* McGraw-Hill Book Company, New York, 1966, for an extensive discussion of planning in marketing.

uct versus an expanded advertising outlay to promote an existing product. Or, it might desire to choose a best new product. In either case, strategic decisions—those directly related to corporate goals—are being made.

Other levels of decision making are inherent in this illustration. For example, if a new product is chosen, the price, package design, and many other product characteristics must be decided on before it can be marketed. If additional promotion of an existing product is a chosen alternative in the strategic decision, the media to be used to promote the product must be determined and an allocation of expenditure must be made. A "chain of decisions" is thus initiated by each of the alternatives which may be chosen at the strategic decision level.

A repetition of a standard word of caution is in order at this point. As with most useful models and classification schemes, the strategic-tactical dichotomy is useful as a device for channeling one's thinking and organizing his understanding of the world around him. No unique distinction exists between the two types of decisions, and one should take care that he does not take a model, such as the strategic-tactical one, and try to impose it on the world by requiring that every decision clearly fall in one and only one category. Such models are a good way of attempting to understand a complex world. They emanate from the world, and they should be used as far as they are useful, but they should not be used in an attempt to impose order on the world by requiring the world to be a perfect representation of the model.

DECISION PROBLEMS

Each of us knows that he faces many decisions in his life—whether or not to marry, whom to marry, which school to attend, which job offer to accept, and so forth. Yet, few of us think in terms of the salient elements of which such problems and situations are composed, and in failing to do so, we often fail to recognize important distinctions among the various problem situations.

The Elements of Decision Problems

The most obvious element of a decision situation is a decision maker—the individual or group who is faced with the problem.[4] The term "decision maker" as used here should not be interpreted to mean a forceful, dynamic

[4]Similar problem elements have been discussed by C. W. Churchman, R. L. Ackoff, and E. L. Arnoff, *Introduction to Operations Research*, John Wiley & Sons, Inc., New York, 1957, chap. 5.

activist as opposed to one who procrastinates. The meaning of the term implies nothing about the personal qualities of the individual who may fill the role. In the formal sense, *a decision maker is an entity, either an individual or a group, who is dissatisfied with some existing state or with the prospect of a future state and who possesses the desire and authority to initiate actions designed to alter this state.* For example, the marketing vice president who is dissatisfied with a downward sales trend is potentially a decision maker in this sense if he actually possesses both the desire and the authority to alter promotional expenditures or to take other actions designed to increase sales.

The decision maker's desire to achieve some state of affairs—his objectives—is the reason for the existence of a problem. Often, objectives are expressed as a wish for the attainment of some new state, such as higher profits, or the retention of an existing one, such as "our image of the industry's leader." Usually, however, the objectives of the decision maker are expressed as some combination of achievable goals and retentive constraints on the pursuit of those goals—e.g., to maximize profits while simultaneously maintaining an image level.

Of course, the determination of appropriate objectives is itself a part of the planning process. All other levels of decision making consider the basic objectives to be predetermined—i.e., that the basic decision concerning objectives has already been made. The setting of objectives is the highest-level decision problem in the organization. And, since it provides the basis for all other decision making in the organization, it is obviously of great importance. It is therefore usually surprising to the novice to find that many high-level managers cannot explicitly express what their organizational objectives are. Indeed, when we take up this question of setting objectives in a later section in this chapter, we shall see that one of the important values of an explicit planning process is that it requires managers to focus on and explicitly state their objectives.

To pursue a given set of objectives meaningfully the decision maker must have available to him *alternative actions* which can promote the state of affairs he wishes to achieve. These available alternatives, together with a *state of doubt* as to which one is best, constitute the heart of any decision problem.

Of course, all these elements must exist in some context—which we have referred to as the *environment.* In a formal sense, the environment represents all of those *uncontrollable* aspects of the world which may or may not have any influence on the decision situation.

Solving Decision Problems

To solve a decision problem, it is necessary that the decision maker choose the *best* of the available alternatives. In a very simple situation, this is equivalent to saying that he should choose an alternative which leads him to a future state which is at least as good as all other possible states. In more complex problems, the idea of a best alternative is somewhat more subtle. We shall take this up in the next chapter, but for the moment we may define a *problem solution as the best of a set of available alternative actions.*

This apparently simple and straightforward statement has tremendous practical ramifications. What is meant by "best," for instance? How is the best alternative to be found? Is the alternative I consider to be best necessarily the same one that my superiors would so consider? Again, we shall take up each of these questions in the succeeding chapter.

Here, we wish only to point out that the formal concept of a problem solution is itself subject to controversy at the practical level. Professor Herbert Simon of Carnegie Mellon University has proposed, in his *principle of bounded rationality,* that people seldom attempt to find the best alternative in a decision situation. Rather, they select a number of good enough outcomes and an alternative which is likely to achieve one of them. In searching for a new product, for instance, the marketing executive makes no attempt to enumerate *all* possible products, so that he can select the best one; rather, he decides what he wishes to achieve with the new product and selects one which is likely to satisfy these desires.

This descriptive concept (how people *do* act) has normative implications (how people *should* act), for one might argue the irrationality of complete rationality. In other words, a completely rational man should evaluate *all* alternatives and choose the best one, yet it would be irrational to do so if this would involve the expenditure of vast amounts of time and money.

Of course, the idea of selecting the best alternative is questionable in another aspect. Not only might it be irrational to try to investigate every alternative, but in more complex problems, it is actually impossible to do so. Our understanding of the underlying structure of most complex systems is incomplete, and we are often unable to understand the interrelationships of all of the factors bearing on any particular decision problem. To expect optimization (the selection of the best alternative) in such a state of knowledge would be utter folly. And, as we shall illustrate subsequently, the decision maker who recognizes this is likely to achieve the greatest benefit from an explicit planning and decision-making procedure.

The Decision Process

To gain further insight into decision problems, we shall briefly discuss the *process* which one must go through—either explicitly or implicitly—in solving a decision problem. This process serves as a focal point for the discussion of a conceptual framework for decision problem analysis, to be presented in the next chapter.

Figure 9-1 outlines a series of activities, beginning with the collection and analysis of the input data, which describe the general decision-making process. After the basic data have been collected and assembled, the decision maker predicts the outcome which may result from each of the various alternatives available to him. Having done this, he evaluates the outcomes in terms of their worth and compares the alternatives on the basis of the outcomes to which each may lead. The best alternative is selected and action is taken which impacts on the "world"—the uncontrollable factors which affect the actual outcome of the decision situation. When the consequences of his choice are known, the decision maker measures the results and compares them with the predicted outcome. The process can then begin anew with new predictions which incorporate the lessons learned in comparing the previous predictions with the actual outcome.

Of course, a "one-shot" decision problem requires only one such sequence of steps. However, few problems are really so unique that the sequence is not repeated in some general way. For example, other merger opportunities may arise, so that the information gained in one merger decision is useful in another. Most strategic decisions are recurring at least in this sense; therefore, they require that the "loop" be navigated again and again, each time with (it is hoped) better input information and predictions.

In any one of the "tours" through this decision-making process, the decision maker is called upon to choose an alternative which appears to him to be best. Of course, as we have said before, the selection of best alternatives need not actually be operationally implemented. We shall see, however, that it is possible to think in terms of attempting to achieve the best alternative while actually recognizing that the manager can be well satisfied with something less than the best in any particular decision situation.

FIGURE 9-1 *The decision-making process.*

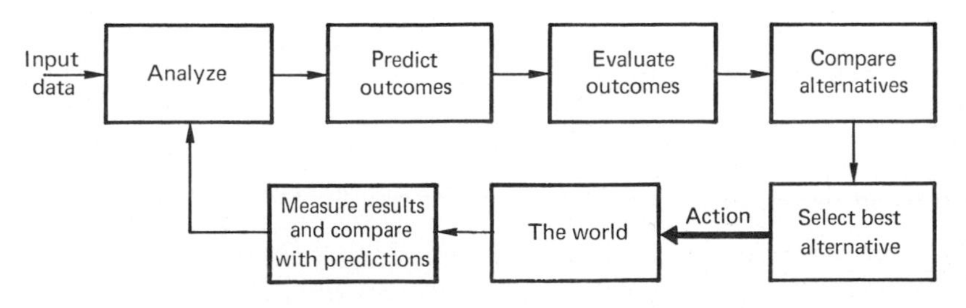

THE ESTABLISHMENT OF OBJECTIVES

The highest-level and most significant decision problem involved in the planning process is that of determining appropriate objectives for the organization. Previously we have extensively discussed the various claimants to the organization and the problems which their conflicting objectives impose on the setting of organizational objectives. In effect, the organization's managers must determine the answers to two basic questions before they can appropriately entertain the question of appropriate organizational objectives. The first of these questions involves considerations of the various claimants and the second is that of determining which "business" the organization is really in.

Who Is Really Boss?

The question of who is really the boss of the business organization does not seem to many to be anything more than academic. Those who are skeptical about the operational significance of our prior discussion of the various claimants to the organization would propose that the owners are the boss. In the case of the corporation, the owners are the shareholders.

A little thought quickly leads one to the realization that this concept is not as simple as it might sound. Even if one were willing to ignore the claims of anyone save the shareholders, the fact that most large corporations have thousands of shareholders indicates the magnitude of the problem of catering to them. Each of the shareholders has different objectives. For example, the typical top-level executive has short-term speculators looking over one shoulder and long-term estate builders over the other. In other words, one major group of stockholders is particularly concerned with short-term increases and the market price of the corporate securities, while another group is more interested in long-term appreciation and income through dividends.

Of course, it is apparent that management's overwhelming concern must be with the long-term welfare and destiny of the corporation, but to imply that long-term considerations are important to every shareholder is obviously preposterous. At any given point in time, a large portion of the total shares of any given corporation are probably held by those who have bought them in the hope of short-term price appreciation, and, of course, it is impossible for management to distinguish between long-term and short-term shareholders. Thus, it is practically impossible for management to give consideration to the goals of shareholders as a body in the hope of using the shareholders' goals as a basis for setting the organization's objectives.

Indeed, there are those who would argue that other claimants to the organization have a more basic interest in its long-term destiny than do the owners. As corporations have become more successful as instruments of innovation, they have gained the public's admiration through their demonstrated ability to convert ideas very rapidly into useful products and services. Andrew Shonfield, Director of Studies at the Royal Institute of International Affairs in London, is quoted by *Business Week* as expecting that there will be ". . . increasing numbers of consumers and employees with a long-term interest in the policies of the giant corporation to which many of its legal masters will be indifferent."[5] And, as things are developing, ". . . with the different groups crowding in on company boards and asserting their right to a say in management decisions . . . ," the shareholders ". . . may have to struggle to be heard above the din of competing claims."[6]

Nowhere is this so apparent as in the largest American corporation—General Motors. In discussing the current role of the GM chairman, a brokerage firm's house organ says:

> *The chairman must simultaneously consider the interests of 1,300,000 stockholders, 794,000 employees, 15,000 independent GM auto dealers and some 7,000,000 car and truck buyers. Add to these the public at large.*
>
> *Add also the critics such as Ralph Nader and the promoters of Campaign GM. There is the ever watchful eye of Government, particularly in the Antitrust Division. And, then, because what's good for the country is good for GM, other considerations are the general health of the economy and business. There are also women's lib, anti-war constituents and just plain pranksters—like those who stole the chairman's Cadillac right from the company garage last year.*[7]

What Business Are We Really In?

The second basic question which must be answered by the planner in determining organization objectives is: "What business are we really in?" The answer to this question both imputes goals for the enterprise and serves to define the scope of its activities. This is another one of those obvious points which are often overlooked. Levitt has pointed out, for example, that many of the difficulties of the United States railroads could

[5] "Management Outlook," *Business Week,* Dec. 14, 1968, p. 13.

[6] *Ibid.*

[7] "Managing GM—A Big Responsibility," *Investor's Reader,* Merrill Lynch, Pierce, Fenner, & Smith, Inc., Feb. 3, 1971, p. 21.

have been avoided had their business been defined as that of ground transportation rather than railroad activities, as it implicitly was. Although such an indictment of an entire industry may be viewed as a frivolous exercise in the 20/20 hindsight vision of which even the least perceptive of us are abundantly possessed, it is precisely the preclusion of the opportunity for such retrospective judgments which is the reason for planning. In this vein, we see today that many businesses are broadening their horizons and opportunities by asking this vital question. Thus, oil companies have begun to view themselves as in the travel business rather than in the oil business. The manifestations of this are already being seen—e.g., in the marketing of travel insurance via correspondence with credit card holders. Similarly, insurance companies are beginning to view their role as one of providing a broad range of financial services rather than just insurance.

If any organization should know the business that it is really in, one would think that it would be the federal government. Indeed, the federal government actually has traditionally behaved as though it were uncertain as to what business it is really in. The broad objectives of the federal government are stated in the Preamble to the United States Constitution—to "provide for the common defense" and to "promote the general welfare." In modern terms, this is interpreted to mean defense, the maintenance of order, the promotion of health, education, welfare, economic development, and the conduct of essential services such as the post office (although it has now formally divested itself of that business). Each modern government does all these things to one degree or another and, in this sense, each knows what business it is really in.

However, no government attains all of these objectives to the ultimate degree. No one would claim, for example, that the United States' defenses are perfect or perhaps even as good as those of some other nation in the world. On the other hand, the educational level of the United States population is higher than that of most other countries. Thus, the attainment of government objectives is a process involving the allocation of resources—of compromises between money spent in one way to better achieve one objective than that spent in another way to achieve another objective. The conclusion that the federal government has not traditionally known what business it is really in rests on the fact that it has not always made decisions as though it recognized the basic trade-offs between these objectives. Indeed, for a long period of time the federal government's budget was input-oriented—rather than output-oriented. In other words, it focused on the resources being consumed (personnel, pay, construction) rather than on programs oriented toward the achievement of basic objectives. The federal government was thereby unable to contrast rationally the relative benefits associated with spending a dollar on "promoting the general welfare" versus the results of a dollar spent on "providing for the common defense." In this sense, then, the federal government did not really recognize which business it was in, for although its basic objectives are stated in the United States Constitution, the ways in which it might rationally pursue those objectives were obscured.[8]

In many businesses, the question of determining the nature of the business is equivalent to determining the class of products and services which the firm will produce. In many companies, products which fit into the existing raw material, production, and distribution scheme would appear to define the business. Hence, an oil company might consider only products which use oil as a raw material, which require refining, and which may be sold through service stations. In fact, the list of products meeting all of these criteria is very limited. As a result, major oil companies have broadened this limited view of raw materials, production facilities, and the distribution organization to include their clerical and managerial resources and correspondence with credit card customers. The previously noted marketing of travel insurance policies has been one new-product result. This new product utilizes many of the human resources already available to the company as "raw material" and "production capacity." In addition, the sale of these policies through existing postal correspondence with credit card customers represents a much broader view of promotion and distribution of facilities. Indeed, the possibility of direct marketing of insurance through the standard outlets—the service stations—has not been omitted from consideration.

Thus, by asking about the definitions of the existing organization and whether they are sufficiently broad to encompass those things which the organization might wish to do, those things which the organization might be capable of doing, and those things which the organization might have an opportunity to do, the organization is "defining what business it is really in." Only after it has done so is it possible for it to take up the first decision aspect of planning—the setting of objectives.

Values and Objectives

The question of establishing the objectives of the organization is basically one of implementing the values of the managers of the organization. "Value" is the term which is used to mean the assessments of worth which are made of various things and states by people. Thus, the fact that I

[8] Fortunately, the introduction of program budgeting, as discussed in Chapter 8, has resolved many of these ills to a large degree. See David I. Cleland and William R. King, *Systems Analysis and Project Management,* Mc-Graw-Hill Book Company, New York, 1969, chap. 6, or David Novick (ed.), "Program Budgeting, Program Analysis and the Federal Budget," U.S. Government Printing Office, Washington, 1965.

choose to drive a bright red Corvette rather than a dark four-door sedan reflects something about the values which I place on automobiles and the image which they convey. So, too, with the managers of an organization. No two managers or organizations will have the same values. Thus, no two will prescribe the same objectives for the organization.

In truth, when top management determines the objectives of the organization, they are setting the goals, and implicitly the values, of the organization. In doing so, their personal values play the most important role. This is so because the establishment of objectives is necessarily approached on a much more subjective basis than are any of the other decision aspects of planning. There is no formal structure to tell the manager which values are right and which are wrong, or indeed which are more important than others. Values are a personalistic concept. Each of us has his own, and presumably the values of each of us are somewhat different.

This means that the decision problem involved in the setting of objectives is extremely complex. It is not subject to the same degree of rational analysis as are other lower-level decision problems in the organization. This complexity is further amplified by two additional aspects of the problem of setting objectives. These aspects have to do with the question of *group values* and with the *multiple value systems* of each individual.

GROUP VALUES One of the great difficulties in the rational consideration of decision problems of all kinds is the generally accepted difficulty associated with interpersonal comparisons of values. Thus, my values are mine, yours are yours, and the two are not comparable.

Since most organizations' objectives are not set by one individual, but rather by a group of individuals, the question of determining an aggregate value for the group in the face of the differing values of the individuals who form the group is an extremely difficult one. In general, there is no way of taking the known values of individuals and developing from them the values of a group.[9] The difficulties associated with the construction of some procedure for passing from a set of known individual values to a pattern of group values are illustrated by the well-known "voting paradox." For example, suppose that the democratic concept of "majority rule" is to govern the deliberations of three persons for three alternatives labeled A, B, and C. Suppose that Table 9-1 gives the preference order of each of the three persons for the three alternatives. It is clear that A will be preferred to B by the group, since both individuals 1 and 2 prefer it to B, and

[9] See, for example, Kenneth J. Arrow, *Social Choice and Individual Values,* John Wiley & Sons, Inc., New York, 1963, for a wide-ranging discussion of this and associated issues.

"the majority rules." Similarly, B will be chosen by the group in preference to C by a vote which goes against the preferences of the middle individual. As a result, since A is preferred to B and B is preferred to C by the group, one might logically infer that alternative A would be preferred to alternative C by the group. After all, if one likes chocolate ice cream better than vanilla and vanilla better than strawberry, the logical conclusion is that chocolate is preferred to strawberry. A glance at Table 9-1 indicates that this apparently logical deduction is not valid for this particular group. In fact, two of the three individuals prefer C to A, and hence the group, by virtue of majority rule, prefers C to A. This potential violation of basic logic in determining group values under the rule of the majority is one reason why most informally organized groups actually make decisions in more subtle ways. More importantly, it illustrates that this obvious method of passing from individual to collective values fails to satisfy the basic condition of rationality as most people understand it. Indeed, this is the case with most other bases for developing group values.

Thus, the basic difficulties associated with the inherently complex problem of establishing objectives are complicated by difficulties involved in interpersonal comparison and aggregation of values. It is even further complicated by the multiple value systems which exist within each person.

MULTIPLE PERSONAL VALUES There are at least two value systems wrapped up within every individual. Each individual has, first, a set of preferences and values which are "his own"—his favorite color is green, he likes redheads, and a Mustang is just his speed. In addition to this, he has a set of values which emanate from the organization. He would be less than human in setting organizational objectives and in making organizational decisions if both value systems did not come into play. It is not uncommon, for example, for most of us to consider the possible influence of our organizational action on our own future promotions, remuneration, and so forth, as well as its influence on the purely organizational outcome.

Thus, every decision that is made by a manager reflects both his personal and his organizational value systems to one degree or another.

TABLE 9-1 Voting paradox

	PREFERENCE ORDER		
Individual	*Most Preferred*	*Second most preferred*	*Least preferred*
1	A	B	C
2	C	A	B
3	B	C	A

This has always been recognized and put to use by successful executives in their selection of managerial personnel on bases other than "purely technical" ones. An extremely successful friend of one of the authors says, for example, that he chooses people "by the gleam in their eyes." Although this is clearly a subjective and unscientific criterion for personnel selection, it has the virtue of giving some kind of consideration to the personal values of the individual and the way in which he "reflects" them.

The Value of Objectives

The seemingly straightforward establishment of objectives is intrinsically confounded both by the multiple value systems within each individual and by the difficulties in making consensus value judgments on the part of the management group who must set the objectives. There is a good deal of empirical evidence that both of these conflicts are pervasive. However, there is no reason to be greatly concerned with this unless one is seeking neat rules and formulas for setting objectives.

The problem of setting objectives is a complex one, and as a result it has no easy resolution. However, most important problems faced by managers are of this variety. The important point for the manager to recognize is that the problems do exist and that *they do not exempt him from the responsibility for formally establishing objectives.*[10]

It is undeniably true that a great many organizations, and, hence, the managers who control the operations of the organizations, have skillfully avoided the complex problem of formally establishing objectives. This is vividly exemplified when one queries managers as to their objectives. The stated objectives of increasing sales, profits, and market shares are often found to be complemented by unstated, but natural, desires relating to the stability of management, market leadership, and the organization's image.

As an illustration, one of the authors once spent several weeks attempting to formulate a decision problem in which the specified alternatives involved either increasing promotional effort, discontinuing sales, or making no change in each of a number of market areas, only to find that the company's president would not consider withdrawing from any market area regardless of the consequences. As the issue was pursued, it was found that although it had never been openly stated, the president's view of his product as the industry's leader and his estimate of the loss of face inherent in pulling out of an area were of such overriding impor-

[10] For further discussion of both theoretical concepts and empirical studies of values, see Florence R. Kluckhorn et al., *Variations in Value Orientation,* Row, Peterson, and Company, Evanston, Ill., 1961.

tance as to negate considering mere sales and profits. It is true, of course, that it may be argued that the short-term image objective of the president was really a long-term profit objective and that the president's objectives were perfectly consistent and rational. This is possibly valid; however, the point to be made is that the image objective had to be ferreted out at great effort by the analyst who was involved in the problem. The implications of this to the rest of the people in the organization and to the organization's operations are quite clear. Since this objective had never been clearly stated, presumably many people in the organization knew nothing of its existence. Hence, they were working under an inappropriate set of implied objectives rather than an explicit set of objectives. In this case, the implied objective set was obviously different from that set of objectives which could have been promulgated by top management.

The existence of clearly defined objectives for an organization has two primary values. First, *people know what they should be trying to do, and it is thereby possible for all to work toward the common goal,* thus fulfilling one of the basic tenets of organizations.

Secondly, *clear objectives serve as a basis for performance evaluation and control.* They enable us to obtain answers to the basic question—How well are we doing? Thus, objectives are a standard or idealized model against which actual performance can be gauged and compared.

The "Boss" and the "Business"

The tie-in between the questions, Who is really boss? and What business are we in? is through the stated objectives of the organization. The organization's objectives tell the world about the relative importance it places on the interests of its various clientele and how it intends to define its activities to pursue those interests.

Perhaps this is best summed up by a statement of broad objectives made by the president of a major United States corporation (here unidentified):

> *The XYZ Corporation, as a broadly owned multinational company, is committed to four fundamental, interdependent objectives, all of which are essential to its long-term success. XYZ intends to excel in all of these:*
>
> *One, provide for shareholders a return superior to that available from other investments of equal risk, based on reliable long-term growth in earnings per share.*
>
> *Two, provide for employees a rewarding and challenging employment environment with opportunity for economic and personal growth.*

Three, provide worldwide customers with products and services of quality.

Four, direct its skills and resources to help solve the major problems of the societies with the significant benefits of its other essential objectives.

SUMMARY

Planning is the process of establishing goals, allocating resources, and providing the guidance necessary for the operation of an organization. It is important that planning be explicitly undertaken as a major management function if the organization is to do more than to attempt to react to its changing environment. If it desires to *influence its environment,* to utilize change to advantage, and to take advantage of opportunities, it must plan.

Planning is intrinsically intertwined with decision making. The decision process is one of continually evaluating information, predicting the future, evaluating and comparing alternative courses of action, and selecting the best of those which are available for implementation. All of this is done on the basis of *predictions* of the future.

Strategic decisions are those on which a commitment is made in the current time period, but whose impact will not be felt until future time periods. Tactical decisions are made within the framework of strategic decision in order to implement strategy.

Decision problems require that the decision maker exist in an environment with which he is dissatisfied. If he has alternative courses of action available to him and a state of doubt as to which is best, a decision problem exists. To solve that problem, the decision maker must choose the best of the set of available alternatives.

The highest-level and most important organizational application of this formal description of a decision problem is the planning decision regarding the establishment of objectives. This decision is confounded by the practical problems of determining to whom the organization owes it loyalty, what the nature of the business really is, and the conflicting value systems within the organization.

The guiding philosophy behind the establishment of objectives as with all of planning is the same. Although the problems involved in performing these activities explicity and in detail are significant, the *cost of not doing so is even more significant* since a failure to plan means that the organization will proceed into the future under some implicit and unchosen plan rather than under an explicit carefully selected one.

EXERCISES

1. Relate planning to *(a)* present decisions, *(b)* future decisions, *(c)* information requirements.
2. "Management is nothing more than decision making." Refute or defend.
3. What are strategic decisions and tactical decisions? How are they the same and how are they different?
4. Describe the "chain of decisions" related to the selection of a new product by a business firm.
5. What constitutes a decision problem?
6. Contrast a decision problem with a state of dissatisfaction.
7. Why is a state of doubt essential to the existence of a problem?
8. Define a *solution* to a decision problem.
9. Discuss the previously-noted comment, "To decide not to decide is to decide," in terms of the decision framework of this chapter.
10. What is the principle of bounded rationality? How would it apply to your search for a mate?
11. Is the principle of bounded rationality descriptive or normative?
12. Consider one of the major personal decisions which you made recently in light of the decision process of Figure 9-1. Describe each step.
13. Relate the problem of establishing organizational objectives to the concept of organizational claimants.
14. "The question of what business we are really in is trivial. We know that. Let's get on to the important issues." Comment on this statement made by a business leader.
15. How do personal values influence organizational objectives?
16. Discuss the group value problem reflected by the "voting paradox."
17. Figure 9-1 describes the decision-making process. Is this used both in strategic and tactical decisions, or is it a process only useful in strategic decisions? Why? Defend your answer.
18. Many organizations fail to set objectives. Why?
19. Managers often state objectives in such terms as:

    "Our objective is to be number one in our industry."

    "Our objective is to serve our customers."

    "The objective of this organization is to serve our customers and to earn a reasonable return on our invested capital."

    "The goal of this organization is to meet our responsibility to stockholders, employees, customers, and the public."

    How operational are such objectives? Explain.
20. "Objectives, particularly those which are of a strategic nature, are guides to the executive's thinking. Such guides should clearly be changed as necessary as the executive better comprehends the resources he has, the competition he faces, and the dynamic state of change in the market he attempts to meet." Comment on this statement in the light of objectives such as those in exercise 19.

21. There is not universal agreement that top management should spell out detailed objectives for the organization. Wrapp[11], for example, believes that the general manager should not do master planning nor make clear-cut, forthright statements of policy; rather he should function as an opportunist. Under this philosophy, the manager should not commit the organization to a specific set of objectives, but rather he should provide the organization with a sense of direction. Is this view compatible with that given in this text?
22. Wrapp[12] has also said, "The more explicit the statement or strategy, the more difficult it becomes to persuade the organization to turn to different goals when needs and conditions shift." Comment on this.
23. It has been proposed[13] that businesses should submit to "social audits" as well as to fiscal audits of their activities. What do you think is meant by this? Do you feel that the proposal has validity?

REFERENCES

Archer, S. H., "The Structure of Management Decision Theory," in W. A. Hill and D. Egan (eds.), *Readings in Organizational Theory: A Behavioral Approach,* Allyn and Bacon, Inc., Boston, 1966.

Delbecq, Andre L., "The Management of Decision-Making within the Firm: Three Strategies for Three Types of Decision-Making," *Academy of Management Journal,* December, 1967.

Dickson, G. W., "Management Information-Decision Systems: A New Era Ahead," *Business Horizons,* December, 1968.

Feldman, Julian, and Herschel E. Kanter, "Organization Decision Making," System Development Corporation, Santa Monica, Calif., Dec. 30, 1963.

Findler, Nicholas V., "An Information Processing Theory of Human Decision-making under Uncertainty and Risk," Carnegie Institute of Technology, Pittsburgh, July, 1964.

Hartley, Harry J., "Administrative Decisions and Functional Analysis," *Education,* January, 1966.

Holder, Harold D., and William P. Erling, "Construction and Simulation of an Information Decision Model," *Journal of Communication,* December, 1967.

Jones, William M., "Decision Making in Large Organizations," RAND Corporation, RM 3968-PR, Santa Monica, Calif., 1964.

King, William R., "Human Judgment and Management Decision Analysis," *Journal of Industrial Engineering,* December, 1967.

[11] Edward Wrapp, "Good Managers Don't Make Policy Decisions," *Harvard Business Review* (Sept.–Oct., 1967), pp. 91–99.

[12] *Ibid.*

[13] Robert Reinhold, "Executive Foresees Social Audits for Businesses," *New York Times,* Feb. 14, 1971, p. 8.

———, *Quantitative Analysis for Marketing Management,* McGraw-Hill Book Company, 1967, New York, chaps. 1–4.

Meinhart, W. A., "Project Management: An Incentive Contracting Decision Model," *Academy of Management Journal,* December, 1968.

Miller, D. W., and M. K. Starr, *Executive Decisions and Operations Research,* Prentice-Hall, Inc., Englewood Cliffs, N.J., 1960.

Mockler, Robert J., "The Systems Approach to Business Organization and Decision Making," *California Management Review,* Winter, 1968.

Panos, Robert J., and Alexander W. Aslin, "Using Systematic Information in Making Education Decisions," *Education Record,* Spring, 1967.

10

the analysis of management decisions

In Chapter 8, it was argued that the planning aspect of the manager's job is of basic importance to the organization. Planning was shown to be intrinsically intertwined with the making of decisions, for in making the basic decisions concerning the organization's objectives, its makeup, and its allocations of resources, the manager is to a large degree determining its future.

Of course, it is not just the choices made by the manager in the planning process which influence the future. Other things—many of which are beyond the control of the manager—interact with his choices to jointly produce effects on the organization.

To begin to understand the planning aspect of the manager's job requires that we understand decision problems and the way in which decision problems may be analyzed and solved. Having developed the rudimentary elements of a decision problem in the previous chapter, we begin here to develop the methods which have been applied to analyzing and solving decision problems.

In placing emphasis on the use of formal methods of decision problem analysis, we are quick to point out that in no way do we imply that all decision problems are (or should be) approached in this fashion. The conceptual framework to be developed provides *a basis for thinking about problems.* Whether a manager formally uses them by performing paper-and-pencil analyses, or whether he simply uses them as a way of thinking about problems, is irrelevant to this primary value.

Neither should the emphasis on formal problem analysis be taken to imply a relegation of a manager's experience, judgment, and intuition to a relatively minor role in resolving decision problems. With all of the aid which a formal analytic basis for problem solving can be to the decision maker, his judgment, intuition, and experience still play a primary role. In effect, the conceptual framework for problem analysis fulfills a role complementary to the more subjective aspects which have traditionally been used to define a "good manager."

THE STRUCTURE OF A DECISION PROBLEM

The elements of a decision problem are structured to produce results, or *outcomes,* in problem situations. The outcome of a decision problem is a state of affairs which results as a consequence of the choice of a particular alternative and the occurrence of a particular set of environmental conditions. Outcomes are best assessed in terms of the degree of attainment of the decision maker's objectives.[1] Thus, if the basic objective is one of profit, an amount of profit represents the degree of attainment of the profit objective. Hence, the dollar profit amount would be a valid *outcome descriptor* in such a case. If the problem is one involving multiple objectives—e.g., to make a profit *and* to provide a specified level of service *and* to be the industry's leader—multiple outcome descriptors may be necessary in order to describe individual outcomes.

The alternatives available to the decision maker are composed of basic building blocks which are *controllable* to him. In a decision problem concerning the establishment of a project team to be set up to evaluate new-product ideas, for example, the controllables are the size of the team, the people to be assigned to it, its organizational structure, the person to be designated as the team leader, etc. In the decision problem concerning which new product to produce, the controllables are the physical characteristics of the product, its packaging, price, etc. (Of course, all of these controllables need not be considered simultaneously by the decision maker.)

Every aspect of a decision problem is either controllable or uncontrollable to the decision maker. Usually a few elements are controllable and a great many are uncontrollable. For instance, the quantity of a product to be produced is controllable to the production manager; so, too, is the scheduling of production. However, the demand for the product, the world situation, the price of tea in Hong Kong, and a multitude of other things are uncontrollable to him.

Only some of these uncontrollables are *relevant* to any particular decision problem, of course. The demand for the product obviously is relevant to a decision involving the scheduling of production; the world situation is undoubtedly relevant, but in some tenuous, roundabout way which probably defies explicit definition; the price situation in the Far East is probably totally irrelevant.

Clearly, it is the province of the manager or analyst to determine which uncontrollable elements are relevant and which are not. In doing so, he is constructing a model of the real world, since he will not be able to consider everything which may be in some way relevant. For instance, he would undoubtedly conclude that the demand for the product is a rele-

[1] In the discussion in this chapter, we shall assume that the objectives are "given." In Chap. 9, we discussed the difficulties involved in developing such objectives.

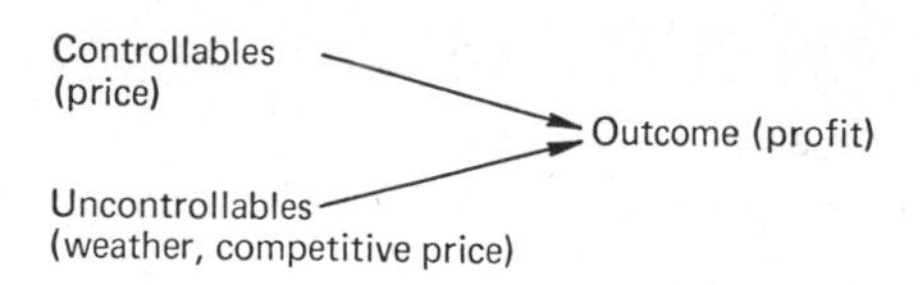

FIGURE 10-1 *Decision outcomes produced through the interaction of controllables and relevant uncontrollables.*

vant factor to be included, and that the Hong Kong tea price is not. The world situation would probably be left out of his model because of its tenuous relevance and because the manager would be hard pressed to know how to include it if it were relevant.

The controllables and relevant uncontrollables interact to produce the decision problem outcome. The situation may be described in the simple fashion of Figure 10-1, where the single controllable in the model is taken to be the price to be charged for a product, the uncontrollables of the model the weather and a competitor's price, and the outcome is described in terms of a single profit objective.

Note that in this situation only one of the many controllable aspects of the new product is being considered. We shall find that this is a pervasive situation in management decision analysis. It is often impossible or impractical to simultaneously consider all of the controllables in a problem situation, in which case one or a few may be considered at a time. In this illustration, the two uncontrollables considered by the model are of different basic natures. The weather is an aspect of the environment which is completely outside the control of the decision maker even though it presumably (since it is included in this simple model) has a significant effect on the profit outcome. The competitive price uncontrollable is subject to the control of some rational being, but not the decision maker. It is subject to the control of a competitor who has objectives which are in conflict with those of the decision maker.

The interaction of the price, weather, and competitive price—the controllables and uncontrollables—produces the outcome of the decision problem which is measured in terms of profit. If the price is high, the weather is hot, and the competitive price is low, the profit outcome would generally be different from what it would be if the price were high, the weather were cold, and the competitive price were low. Indeed, *each possible combination* of the various levels of the controllables and uncontrollables would generally produce a different outcome.

THE OUTCOME ARRAY

The outcome array is a device for conceptualizing the structure of a decision problem in terms of the interaction of controllables and uncontrollables in producing outcomes. In constructing such a conceptualization, the rows of the array represent the various levels of the controllables and the columns represent the various levels of the uncontrollables. Thus, the array is a simple pictorial model of a decision situation.

Suppose, for example, that the decision problem is one in which a salesman must choose between spending the next day calling on customers in a rural area and calling on those in an urban area. The single controllable (where he will go) has two levels—urban and rural. The salesman's home is located so that he may go to only one of these areas on the coming day. He knows that the sales results he can expect to achieve depend on what sort of weather he will encounter. On a sunny day, the farmers will be in their fields, making it difficult to contact them; whereas on a rainy day, they will be working near the farmhouse where they can be contacted easily. On the other hand, city traffic will be so congested on a rainy day that it will preclude his making many calls. The only relevant uncontrollable is therefore the weather, which can be considered as either "rainy" or "not rainy."

The outcome array for this problem is shown as Table 10-1. There the controllable levels "urban" and "rural" are used to designate the rows and the uncontrollable levels "rainy" and "not rainy" are used to designate the columns.

TABLE 10-1

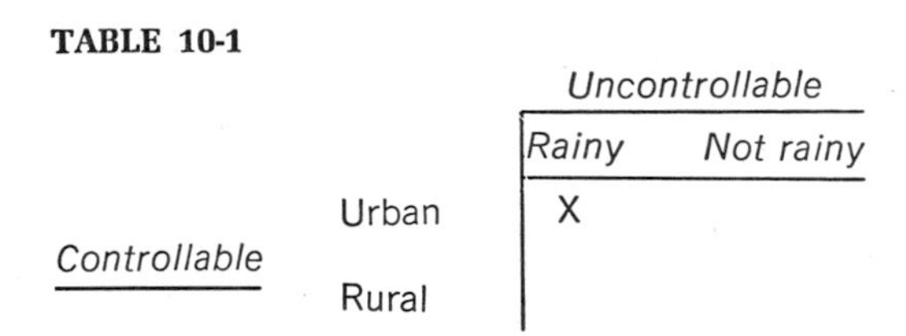

		Uncontrollable	
		Rainy	*Not rainy*
Controllable	Urban	X	
	Rural		

Each of the four positions in the body of the table represents an outcome which results from the interaction of a particular level of the controllable and a particular level of the uncontrollable. For example, the outcome indicated by the "X" is the one which would result *if* the salesman chosen to go to the urban area and *if* the weather were rainy.

The simplest outcome descriptor in this situation might be "profit," since each of the four possible consequences (outcomes) which might ensue would result in a different sales level. Suppose that he anticipates $100 in profit if he goes to the urban area and the weather is good, but only $40 if it is rainy. Conversely, he anticipates making $75 profit in the rural area if the weather is rainy but only $50 if it is not. This information is concisely conveyed in the outcome array shown as Table 10-2.

TABLE 10-2

	Rainy	*Not rainy*
Urban	$40	$100
Rural	$75	$ 50

The decision problem here is the answer to the question: Should he go to the urban area or the rural area? Clearly, *if the salesman knew what the weather would be on the next day, the choice would be simple.* If he knew that it would be a sunny day, he would go to the urban area, since he can make more profit there in good weather. On the other hand, if he were certain that the weather would be bad, he would go to the rural area, since he would make greater profit there.

In the absence of any certain knowledge about the sort of weather which will ensue, the decision maker needs additional information in order to arrive at a rational choice. Before we go on to discuss the required information, however, let us stop to consider the value of the outcome array to the decision maker.

The primary value of the outcome array is that it lays out the interaction of the controllable and uncontrollable in terms of the entire range of outcomes which they *may* produce. Assuming that the levels of the controllable and the uncontrollable are in some sense exhaustive, every conceivable outcome of the decision problem is laid out before the decision maker in the outcome array. In a realistic problem involving many levels of both quantities, for the decision maker to attempt to consider such a range of outcomes in any less explicit fashion would inevitably lead to the omission of some, possibly important ones. Thus, the outcome array provides a framework within which a problem can be rationally considered with the assurance that, within the assumptions of the model, nothing has been overlooked or omitted from consideration. It is in this simple virtue that the outcome array most frequently plays its greatest role as an aid to decision making.

THE STRATEGY CONCEPT IN DECISION ANALYSIS

The analysis of decision problems in an outcome array format is often termed "decision theory."[2] The parlance of decision theory has developed along lines which are quite independent from those of management thinking. It is not, therefore, surprising that some confusion exists in the terminology used in the two fields. Such is the case with the term "strategy."

[2] See Robert M. Thrall, C. H. Coombs, and R. L. Davis, *Decision Processes,* John Wiley & Sons, Inc., New York, 1954.

Previously (in Chapter 8), we introduced the term "strategy" to mean a complex of plans. Decision theorists use the term in what appears to be a different way. Actually, the two uses are remarkably similar in their salience.

Decision theorists define a *strategy* as a *complex alternative which is available to a decision maker. Strategies are composed of the various controllable elements of the decision model, and they may encompass sets of responses to contingencies which might arise.* Thus, strategies are sophisticated kinds of alternatives—the two ways in which the sophistication is achieved are by allowing strategies to encompass more than one controllable element and by providing with the strategy the basis for different responses to various contingencies.

Mutiple Controllables in a Strategy

The simplest variety of strategy is an alternative which specifies a single level for a single controllable; for example, an advertising expenditure of $1 million on TV "spots" could be such a simple alternative. Other alternative strategies to be considered in such a situation would each specify a different expenditure level—e.g., $1.1 million, $1.2 million, etc. This is the sort of strategy which we dealt with in the simple decision problem of Table 10-2, since each alternative consisted of a specified level of a single controllable.

In more complex decision models, more than a single controllable needs to be given consideration. For example, a new-product decision involves controllable elements of price, design, packaging, promotional expenditures, distribution channels, etc. Although it may not always be necessary to consider all of these elements simultaneously, many decision situations are sufficiently complex to require that more than a single controllable be incorporated into any useful model. (We shall subsequently discuss the question of deciding which elements need be considered concurrently and which may be deferred in a section on "Suboptimization.") Here we assume that it is necessary that more than one controllable be included in each strategy.

When more than one controllable element is included in a strategy, *each strategy consists of a specified level of each controllable.* Thus, if the relevant controllable aspects of a decision problem were the selling price and the amount to be expended on advertising, a *single* strategy might be:

Strategy #1: (price: 50¢; advertising: $1 million)

Other alternative strategies would be composed of other combinations of levels of the two controllables—e.g.,

Strategy #2: (price: 55¢; advertising: $1 million)
Strategy #3: (price: 55¢; advertising: $1.5 million)
Strategy #4: (price: 59¢; advertising: $1.3 million)
•
•
•
•
etc.

The Adaptive Aspects of a Strategy

The other important consideration which distinguishes a strategy from a simple alternative is that a strategy may involve predetermined responses to a set of contingencies which may occur. Perhaps this is best illustrated in the context of a pricing situation involving the consideration of a competitor's (presently unknown) price. A *single* strategy might be:

Set initial price at $1.95.
If competitor's price is lower, reduce price to meet it.
If competitor's price is $1.95 or higher, retain price at $1.95.

Thus, with such a strategy the decision maker is determining in advance what actions he will take in response to various contingencies (competitive prices) which might arise. Alternative strategies would involve other possible responses to each of the contingencies.

It is clear that the potential complexity of general strategies of this form is virtually limitless. Also, the number of possible strategies of this kind which need to be considered grows very large as the number of elements in each strategy increases. Practically speaking, it is usually not feasible to consider strategies which are extremely complex—either from the analytic viewpoint or in terms of the degree to which it is possible to delineate *meaningfully* the possible combinations of circumstances and events which may ensue. Thus, most such strategies are limited to four or five essential elements.

The Semantics of Strategies

The two uses of the term "strategy" in this text appear to be quite different. In the one case (Chapter 8) a strategy is said to be a "complex of plans." Here, the decision theoretic viewpoint views a strategy as a "complex alternative in a decision problem." However, the similarities are also obvious. The word "complex" is used in both instances, for example. Thus, in both uses of the term, a strategy is something more than a basic building block; it is a complex aggregate of building blocks. In the one case, the building blocks are plans, and in the other, they are controllable elements of a decision problem.

The other major similarity is the adaptive nature of both uses of the term. With regard to the primary management view of "strategy," Koontz and O'Donnell say:

> *Plans and policies are made with the realization that they must be adjusted in accordance with the reactions of competitors, customers, suppliers, employees, subordinate managers, and others inside and outside the enterprise. The policy of a business firm may be clear and its plans well developed, but strategy may require some adaptation to meet the plans and policies of others.*[3]

Thus, the adaptive element in the Koontz-O'Donnell concept of a strategy is very much like that aspect of the decision theoretic concept. In both cases the power and flexibility gained by incorporating contingency planning into strategies are apparent.[4]

A DECISION ANALOG

As an aid in understanding the concept of a strategy and its application to the analysis of management decisions, let us consider a fictitious decision problem analog which serves to strip away the unnecessary complexity from the concept. The situation to be used has no significance or realism whatever, but it does serve to illustrate the *structure* of problems and strategies and the makeup of outcome arrays.

Consider a laboratory situation in which you and some other person —your "competitor"—are seated on opposite sides of a "black box." Each person has two dials before him—one red and one blue. Each dial can be set in either of three positions—left, middle, or right. When so directed, each individual sets both of his dials to some position. Then, the black box is activated and some outcome is generated. For example, a monetary reward *for you* may be the result.

In each case, you have two controllable elements—the red and blue dials. Each of these controllables has three possible levels—left, middle, and right. Since both dials must be set in order for you to achieve a re-

[3] H. Koontz and C. O'Donnell, *Principles of Management,* McGraw-Hill Book Company, New York, 1964, pp. 175–176.

[4] The notion of a strategy as used here is that used in the game theoretic analysis of conflict situations. For example, see J. C. C. McKinsey, *Introduction to the Theory of Games,* McGraw-Hill Book Company, New York, 1952. In a managerial context, the idea of a strategy is pursued further in J. Thomas Cannon, *Business Strategy and Policy,* Harcourt, Brace & World, Inc., New York, 1968.

TABLE 10-3

Strategies	*Level of* Controllable #1 (Red dial)	*Level of* Controllable #2 (Blue dial)
S_1	Left	Left
S_2	Left	Middle
S_3	Left	Right
S_4	Middle	Left
S_5	Middle	Middle
S_6	Middle	Right
S_7	Right	Left
S_8	Right	Middle
S_9	Right	Right

ward, the alternatives open to you are "complex" combinations of the two controllables. You have, in this case, nine possible *strategies,* as shown in Table 10-3.

Note in Table 10-3 that each possible combination of levels for the two controllables is enumerated once and only once. Hence, these are different strategies.

There is no reason for you to consider any simpler alternatives, since each action that you take must encompass a setting on *both* dials. Your outcome array should therefore be made up of these nine strategies in interaction with the uncontrollables.

The person sitting opposite you, together with the black box, represents the uncontrollable aspects of the situation. Obviously, his strategies are identical with those which you have depicted in Table 10-3. Hence, the format for an outcome array whose rows designate complex strategies and whose columns designate the uncontrollable actions of another person (labeled $C_1, C_2, \ldots C_9$) is that shown in Table 10-4.

The entries in the outcome array of Table 10-4 represent the reward which you obtain in the case of each combination of strategies selected

TABLE 10-4

	C_1	C_2	C_3	C_4	C_5	C_6	C_7	C_8	C_9
S_1	$1	0	0	0	0	0	0	0	0
S_2	0	$1	0	0	0	0	0	0	0
S_3	0	0	$1	0	0	0	0	0	0
S_4	0	0	0	$1	0	0	0	0	0
S_5	0	0	0	0	$1	0	0	0	0
S_6	0	0	0	0	0	$1	0	0	0
S_7	0	0	0	0	0	0	$1	0	0
S_8	0	0	0	0	0	0	0	$1	0
S_9	0	0	0	0	0	0	0	0	$1

one each by you and the other person. Depending on your level of understanding of the way in which the black box operates, these entries may or may not be simple to develop. In that table, it is assumed that you know that *your* reward is $1 if both you and your competitor select correspondingly numbered strategies, and that you receive nothing otherwise.

In this trivial analog, the "other person" represents another rational being whose actions, in part, affect the outcome with which *you* will be presented. In the real world, a competitor who can promote his product and thereby lure potential buyers away from yours is such an entity. Of course, his actions are uncontrollable to you in the typical case.

The black box in the analog represents other environmental conditions which influence the outcome. In the case in which your understanding of the operation of the black box is that described in Table 10-4, there is only one level of this uncontrollable being considered.

To summarize the black box analog, let us consider the roles played by the various elements. You, the decision maker, must choose a strategy from a set of available ones. Each of these strategies is made up of one of three specific levels of two basic controllable quantities. The other person, who also has a set of strategies from which he must choose, represents uncontrollable competitive elements in the environment. The operation of the black box sums up the environment's uncontrollable effect on the outcome. Thus, *three* entities (the decision maker, the competitor's actions, and the environment) of *two* different varieties from the viewpoint of the decision maker (controllable and uncontrollable) interact to produce the outcomes of the decision problem. Figure 10-2 sums up this interaction in the same general way as was done previously.

Considering the interaction depicted by Figure 10-2 in a realistic decision situation might be useful. In the United States government's decision problem concerning the development of a supersonic transport (SST), for example, the significant controllable elements were the design and the manufacturers (various combinations of which defined the strategies), the relevant environmental factors were the demand for fast long-range travel and the "sonic-boom" characteristics of supersonic flight,

FIGURE 10-2 *Decision outcome produced through the interaction of strategy, competitive action, and environmental factors.*

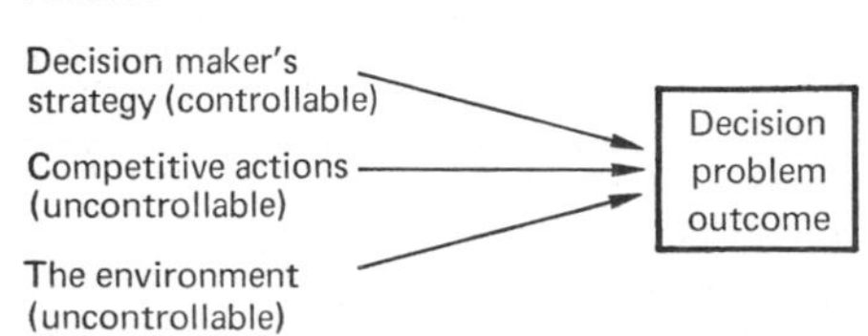

and the competitive actions which needed to be considered were the design and usage of competitive airplanes made by France, Britain, and the U.S.S.R. All of these factors interact in an outcome table model to produce a set of outcomes of the decision problem.

Of course, *only one of these outcomes is ever actually realized in any decision situation.* Before the outcome becomes fact, however, the entire range of outcomes which *may* result need to be considered by the decision maker in some explicit fashion. In the next section, we shall consider the various ways in which this can be done.

A TAXONOMY OF DECISIONS

Decision problems may be modeled from either of two major standpoints, depending on the basic nature of the real-world problem, the analyst's level of information about it, and the degree to which he requires that his model be "accurate." In this section we shall first describe these two major viewpoints and then discuss those elements which determine which is to be used in any particular problem analysis.

Decisions under Certainty

The simplest model of a decision situation is that termed "decision making under *certainty.*" In such a case, *the known set of environmental conditions is presumed to be presented to the decision maker; each possible strategy, therefore, leads to a single known outcome.* The array of Table 10-5 describes a certainty situation in which each possible strategy (each strategy involving a production quantity and an interval between production runs) leads to a known profit outcome. The right-hand column of the

TABLE 10-5

		Known environmental state
S_1	(Produce 300 units at 1-week intervals)	\$100,000 profit
S_2	(Produce 600 units at 2-week intervals)	\$ 78,000 profit
S_3	(Produce 900 units at 2-week intervals)	\$106,000 profit
S_4	(Produce 700 units at 1-week intervals)	\$121,000 profit

table represents the known state of the environment, which is here not described. (For instance the environmental state might be a known level of demand.)

SOLVING CERTAINTY DECISION MODELS In such a situation as that described by Table 10-5, the criterion to be applied for the determination of the best strategy is deceptively simple. Since each strategy leads to a single outcome, the decision maker should choose the strategy which gives the best outcome. Thus, the problem of choosing among strategies is transformed into one of choosing among outcomes. In this case, this is indeed simple, since each outcome is described by a single outcome descriptor—profit—and that measure is one that everyone holds the same preference for—i.e., more profit is preferred to less profit. Therefore, in the case of Table 10-5, since S_4 leads to $121,000 in profit, and since no other strategy leads to an outcome involving this much profit, the decision maker should choose S_4—i.e., he should produce 700 units at one-week intervals.

The apparent simplicity of this criterion is misleading in the context of real-world problems, for a number of reasons. First, in many instances the idea of making a choice between outcomes rather than strategies is not a trivial problem. Also, the number of strategies which need to be considered in decision models under certainty may be so great as to preclude straightforward analysis of them.

EVALUATING AND COMPARING OUTCOMES The complexities of expressing preferences among outcomes can be readily illustrated by considering a marketing decision maker who has objectives whose degree of attainment can be assessed in terms of two measures—"market share" and "profit." Two possible outcomes which might evolve are:

Outcome O_1	Market Share = 10% Profit = $50,000
Outcome O_2	Market Share = 15% Profit = $40,000

There is no clear basis for choice between these two simple outcomes. In O_1, a larger profit is achieved together with a lower market share than in O_2. It is not at all obvious which outcome a particular decision maker or organization might prefer. In fact, it is apparent that two different individuals or organizations might have different preferences regarding these outcomes. The marketing manager associated with a newly introduced product might be very concerned with penetration of the market and might, therefore, prefer O_2 over O_1, even though less immediate profit is achieved. The manager whose product is mature might prefer O_1.

Economists have long concerned themselves with a concept called *utility,* which may be thought of as the capacity of an event, object, or state of affairs to satisfy human wants. The idea of a basic measure of the degree of human satisfaction which is derived from a state of affairs leads naturally to consideration of using this measure as a way of evaluating and comparing outcomes. If, for example, O_1 is preferred to O_2, O_1 possesses greater utility than O_2. If one had a numerical measure of the utility of each outcome, he could easily select that outcome which had the greatest utility as the one that is most preferred.

However appealing the concept of utility as indicative of human satisfaction might be, the operational measurement of utility is not well developed.[5] In practice, utility measurements are seldom made by managers in their problem analyses because of the deficiencies and operational impracticalities of these measurement techniques.

The way in which practical managers and analysts often handle this difficulty is by using the outcome descriptors themselves as proxy measures of utility. It is often possible to do this because the outcome measures are all of one or two general varieties—resource costs or benefits.

The *resource costs* are the resources expended in attaining an outcome. *Benefits* are the returns associated with an outcome, in terms of either the resources gained or the psychological, sociological, or other intangible values derived from the state. Examples of such benefits are the profit earned, the time saved, and the number of maternal deaths prevented, as are the intangibles "greater freedom enjoyed," "freedom from want," "higher quality of life," etc.

Outcome descriptors—whether of the resource cost or the benefit variety—are measures of the degree of attainment of objectives, and since most complex decision problems involve multiple objectives, it is common to find each outcome in an outcome array described by a number of such measures. We have already demonstrated, for the situation involving "market share" and "profit," that multiple outcome descriptors may complicate the problem of evaluating and comparing outcomes.

A key method for obviating the problem of multiple-outcome descriptors, without resorting to formal utility measurement, is to reduce them to an overall single measure which reflects their aggregate worth. The concept of aggregate worth is basic to utility measurement; however, as we have already pointed out, the problems involved in utility measure-

[5]The interested reader may refer to P. C. Fishburn, *Decision and Value Theory,* John Wiley & Sons, Inc., New York, 1964, for a summary of utility measurement techniques and a discussion of the applicability of the concept in problem solving.

ment are so great as to impair the operational usefulness of attempting utility assessments.

To demonstrate how we may assess the aggregate worth of outcomes described by cost and benefit measures without resorting to formal utility measurement, consider a decision model involving two strategies (S_1 and S_2) and a single environmental state. Table 10-6 shows two outcomes described in terms of revenue (benefit) and cost. Note that these outcomes have the same characteristics as those described previously in terms of market share and profit. The outcome for S_2 is better in terms of the benefit measure—revenue—while S_1 is better in terms of the other measure—cost—since it has the lower cost.

However, few would hesitate in this situation to declare that the outcome associated with S_1 is better.[6] Why? The obvious answer is that the *profit* there is $100,000 minus $40,000, or $60,000, while the profit associated with the outcome for S_2 is only $50,000.

The critical point illustrated by Table 10-6 is that since the resource costs and benefits are expressed in the same terms—dollars—the *difference* between benefits and costs seems to be a valid measure of the *net worth* of each outcome. Additionally, of course, this difference has an accounting significance which further enhances its intuitive appeal.

TABLE 10-6

	Environmental state
S_1	Revenue = $100,000 Cost = $ 40,000
S_2	Revenue = $110,000 Cost = $ 60,000

Decisions under Uncertainty

The outcome array of Table 10-7 describes a decision situation under *uncertainty.* A comparison of this table with Table 10-5 shows that the primary difference is that *more than one set of environmental conditions is considered here.* This is the key to the decision model under uncertainty. In an uncertainty model *each strategy may lead to any one of a number of possible outcomes.* That Table 10-7 describes such a model is made clear from its *two* columns (as compared with the one column in Table 10-5). Since each column in an outcome array describes a particular set of environmental conditions, it is clear that this model contemplates a situation in which *the single environmental condition that will eventually occur is*

[6]This statement implies some assumption about the decision maker's availability of funds, etc. If the alternatives were presented to a large, prosperous corporation, the statements made here would invariably be true.

unknown in advance. Therefore, a number of possibilities must be considered.

TABLE 10-7

		Environmental states	
		N_1	N_2
Strategies	S_1	$50	$100
	S_2	$48	$ 96

Each of these possible environmental states in combination with a given strategy will lead to a different outcome. For instance, in Table 10-7, strategy S_1 will lead to a $50 outcome *if* environmental state N_1 occurs *or* to a $100 outcome *if* environmental state N_2 occurs. *Since the decision maker does not know with certainty which of the two states will take place, he must consider both possible states when he seeks to evaluate the relative worth of the strategies.*

SOLVING UNCERTAINTY DECISION MODELS The determination of the best strategy in a problem such as that described in Table 10-7 is a relatively simple matter. The reason is that regardless which of the environmental states occurs (and recall that only one of them will), S_1 leads to the better outcome. Since S_1 leads to a better outcome than S_2 for every possible contingency, it is clearly the better strategy.

Of course, most uncertainty decision problems are more complex than this. Consider, for example, the model of Table 10-8. There, S_1 is better if N_1 occurs, but S_2 is better if N_2 occurs. Which is better—S_1 or S_2? The answer is: "It depends." It depends on the relative likelihoods associated with N_1 and N_2, for one thing. For example, if one knew that N_2 was 1,000 times as likely to occur as N_1, he would probably choose strategy S_2 because it is best in the event that the extremely likely environmental state occurs. On the other hand, if the likelihoods are not so extreme—say,

TABLE 10-8

		Environmental states	
		N_1	N_2
Strategies	S_1	$50	$ 96
	S_2	$48	$100

that there is a 50:50 chance for each of the environmental states—no strategy is clearly best.

In decision models under uncertainty, it is clear that *two* things must be considered in order to arrive at a rational determination of which strategy is best. First, the relative *worth* of every possible outcome must be considered. This worth is described in terms of evaluations of the outcomes in the outcome array. Secondly, the relative *likelihood* of each of the environmental states must also play a role.

We have already discussed evaluating the worth of outcomes in terms of utilities or some proxy for them. Assessments of the likelihood of various environmental states are made in terms of probabilities.

LIKELIHOODS AND PROBABILITIES[7] The basic ideas relating to probability are familiar to most of us. Any event whose outcome is at least partially determined by chance, such as the flip of a coin, may be described in probabilistic terms.

The probability of an outcome of an event is most easily thought of as the long-term percentage of times the outcome would occur if the event were repeated again and again. In terms of the coin flip, after a long series of flips, one might divide the number of occurrences of the outcome "heads" by the total number of flips and call the resulting decimal the "probability of heads." As such, one would likely determine the probability of a head to be near ½. Similarly, the probability of a 6 on a single throw of a die would be about ⅙. The concept of probability may also be applied to sequences of events. The knowledge that the probability of a head is ½ will in no way help one to predict what the outcome of a particular flip of a coin will be, for it will be either heads or tails, and one is either right or wrong. But this knowledge does permit one to predict that in a long sequence of flips, the relative frequency of heads will be close to 50 percent.

To apply probabilities to the analysis of managerial decisions, one must recognize that the basic idea is applicable to the uncertain events which can influence these decisions—the environmental states. To conclude that the probability is ⅓ that "June rainfall in Cleveland will exceed 2 inches" implies that an investigation of weather bureau records for many past Junes has indicated a relative frequency of ⅓ for such a state. If this is so, and if there is no reason to believe that weather patterns now and in the future will differ from those in the past—i.e., if the pattern exhibits stability over time—one might conclude that the future relative frequency of occurrence of rainy (over 2 inches) Junes will also be about ⅓.

[7]Material in this section is adapted from William R. King's *Probability for Management Decisions,* John Wiley & Sons, Inc., New York, 1968, with permission of the publisher.

Several properites of the probabilities for environmental states which we shall make use of are implied by the foregoing statements.

1. Each environmental state must have an associated probability greater than or equal to 0 and less than or equal to 1.
2. The environmental states must be defined so that they are disjoint, i.e., the occurrence of one precludes the simultaneous occurrence of any other. (For each strategy-and-environmental-state combination, there must be one and only one clearly defined outcome.)
3. The environmental states to which probabilities are attached must be an exhaustive listing. (No environmental states which might possibly occur can be omitted from our outcome table.)

Taken together, these properties imply that the probabilities attached to all of these environmental states should sum to unity. This is equivalent to saying that it is certain that one of the possible states will occur. Correspondingly, in terms of outcomes, exactly one of the disjoint and exhaustive outcomes will occur for each strategy, and hence the probabilities associated with the possible outcomes for each strategy should also sum to 1.

Probability, then, is simply a way of dealing with our uncertainties about the future. In attaching probabilities to environmental states, we evaluate the likelihood of them occurring, and we thereby synthesize our information about future likelihoods into a single number. We can then utilize this information in determining the best strategy in an uncertainty decision model.

To use the likelihood assessment—probabilities—and the worth assessments—utility or some proxy for it—in solving uncertainty decision problems requires that some criterion be developed which synthesizes these two elements. The criterion, which is generally accepted, is that of *maximum expected utility or maximum expected net benefit.*[8] Thus, that strategy is chosen as best which has associated with it the largest expected utility (or net benefit).

The term "expected" warrants further explanation. In its usage here, the expected value (or "expectation") of some measure such as utility or net benefit is analogous to the familiar arithmetic average. If, for example, a class of five were given a test in which three people scored 90 and two scored 80, the instructor would calculate the average score to be

$$\frac{3(90) + 2(80)}{5} = 86$$

[8]"Net benefit" is simply the benefit attained minus the cost of the resources used to attain it.

Written slightly differently, this is $\frac{3}{5}(90) + \frac{2}{5}(80) = 86$, where $\frac{3}{5}$ is the relative frequency of occurrence of the score 90 and $\frac{2}{5}$ is the relative frequency of occurrence of the score 80.

The *expectation* (or "expected value") of a measure is analogous to an average. *If a measure takes on a number of values with known probabilities, the expectation of the variable is the sum of the products of the probabilities and the associated values of the measure.* Thus, if a new product is predicted to return either \$1 million, \$2 million, or \$3 million with equal likelihood (probabilities of $\frac{1}{3}$ for each), the expected return is:

$$\tfrac{1}{3}(1 \text{ million}) + \tfrac{1}{3}(2 \text{ million}) + \tfrac{1}{3}(3 \text{ million}) = 2 \text{ million}$$

Here "return" is a measure which takes on three different values (\$1 million, \$2 million, or \$3 million), each with a known probability (in this case, $\frac{1}{3}$ for each). The probabilities play the same role as do relative frequencies in a calculation of a simple average.

The expected net benefit (or utility) measure is, then, a weighting of the possible outcomes by the probabilities associated with each outcome. The criterion for selection of a best strategy determines that this quantity should be maximized.

In the illustration previously given and repeated here as Table 10-9, if it were determined that each environmental state was equally likely to occur, a probability of $\frac{1}{2}$ would be imputed to each. The expected net benefit (expected profit) associated with strategy S_1 is, therefore, $\frac{1}{2}(\$50) + \frac{1}{2}(\$96) = \$73$, and the expected net benefit for S_2 is $\frac{1}{2}(\$48) + \frac{1}{2}(\$100) = \$74$. Therefore, strategy S_2 is the better of the two, since it has the higher expected net benefit.

Certainty versus Uncertainty

Both the certainty and the uncertainty formulations which we have discussed are idealized models of real-world decision situations. The question

TABLE 10-9

		Environmental states	
		N_1	N_2
Strategies	S_1	\$50	\$ 96
	S_2	\$48	\$100

of whether the manager should use the certainty or the uncertainty model in a particular instance depends on the *nature of the problem* with which he is dealing, the *level of information* available to him, and the *"accuracy"* required.

THE NATURE OF THE PROBLEM There are few meaningful decision problems with which a manager may be faced which are inherently of the certainty variety. Most meaningful decision problems require that the future be *predicted* in order for an outcome array to be developed. In other words, *every outcome in such a table is a prediction of something that will happen in the future if a specified strategy is adopted.* Obviously, such predictions are difficult to make with the perfect accuracy assumed by the certainty model because various future environments may generally produce various outcomes for a given strategy. Of course, such uncertainty is explicitly taken into account in the uncertainty model.

However, even if the basic problem situation is not really of the certainty variety, the certainty model may serve as a useful approximation to the real world. For example, the decision relative to scheduling production facilities depends on the anticipated level of sales for the products in question. Future sales are inherently uncertain; yet many products have such a stable sales pattern that the future sales level can be treated as though it were known. The use of the certainty model in such a case allows the decision maker to concentrate on the important aspects of the problem and to omit the minor uncertainty about the sales level. Thus, since no model is a perfectly accurate representation of the real world, the use of the certainty model to describe an inherently uncertain situation is simply one of the ways in which the model deviates from reality.

THE LEVEL OF INFORMATION The information which is available to the decision maker and analyst is another element in the determination of an appropriate decision model. Often, likelihood information of the kind necessary for the assessment of probabilities is unavailable. In such a case, one of the basic elements of any analysis under uncertainty—the probabilities of the various environmental states—is missing and the variety of uncertainty analysis which we discussed previously is inapplicable.

Such a lack of likelihood information might be the case if no experience with a similar situation is available—e.g., a decision concerning a new product to be placed on the market—or if the principal relevant uncontrollable involves the conscious acts of a competitor. In the latter instance, it might not be deemed meaningful to treat a competitor's possible acts in the same way as we would treat other uncontrollable elements of the environment.[9]

If appropriate likelihood information is not available, the manager has two basic alternatives. He can model an uncertain situation in terms of certainty as we have just discussed, or he can treat the uncertain situation directly. The way in which he should proceed in the latter case is not completely clear, but there are a variety of logical criteria which have been proposed for this situation.[10]

For all practical purposes, the manager has only the two basic alternatives:

1. Use the certainty model.
2. Find some way of assessing likelihoods.

The latter course is not as undesirable as it might appear to be, even for the situation in which little or no history is available to provide a basis for assessing probabilities. This is so because *precise* assessments of probabilities are not always essential to the solution of a problem,[11] and because in most real-world circumstances, some information, however vague, is available concerning such likelihoods. Only the most obscure circumstances which one can hypothesize, such as "war versus peace," do not involve *some* information as to the likelihood of the environmental states on which probability estimates may be based.

THE ACCURACY OF THE MODEL Another of the basic elements in the decision concerning whether a certainty or uncertainty approach will be taken to the development of a decision model is the accuracy of representation which is necessary. The overall question of selecting an appropriate model is itself one of balancing the *cost of accuracy* and the *cost of error.* Presumably, a very accurate model is costly to construct and manipulate relative to a less accurate one. Conversely, because it is more accurate, it is also likely to introduce fewer errors into the decision process.

Much of the determination which the manager must have concerning the required accuracy must be done subjectively. Although quantitative procedures are available for aiding in this determination,[12] it is the mana-

[9] The body of knowledge termed "game theory" is applicable to this variety of situations. See A. Rappoport, *Two-Person Game Theory—The Essential Ideas,* The University of Michigan Press, Ann Arbor, 1966.

[10] See William R. King, *Quantitative Analysis for Marketing Management,* McGraw-Hill Book Company, New York, 1968, pp. 54–61.

[11] *Ibid.,* pp. 61–64.

[12] *Ibid.,* pp. 100–104.

ger who knows how accurate he needs to be in order to arrive at the best strategy for any particular decision situation. Analytical personnel can readily provide him with estimates of the cost of achieving greater levels of accuracy, and by combining these with his assessment of the cost of making mistakes (cost of error), he can determine how accurate he must be.

VALUES OF MANAGEMENT DECISION ANALYSIS

To address the question of the value of formal decision problem analysis, one must recognize that *formal analysis is not a substitute for the intuition, judgment, and experience of the manager. Rather, it complements them.* Thus, in stressing the value of such devices as the outcome array and associated formal analytic procedures, we are in no way suggesting that they should supplant the manager's basic role and duty. They are simply tools which can help him to make better decisions.

The role of the outcome array is particularly apparent in this regard. In a typical organizational decision situation, many alternatives are available and many outcomes may result from each. It is virtually impossible for any human being, regardless of his intellect and experience, to give adequate consideration to all of these outcomes without resorting to an analytic device such as the outcome array. Of course, it may be validly argued that it is not always necessary for the decision maker to consider every possible outcome. While this may be so, it is also true that *it is invariably difficult to determine which are the relatively unimportant ones without making them explicit.* The most obvious value of the outcome array, then, is that *it requires that each outcome be explicitly delineated and considered.*

The full impact of this simple point can be made clear by considering the large number of very poor outcomes which are actually attained daily as a result of organizational decisions. For example, it has been established that the failure rates for new products in some industries run as high as 98 percent.[13] The overall economic loss is estimated at $4 billion annually.[14]

Of course, every situation which leads to a bad outcome is not necessarily attributable to a failure by the decision maker to consider the entire range of possible outcomes; in many cases, actions are necessarily taken with full cognizance that bad results *may* be obtained. One suspects,

[13] P. Hilton, "New Product Introduction for Small Business Owners," *Small Business Management Series,* no. 17, U.S. Government Printing Office, Washington.

[14] Estimated by Batten, Barton, Durstine and Osborn, Inc.; quoted in *Saturday Review,* Mar. 11, 1967, p. 128.

however, that many of the bad outcomes could be avoided by giving advance considerations to the range of all possible outcomes *in the context of their relative significance.*

The foregoing italicized phrase provides another clue to the less obvious values of simple analytic presentations such as the outcome array, for not only does the development of such a table require that explicit consideration be given to all possible outcomes, but the associated techniques for analysis of the array require that they be viewed in proper perspective—i.e., in terms of their relative likelihood. The natural tendency of the optimistic manager is to explicitly consider a range of relatively good outcomes in any decision situation and to lump together all of the bad ones into a single unpleasant category. In doing so, his inherent optimism is reinforced by the apparent pervasiveness of good potential results and the relative paucity of bad ones. In fact, of course, this favorable view may be more apparent than realistic. Analyses such as those made possible by the basic uncertainty model tend—in their use of likelihood assessments—to put such factors into proper perspective.

Thus, the overall value of management decision analysis is in putting realism into managerial decisions. Although one may argue that the "toy models" which we have dealt with in this chapter are wholly unlike real-world decision problems, he should remember that so, too, are the "mental models" which individuals use to subjectively assess decision problems. What management decision analysis does is to take important aspects of the problem out of the nebulous domain of the mind and force them into the logical domain by requiring that their constituent elements be made explicit. Then, after all of the explicit objective analysis has been performed, the manager may use the model's output (the solution to the model) as a basis for decision. He need not take the model solution as gospel, since he knows that some elements of the real world have been abstracted out of the model's depiction of the problem. He can, however, take the model's solution and adjust it subjectively to account for these omissions. In doing so, he will probably choose a better course of action than he would have if he had simply chosen to "mull over" the problem and arrive at a choice on a completely subjective basis.[15]

SUMMARY

A formalized conceptual framework for the analysis of decisions provides the manager with a basis for performing organized thought about a decision problem. Without such a framework, he is left to the vagaries of his

[15] In effect, this view of management decision analysis is built on the assumption that the rigorous solution developed from the model will lead him close to the solution of the real-world problem so that he can achieve the best possible outcome, or something close to it, by adjusting the model solution.

memory and his limited ability to simultaneously consider the many complex elements of the problem.

The outcome array—which summarized the interaction of controllables and uncontrollables in producing decision-problem outcomes—is a basic decision-problem model. The outcome descriptions entered into the body of the array represent descriptions of the various outcomes measured in terms which are related to the objectives of the decision maker.

"Strategy"—a concept which is important to the manager, but one with a variety of meanings—is used by decision theorists to mean a complex alternative which is available to the decision maker. Strategies are composed of controllable elements of the problem, and they may include sets of responses to contingencies which may arise. Thus, strategies are complex alternatives which permit the decision maker to "make up his mind" while at the same time retaining some flexibility.

Decision problems may be modeled in either certainty or uncertainty terms. Decisions under certainty presume a knowledge of an existing environmental state. Decisions under uncertainty deal with a range of possible environmental states in terms of the relative likelihood of each.

The determination of which model to use—certainty or uncertainty—is based on the nature of the problem (whether it is reasonable to approximate the environment with a known state), the level of information which is available, and the degree of accuracy which is required.

In any case, the prime value of management decision analysis lies in its role as a complement to the intuition, judgment, and experience of the manager. If formal analysis is viewed in this fashion, it can be of great aid to the manager in the decision problems which he must face.

EXERCISES

1. What is the practical use of having a conceptual framework for dealing with decision problems?
2. Consider yourself as a marketing manager who must express a preference between the outcomes below. What would be your preference? Try to describe the mental process which you went through to arrive at this conclusion.

	Sales revenue	*Total costs*	*Employee morale level*
O_1	\$2.0 million	\$1.5 million	Low
O_2	1.5 million	1.0 million	Medium
O_3	1.5 million	0.9 million	High
O_4	2.0 million	1.6 million	High
O_5	1.4 million	0.9 million	Medium
O_6	1.4 million	0.8 million	Low

3. A particular strategy may result in any one of four possible outcomes. The probability that O_1 or O_2 will occur if this strategy is chosen is ½ and ¼, respectively. Outcomes O_3 or O_4 are so vaguely defined that we find it difficult to get estimates of their relative likelihood of occurrence under the strategy in question. What might we do to facilitate the application of the conceptual framework?
4. A salesman wishes to travel to each of four cities, beginning at his home office, which is not in any of the cities, in a manner which will minimize the total distance traveled. What are his available strategies? Why is this a decision problem under certainty?
5. A decision maker who is faced with a different problem might decide to go and play a rouna of golf. In what circumstances would this constitute a chosen alternative in the problem and in what circumstances would it not?
6. A successful manufacturer of buggy whips began to notice falling sales about the time that the automobile came into common use. Being dissatisfied, he investigated and found that the decreasing use of horse-drawn vehicles was the reason. Did he have a problem? As best you can, formulate his problem by enumerating what you think would be his objectives, alternatives, etc.
7. Is it necessarily true that the alternative which is actually chosen by a decision maker is the solution to his decision problem?
8. How would you apply the "principle of bounded rationality" to your search for a job?
9. As a regional sales manager, you are faced with dealing with a salesman whose performance has dropped off drastically. He is paid on a strict commission basis, and company policy will not allow you to discharge him. Several conversations with him have not proved fruitful. The possibility of transferring him to another region is brought up. You know that if he is away from his home town and family, his performance is likely to drop further. What would be the systems view of this problem? Is it a managerial decision problem? Why?
10. What is meant by the controllables and uncontrollables in a decision problem? Give examples.
11. What is the major value of an outcome array in a real decision problem?
12. Define a "strategy." Give an example. How can a strategy be made adaptive?
13. Distinguish between decisions under certainty and uncertainty. Illustrate the difference in the form of the outcome table for these two cases.
14. "A decision maker needs to have an objective assessment of two things—worth and likelihood." Explain this statement in terms of this chapter. Use an outcome array as an aid in the explanation.
15. Describe the salient aspects of the "model selection decision problem"—i.e., the decision problem concerning which model—certainty or uncertainty—the manager will use to describe a problem situation.
16. Consider the outcome array presented at the top of the next page. Which is the best strategy?

	N_1	N_2	N_3
S_1	$10	$20	$ 8
S_2	$19	$40	$11
S_3	$ 5	$19	$ 4

17. Suppose that dollars are being used as an approximation to utilities and that an individual is faced with the decision situation under uncertainty described below.

	N_1	N_2	N_3
S_1	$ 5	$10	$15
S_2	$10	$20	$30
S_3	$ 5	$30	$10

If the environmental states are equally likely, what should the decision maker do?

18. Consider the situation which many people face when their alma mater asks for contributions, saying that the average contribution last year was $1,000. The alumnus who is considering giving $50 is awed until he realizes that several people have given very large sums, while most have given $100 or less. The weighted sum of the contributions called an *average,* is grossly unrepresentative of one's intuitive idea of an average. How is the expected net benefit concept similar to this?

REFERENCES

Ackoff, R. L., *Scientific Method—Optimizing Applied Research Decisions,* John Wiley & Sons, Inc., New York, 1962.

Arrow, K. J., *Social Choice and Individual Values,* John Wiley & Sons, Inc., New York, 1951.

Churchman, C. W., R. L. Ackoff, and E. L. Arnoff, *Introduction to Operations Research,* John Wiley & Sons, Inc., New York, 1957.

Feller, W., *An Introduction to Probability Theory and Its Applications,* John Wiley & Sons, Inc., New York, 1957.

Fishburn, P. C., *Decision and Value Theory,* John Wiley & Sons, Inc., New York, 1964.

Hurwicz, Leonid, *Optimality Criteria for Decision Making under Ignorance,* Cowles Commission Discussion Paper, Statistics, no. 370, 1951 (mimeographed); cited in Luce and Raiffa (see below).

Luce, R. D., and H. Raiffa, *Games and Decisions,* John Wiley & Sons, Inc., New York, 1957.

Mosteller, R., R. E. K. Rourke, and G. B. Thomas, Jr., *Probability with Statis-*

tical Applications, Addison-Wesley Publishing Company, Inc., Reading, Mass., 1961.

Savage, L. J., *The Foundations of Statistics,* John Wiley & Sons, Inc., New York, 1954.

Schlaifer, R., *Probability and Statistics for Business Decisions,* McGraw-Hill Book Company, New York, 1959.

11

an operational planning system

A "planning system" consists of a *system of plans, a process for planning,* and a *management system for guiding and directing the planning process.* In this chapter, an operational planning system is presented by focusing on these three crucial elements of a planning system.

The operational system described here is a composite of those which have been implemented in a wide variety of organizations. Although it is not directly transferable to any particular organization, a wide range of business, educational, and other organizations would probably find it to be workable and useful.

A SYSTEM OF PLANS

The output of a planning process is a series of documents called "plans." The various plans should be interrelated to form a *system of plans* which can be used to guide the organizational system.

A conceptual model for a system of plans is schematically displayed in Figure 11-1. Much of the terminology used in that figure and the subsequent discussion are directly appropriate to the business firm. However, the functions and purposes of the various plans are appropriate to any organization. Indeed, the authors have themselves applied this system of plans in a wide variety of nonbusiness contexts.

Project plans are the basic building blocks of the system of plans in Figure 11-1. Each echelon of plans receives guidance from prior plans and further specifies such direction by focusing on groups of activities having a common purpose.

The Strategic Plan

The strategic plan of the organization is at the summit of all the plans; it outlines the broad goals and strategies that the organization wants to achieve. It includes the answers to such questions as:

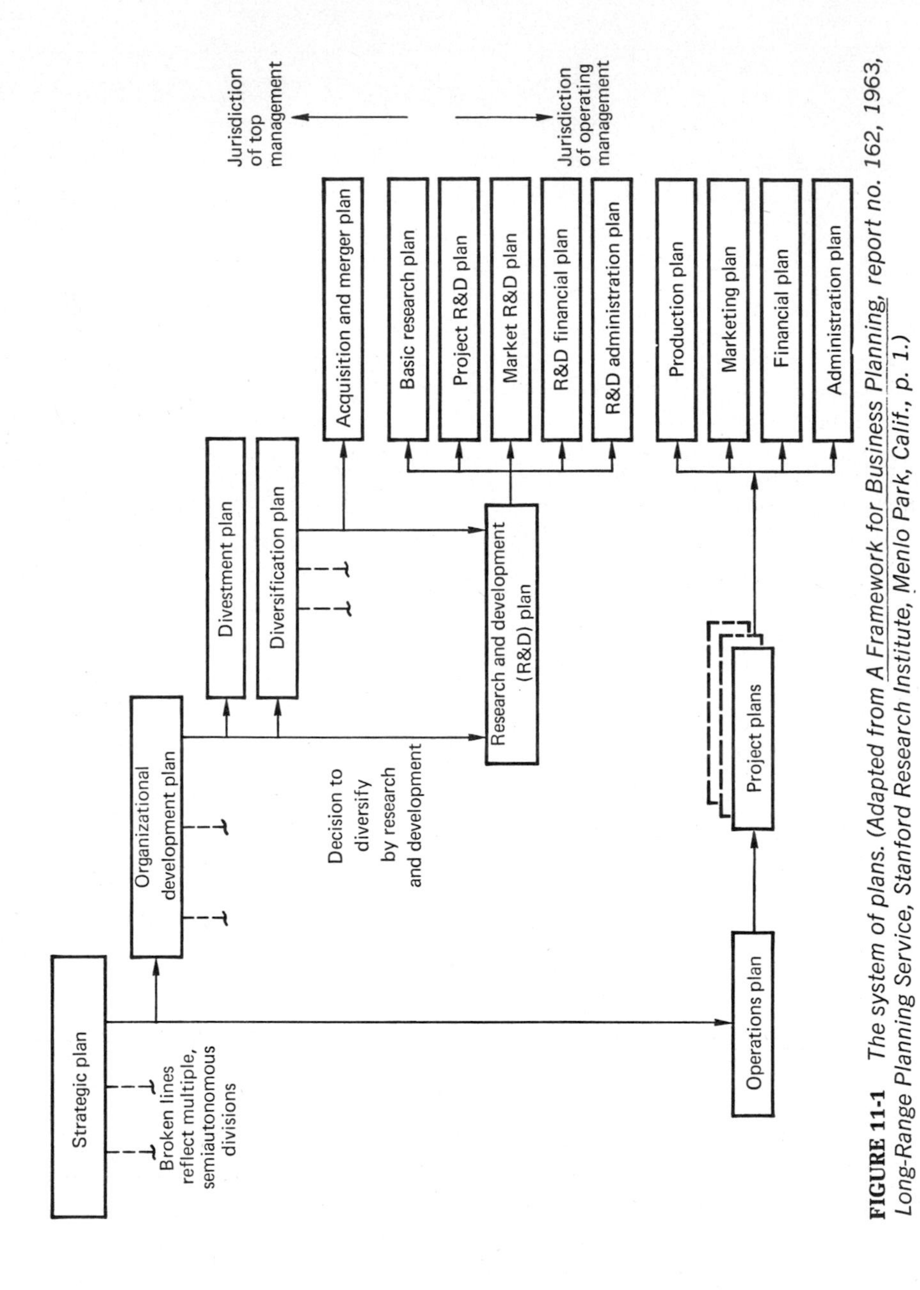

FIGURE 11-1 *The system of plans. (Adapted from A Framework for Business Planning, report no. 162, 1963, Long-Range Planning Service, Stanford Research Institute, Menlo Park, Calif., p. 1.)*

What are the broad missions and roles of the organization?
What strategy is required to move from today's position to a competitive position desired in the future?
When compared with the other plans that are developed in the organization, how realistic is the overall organizational strategy?

What image does the organization want to portray in the greater system (social, economic, political) of which it is a part?

The strategic plan is important because it becomes a standard for deciding what direction the subsidiary plans in the organization should take. It is the overall document used to determine the organizational compatibility of the project plans with the other plans. The strategic plan guides the organization in decisions involving the current product (or service) and the future generation of products and markets.

The Organizational Development Plan

The next echelon in the system of plans is the *Organizational development plan,* which determines the activities necessary for a new generation of products or services. At the same time, the organizational development plan maps in greater detail the route toward the future position of the organization that has been specified by the strategic plan. The development plan answers such questions as:

What will the future environment consist of in terms of demand for our goods or services? What will be expected of our organization in that future time period?
What favorable conditions must be created within the organization in order that new products and new markets can be conceived and defined?
What techniques will be used to screen out poor investments and products and high-risk ventures?
What are the expected resource requirements for the new products or services?

The organizational development plan provides the guidance for three succeeding plans:

THE DIVESTMENT PLAN This plan deals with the divestiture of major elements of the organization. These elements can consist of products, services, property, or organizational entities.

THE DIVERSIFICATION PLAN This plan describes the development of new products, services, and markets to join or replace the current generation of products. It selects new product areas and determines when entry should be made by acquisition of other organizations which possess required capabilities, by merger with another organization, or by conducting in-house research and development which builds on existing competence.

THE RESEARCH AND DEVELOPMENT PLAN This plan specifies action oriented to the creation of new products or processes for existing demand or a new market for existing products or services. It is through this plan that the organization does research to advance the state of the art of what it has to offer. This plan cuts across all elements of the organization to include products, markets, finance, and administration.

The Operations Plan

This plan, one of the two plans supporting the strategic plan, is the blueprint for current business action. It guides activity by which the current generation of products and services is distributed to the existing markets. The operations plan exists within each functional area of effort—production, marketing, finance, and administration—and is a composite of the project plans involved in the effort.

The operations plan specifies the total work to be accomplished in the functional area of jurisdiction. It subdivides the work into logical work units, defines work flow, allocates resources, and establishes authority and responsibility patterns between the principals in the effort. And, of course, it sets gross schedules and budgets. Part of this planning is cyclical, such as plant layout, organization, and routine procedures.

Project Plans

Project plans support higher-level plans; for example, the corporate development plan is supported by a complex of project plans covering the details of undertakings reflected in a corporate development plan. In such a case, project plans might be developed to sustain and guide the following types of actions.

1. Development and installation of a management information system
2. Developing and producing a new product
3. Building and activating a plant
4. Effecting a corporate merger or acquisition
5. Developing a new organization structure
6. Moving an organization from one location to another
7. Marketing a new product
8. Developing, testing, and implementing a specific organizational strategy

The corporate development plan will be supported by a series of ad hoc plans covering the details of a singular undertaking related to the implementation of the corporate development plan. Project plans are the basic building blocks of a planning system. Thus project plans can have short-, intermediate-, and long-range implications. What makes the project plan unique is its concern with an identifiable effort having cost, schedule, and technical parameters. Project plans are found at all levels in the organization. They should deal in sufficient specifics to be useful as a standard for control.

The project plan should go into considerable detail in outlining how the project will be developed and produced. In the management of a project which cuts across different functional lines, the specificity of the plan is a critical requirement since the participating individuals and organizations may have competing and conflicting objectives.

Functional Plans

Additional plans of a supportive nature are also a part of a system of plans. These plans take the form of a functional plan in areas such as production, marketing, and finance. The marketing plan would contain a description of the number of salesmen needed, the nature and magnitude of sales promotion, advertising, and the necessary changes in the marketing philosophy and organization to support the products and services that are offered. Similarly, each function requires plans which pull together the desired future position of the function and the actions required to attain that position.

Relationships among Plans

Figure 11-2 shows the system of plans in a slightly different context from Figure 11-1. The previous figure emphasizes the various subplans. Figure 11-2 emphasizes the *relationships among the various different kinds of plans* in the system, together with *typical time horizons* and the specific *content of the various plans.* The terminology used in Figure 11-2 is applicable to the planning process for a school system rather than for a business firm.

THE PLANNING PROCESS

The "adaptive planning model" discussed in Chapter 8 can now be elaborated into a planning process. The process is discussed here in generic terms so that it can be used as a guide in many different organizational contexts.

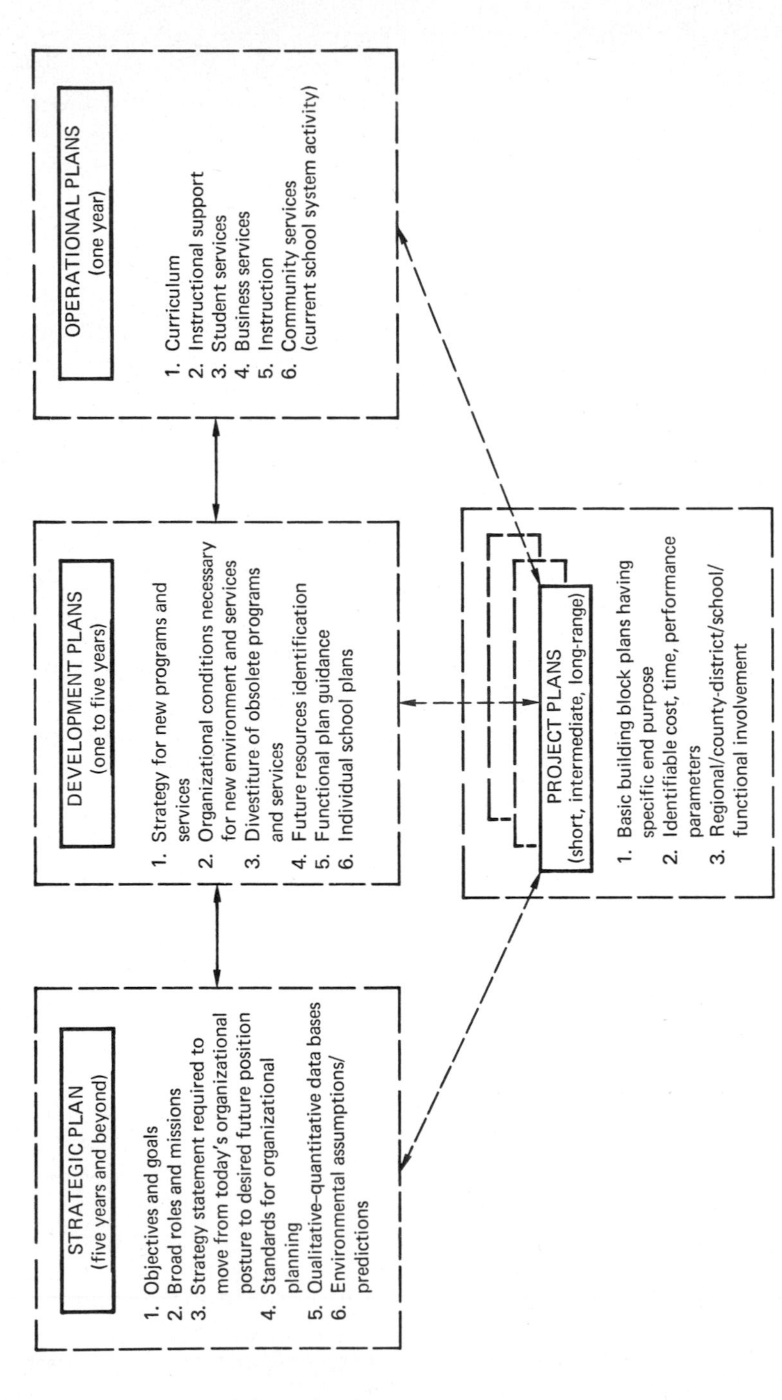

FIGURE 11-2 *The system of school plans.*

The planning process is thought of as consisting of five phases. Usually the phases will *begin* in sequence, but they are not conducted sequentially. At any specific part in a planning process, several of the phases will be being conducted simultaneously—some for the first time while others will be being redone based on feedback information.

The phases of the planning process involve:

1. Establishing general goals
2. Information collection and forecasting
3. Making assumptions
4. Establishing objectives
5. Developing plans

Establishing General Goals

As used here, a *goal* is a broad, general, and timeless purpose.[1] Hence, goals specify the organization's relationship to society, to governments, to the community, and to other elements of its environment, and especially the relationship among the various units which make up the organization.

The process of establishing goals can be more important than are the goals themselves, since the establishment of goals emphasizes where various people and units "fit into" the overall purposes of the organization. It can also be of great benefit in motivating individuals to pull toward the achievement of a set of goals which they have helped establish.

Information Collection and Forecasting

The basic information necessary for a planning process involves a critical assessment of the current status of the organization together with a forward look at the environment which is anticipated for the future. These assessments and forecasts must be done in four major areas:

1. Major products, services, and markets
2. External environment
3. Competitive environment
4. Internal environment

In all instances, *forecasts of the future should be made in terms of ranges of the environmental variables which have reasonable likelihood* rather than in terms of a single numerical quantity. This will permit the

[1] An objective, on the other hand, is a desired accomplishment whose achievement can be measured in a specified time period.

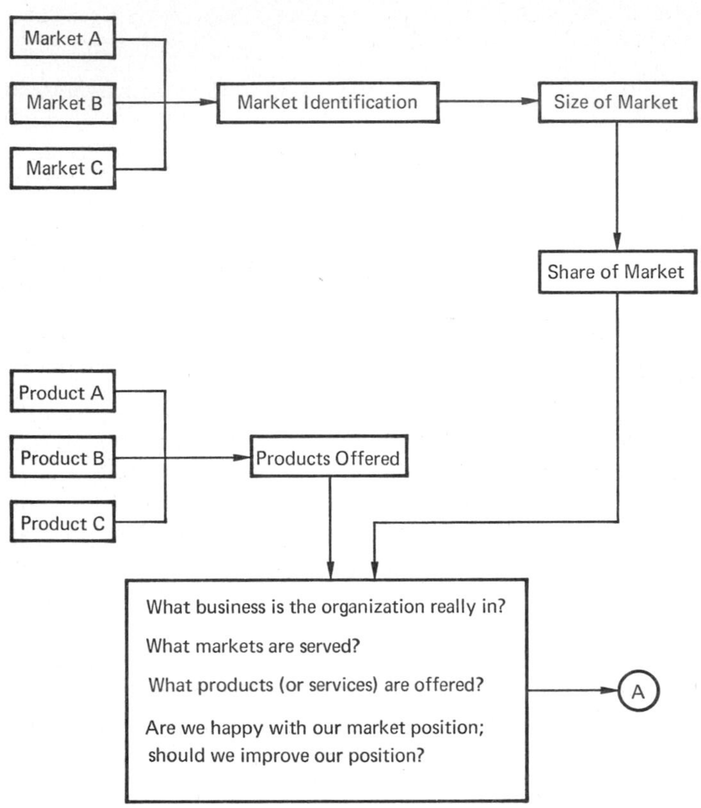

FIGURE 11-3 *Identification of major products, services, and markets.*

explicit consideration of uncertainty (in the manner described in Chapter 10) in the decisions which are to be formulated.

The flow charts of Figures 11-3 through 11-6 describe the general process, variables, and examples of what must be considered in this information collection and forecasting step. (They also provide the crude outline for a management information system which would support the planning process.[2]) The circled "A" on the figures indicates that the answers to all of these questions are fed to a common point in the subsequent planning process.

Making Assumptions

Once the present status of the organization has been assessed and a forward look has been taken in terms of key indicators of the environment, some assumption must be made to serve as a basis for the planning activity

[2] Management information systems will be discussed in Chap. 17.

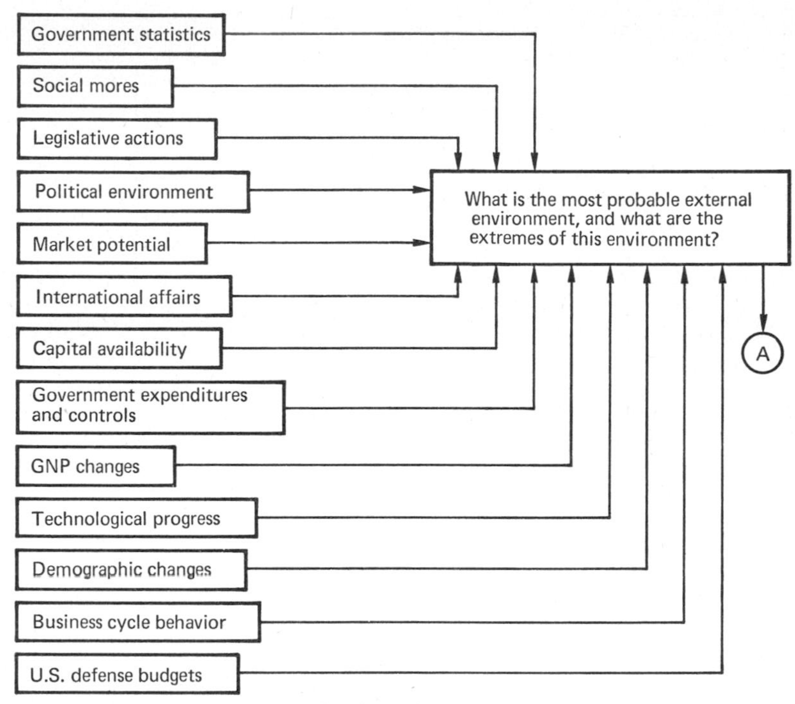

FIGURE 11-4 *Assessment of external environment.*

which will be carried on in the various divisions and centers of the organization. These assumptions will incorporate and extend the forecasts which have been made. Figure 11-7 depicts some of these assumptions. As with the previous figures, the circles "S" on this figure indicates that the information is fed to a common point in the subsequent planning process.

Establishment of Objectives

Based on the assumptions which have been made and the assessments of the current status of the organization, tentative objectives—specific

FIGURE 11-5 *Competitive environment assessment.*

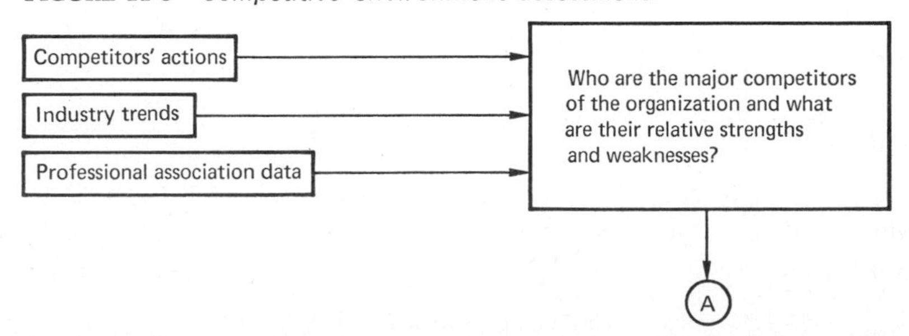

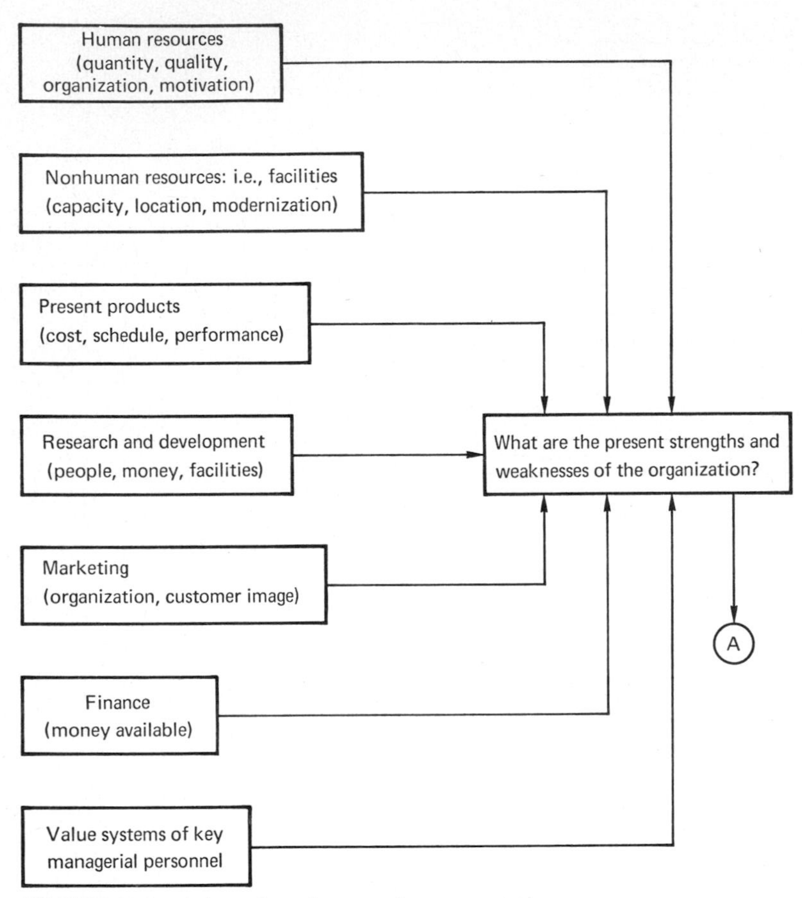

FIGURE 11-6 *Internal envi:onment assessment.*

measurable accomplishments—are specified by the key organizational executives in conjunction with the managers of the various subunits. These objectives provide the managers with a starting point for the development of plans. Since this process has already been described in Chapter 9, we shall devote little additional attention to it here.

The Development of Plans

The development of plans actually involves all of the previous steps. However, because the mechanics of putting plans together are so important, we consider this as a separate step.

In most large organizations, planning must take place over a long time period at various organizational levels. Indeed, planning is probably best accomplished as a continual, year-round process which becomes vocal at periodic reviews. No organization plans effectively if the planning process is compacted into a few days or weeks before a planning review.

FIGURE 11-7 *Making assumptions.*

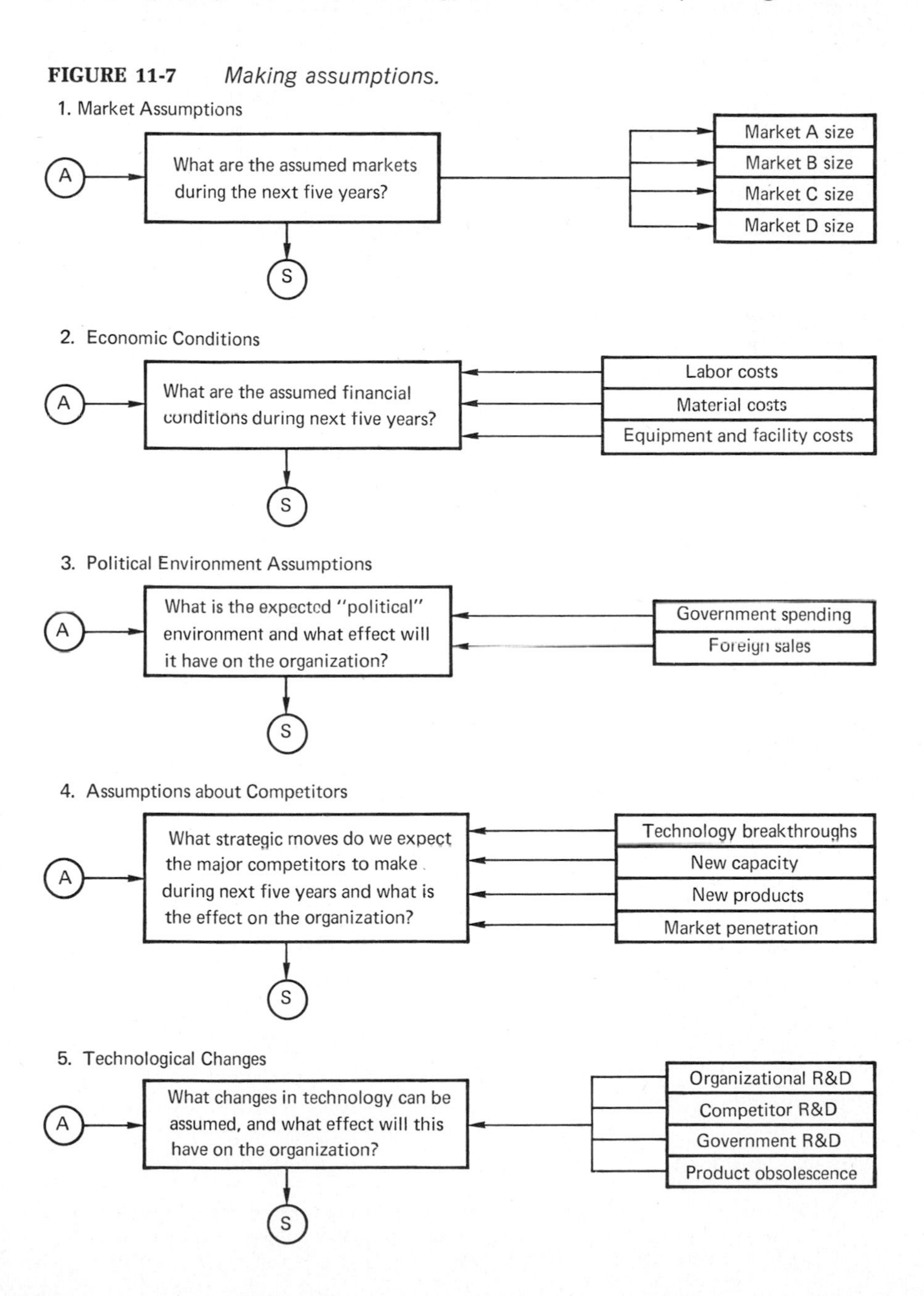

A TYPICAL PLANNING PROCESS IN BUSINESS A typical planning process for a business organization will serve to illustrate the salient features of the planning process and the interactions of the various organization elements in developing plans.

1. Promulgation of corporate objectives, planning guidelines, tentative goals and strategies, key assumptions, etc.
2. Preparation of one- and five-year plans at project level and other levels within major divisions
3. Review, analysis, and consolidation by corporate budget and analysis group
4. Financial review, analysis, and consolidation by corporate budget and analysis group
5. Review by corporate staff
6. Final review by corporate top management committee

The first phase of this process involves a flow from the corporate level of various guidelines and assumptions to be used at lower levels. This ensures some consistency and comparability among plans developed by different units at different levels. Although each subunit should not be compelled to adhere rigidly to each forecast made at the corporate level, deviations in the plan submitted should be justified.

The plans themselves, at all levels, typically involve two key parts. The first is the budget, which summarizes in financial terms what the unit expects to do and what resources will be required to do it. The second portion of the plan is verbal in nature, involving statements of economic, environmental, and operational assumptions, either those taken from corporate-level guidelines or others developed independently,[3] basic objectives of the unit (stated in terms of organizational outputs), and programs designed to accomplish those objectives.

At the major divisional level, each unit's plans are reviewed, consolidated, and summarized. If divisional management has participated in the preparation of plans on a continuing basis, this can be a meaningful review which leads to insights into the goals of the division and how they should be sought. If not, this stage may involve only the shuffling and forwarding of documents.

If an analytic capability—in terms of either professional analysts or analytically oriented managers—is available, a continuing dialogue between divisional unit planners should produce better plans. Thus, while alternative ways of achieving goals should be compared at the lowest

[3] If all guidelines, assumptions, and forecasts passed down from the corporate level are used in the plans, the verbal portion would probably simply summarize these to indicate that they form a part of the plan.

possible level, it is usually most practical to have plans made at one level and to have staff analysts, who may advise and participate in the analysis, review the conclusions at the divisional level. Of course, the level at which formal analysis is done depends heavily on the size of the organization.

If there is time to do so (and the process should be structured so that corporate goals do not become submerged in planning review deadlines), the key goals, assumptions, and forecasts should be circulated among the divisions of the company. This will permit one division to challenge another and to provide information to others. Moreover, it increases the likelihood of joint activities being undertaken by various divisions.

The budget and analysis group of the financial function consolidates financial data and depicts their impact on the overall corporate budget. If the plans are based on analysis and the objective comparison of alternative ways of achieving goals, it is possible for the budget and analysis group to challenge assumptions and forecasts and to "analyze the analysis." This is the most meaningful contribution which can be made at this level. The verbal portion of the plan may be reviewed by relevant corporate staff groups in parallel with the financial analysis. The marketing staff would focus on marketing issues in the assumption, objectives, and programs just as the production staff would give attention to issues within their domain.

Finally, corporate top management is called on to review the plans in the light of their previously determined objectives, assumptions, and guidelines. Their function is to check the plans and ensure broad compatibility with corporate goals and resources. Of course, at this level, judgment is supreme. The executive committee may ask that some plans be revised as a result of the insights gained concerning the lack of validity of the guidelines laid down, the need for greater interaction among divisional plans, the desire to see a wider range of alternatives considered, or any of a wide variety of other reasons.

THE MANAGEMENT OF PLANNING

Planning does not just happen—it must be motivated. An important part of motivation is the attitude that managers create and the climate that exists in the organization. *Since it is people who perform planning, the planning process must itself be structured and managed.*

The Planning Climate

Part of the planning process includes giving attention to the organizational climate necessary for creative planning. An effective way of enhancing the climate for innovating planning is to encourage widespread participation in planning at all levels. Individuals can be encouraged to submit their own ideas for planning in terms of product modification, new products, new organizational arrangements, new strategy for the organization, and so forth. Such ideas should have enough justification and documentation to enable analysis groups to perform an initial appraisal and see if the idea is worthy of further investigation.

An important factor in creating a suitable climate for planning is the stressing of the idea that change is normal and is to be expected as the organization faces a dynamic environment. Top executives must not only be change seekers; they must convince other people in the organization of the inevitability of change. Part of this convincing can be accomplished by drawing up organizational policies which reflect the official attitude toward change and planning, but there is more to it than just the paper work. In this respect Irwin has noted:

> *Top managers must be change seekers. Their leadership role is to provide a climate for rapid improvement toward excellence. The success their business achieves in the future will be in geometric proportion to their understanding of, planning for, dedication to, personal involvement in, and self-motivation toward the implementation of purposeful change. For many companies this demands a reorientation in the thinking of senior executives. It means honest commitment to a new concept. Insincerity or lip service will soon destroy confidence.*[4]

Of course, the only truly effective way of creating a proper climate for planning is to permeate the organization with planning, to demonstrate that it works, and to make use of it. When this pragmatic test of results has been passed, skeptics will be stilled and the organizational climate will be ripe for the institution of planning.

One of the key factors in permeating the organization with a successful planning activity is the personal involvement of key executives. Not only does such personal involvement in and of itself demonstrate the importance of the planning activity—thereby creating a climate in which planning can have payoff—but it ensures that actions are taken based on the strategy which is developed.

The best of plans—logical and complete as they may be—can end up on a shelf unused and unappreciated if individuals do not act to implement them. The function of developing organizational strategy is, in terms

[4] Patrick M. Irwin and Frank W. Langham, Jr., "The Change Seekers," *Harvard Business Review,* January–February, 1966, p. 83.

of basic ideas, a two-step process of *making and executing strategic decisions.* It is not sufficient to have good ideas which have been integrated into the corporate strategy—the strategy has to be integrated into the organizational activities as a continuing way of life. This idea is well put by Learned et al.: "A unique corporate strategy determined in relation to a concrete situation is never complete, even as a formulation until it is embodied in the organizational activities which reveal its viability and begin to affect its nature."[5]

Objective Analysis and Accountability

Ideally, top-level analysis of both the financial and the verbal aspects of plans should be conducted by objective analysts who have no vested interests in the various organizational divisions. This ensures that challenges to overly optimistic assumptions and forecasts will be made. Such objectivity can be acquired by having top-level planning analysts formally involved in the planning process and by having them rewarded in terms of measures related to planning rather than in terms of corporate or divisional results. Thus, while operating managers naturally tend to focus on achieving short-run results (often at the expense of the future), the corporate analyst's personal advancement should depend on his taking exactly the opposite view. If the future is sacrificed for today's results by a division, the corporate planning analyst should bear a portion of the blame. Similarly, if the future is well cared for, the corporate analyst should be rewarded.[6]

Another important aspect of accountability for planning lies in top-level *comparison of present plans with past plans and with actual results.* No one would expect results and past plans to be in perfect agreement, or even expect past plans and present plans for the same time period to agree. However, these deviations should be explained. If goals are changed, top management should be told the reason, or if forecasts have been radically changed on the basis of newly available information, this fact should be brought out into the open and its impact discussed.

The greatest benefit of a review of new plans, old plans, and actual results may well be in the implied accountability for planning. Too frequently the "ideal" planning system deteriorates into a valueless exercise when managers find that their payoff is solely on the basis of short-run

[5]Edmund P. Learned, C. Roland Christensen, Kenneth R. Andrews, and William D. Guth, *Business Policy,* Richard D. Irwin, Inc., Homewood, Ill., 1965, p. 619.

[6]Here we are discussing an organization which is well prepared for today. Our comments would be inappropriate in the case of an organization experiencing current financial problems. Indeed, in such a case, long-range planning should probably be forsaken, or at least reduced, until current problems are attended to and solved.

performance. If some accountability for planning is applied to managers as well as to analysts, the likely result is better planning.

Planning as Organized Activity

The senior executive has ultimate responsibility for planning the organization's future. Although all managers must be involved in planning, this manager will usually find it necessary to have a planning staff operating in a facilitating role.

The planning staff can be made responsible for "quarterbacking" the operation of the planning system—for updating the documents.

The planning function definitely *cannot* be delegated to a staff element. If this is done, planning will be neither implemented nor effective. Rather, the planning staff functions as a *focal* point for the planning effort going on throughout the organization.

OBJECTIVES OF A PLANNING SYSTEM One should begin the design of a planning system by specifying the objectives of the system. A reasonable set of objectives might be:

1. The system of planning that emerges must encompass planning for all organizational activities and all relevant time periods.
2. The initial point of departure must be a set of strategic objectives developed for the major echelons of the organization and for the organization as an entity.
3. The planning system must operate in such a way as to reduce the effect of "organizational parochialism" yet at the same time respect and protect the prerogatives of the key executives within the organization.
4. Responsibility and authority for planning must be delegated to the lowest possible level in the organization. Planning at a headquarters staff level *per se* should be of a *facilitative* nature and deal mainly with the development of the strategic plan and the facilitation of lower-echelon planning, *never as a substitute* for planning by those organizational elements ultimately responsible for the implementation of the plans at a future date.
5. The planning system must operate in such a way as to reduce the effect of "disciplinary parochialism," i.e., the tendency to overemphasize planning in a particular functional area to the neglect of other areas and the overall system.
6. The planning system must achieve total organizational involvement in the preparation of the organization's strategy and long-range objec-

tives and establish focal points of authority and responsibility for the execution of the strategic objectives.

7. The planning system must provide for a better understanding of the problems confronting the organization and the environment under which the organization's operations must be carried out in the future.
8. The planning system should provide each participant with an appreciation of the "other fellow's problems and points of view" and a recognition of the interfaces that must be brought about to have the organization present a posture of a *unitary whole.*
9. The strategic planning system must provide policy guidelines from which each suborganization can develop short-range operating objectives in support of approved overall organizational objectives.

Thus, the planning system should have several important kinds of results. *First,* it should result in a set of documents—a system of plans—which serve to guide the organization and to provide a standard to ensure that the efforts of various elements are in harmony rather than in conflict.

Secondly, and perhaps more importantly, the planning process itself serves as a valuable educational device. It provides the managers of the organization with a way of thinking which may be more significant to the achievement of organizational goals than are the plans themselves.

Thirdly, the planning system, at least in its initial stages of implementation, will serve to define the informational needs of the organization. It is unlikely that the first "run-through" of the planning cycle will take place without the identification of serious informational deficiencies as a result. The resolution of these deficiencies in the form of continuously prepared forecasts, the establishment of staff groups, etc., can be a valuable consequence of planning which could have occurred in no other way.

Initiation of a Planning System

Perhaps the greatest difficulty that will be encountered in the initiation of any planning system is getting the system in motion. Planning is a *continuous* and *dynamic* function. After the initial period of operation, the process should proceed much more smoothly than during the initial start-up period.

Often, the initial process is complicated by the lack of any statement of organizational objectives which form the basis of the plans. The development of a proper set of organizational objectives is dependent upon assessments of the external, competitive, and internal environments which occur as part of the planning process. The solution to this dilemma is the initial establishment of organizational goals, the definition of the organization's relationship to society, governments, other organizations, and other elements of its environment, and the definition of the relationship among the various units which make up the organization.

Figure 11-8 depicts a sequence of events and a flow of information which are used by one company to guide the implementation of a planning system. In subsequent sections each of the elements of Figure 11-8 will be discussed. The sections are numbered to correspond to the elements of that figure.

1. INITIAL STATEMENT OF STRATEGY As an initial step, a Planning Guide should be prepared which contains a preliminary statement of the organization's objectives, describes the anticipated future market and business opportunities, forecasts the desired market participation, defines constraints on resource availability, estimates resource allocations, forecasts financial performance, and presents planning strategies and responsibility as succinctly stated by Learned et al.:

> *A complete summary statement of strategy will in fact say less about what the word means than it does about the company involved. First, it will define products in terms more functional than literal, saying what they do rather than what they are made of. At the same time, it will designate clearly the markets and market segments for which products are now or will be designed, and the channels through which these markets will be reached. The means by which the operation is to be financed will be specified, as will the emphasis to be placed on safety of capital versus income return. Finally, the size and kind of organization which is to be the medium of achievement will be described. It is, of course, more important that the identification of strategy capture the present and projected character of the organization than that it elaborate the categories of purpose just cited.*[7]

2. STUDY GROUPS Two organizational characteristics often operate to the detriment of a total systems approach for the design and operation of a long-range planning system, viz., *organizational* and *disciplinary* parochialism. This parochialism is often reflected in the tendency to overemphasize planning for a particular organizational element and/or a particular functional area to the neglect of the overall organizational goals. If not checked, this parochialism will operate against the development of strategies for the organization as a unitary whole.

[7]Edmund P. Learned et al., *Business Policy, Text and Cases,* Richard D. Irwin, Inc., Homewood, Ill., 1969, pp. 16–17.

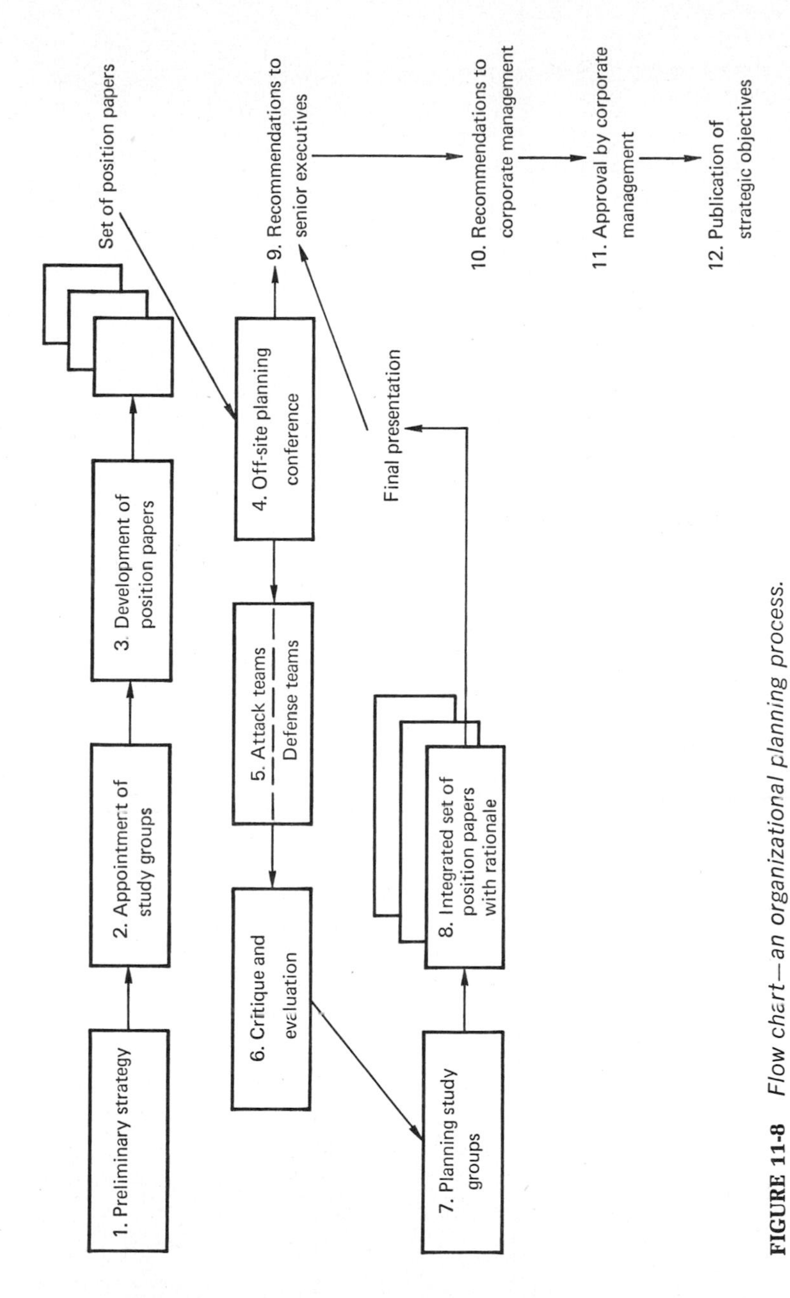

FIGURE 11-8 *Flow chart—an organizational planning process.*

The issue of *disciplinary* and *organizational* parochialism in planning can be alleviated by the use of ad hoc task forces called *planning study groups* established within the organization. The study group's membership is drawn from the various organizations having primary cognizance over the subject of the planning group's effort. The study groups might be active for a three- to four-month period during the time that the strategic objectives and plans are being developed or revised. Study group posture and membership may vary from one year to the next. However, effort should be made to provide for some executive continuity in these groups.

Figure 11-9 reflects a business organization planning system cast in terms of objectives. The summit of objectives within the business organization planning system is the sales and profit relative to investment (return on investment—ROI) derived from the current generation of products flowing through the company. Each product has its profit objective; business area objectives are sustained by emerging projects. Capability development objectives provide the base to support each higher order of objectives. As the product state-of-the-art advances, so must the state-of-the-art

FIGURE 11-9 *Business organization planning system.*

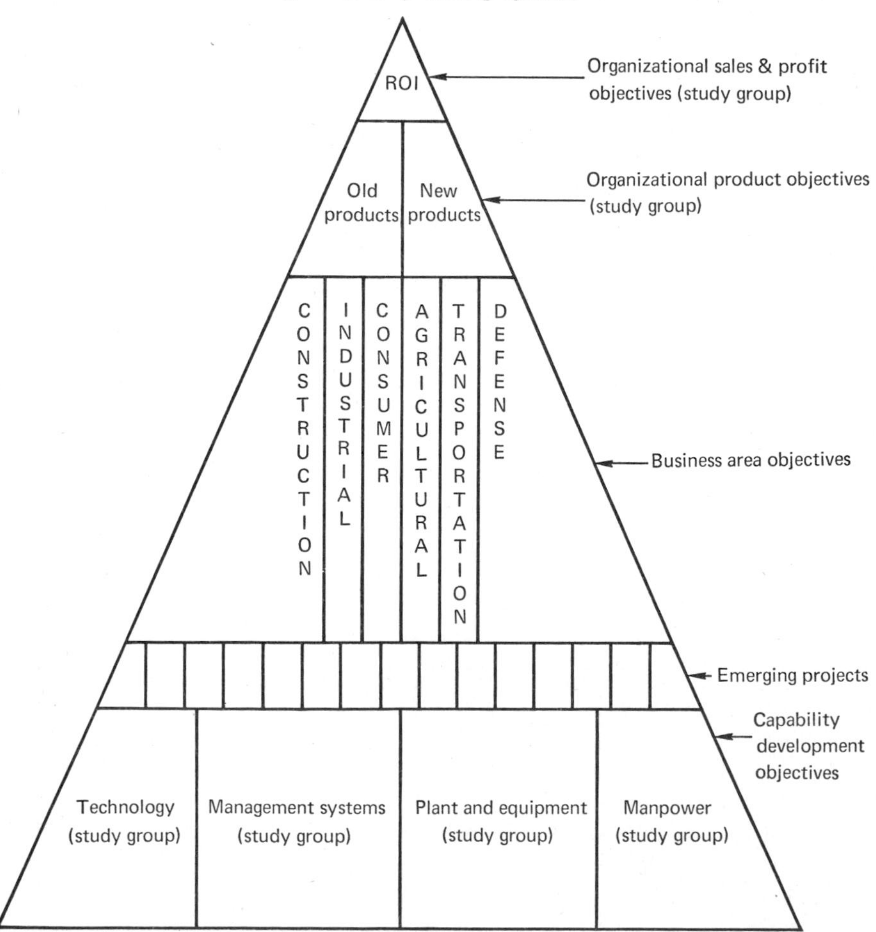

advance in the capability development objectives. The use of the study group approach to the development of capability development objectives will foster the involvement and participation of personnel at different levels in the organization and will contribute to the removal of much of the disciplinary and organizational parochialism which can develop in a planning system. Then, too, the use of the study groups implies built-in "checks and balances" by bringing about a greater degree of involvement.

Such a view of a business organization's objectives might result in study groups defined as:

Type of organizational objective	*Type of study group*
General development	Sales and profit
Product development	Products
Capability development:	
Technology	Technology capability development
Management systems	Management systems capability development
Plant and equipment	Manpower capability development
Manpower	

Each study group should have a minimum of five members appointed by the senior executive of the organization. The chairman of each group should be appointed from the organization having jurisdiction over the subject area to be studied.

3. POSITION PAPERS The final outcome of each study group's effort should be a position paper which evaluates the preliminary strategic objectives developed by the organization and recommends suitable changes based on the study group's deliberations. Since the organizational strategic objectives will ultimately result in a system of plans which will provide a sense of direction for lower organizational elements, it is important that the position papers provide a defensible rationale for the final strategic objectives that are developed. This requires that the study group effort *not* be conducted in a vacuum, but that each study group interface as necessary with other study groups to reach a coordinated position. Factual data, assumptions, logical sequence, and such matters must be coordinated across study group lines. Guidelines to be followed by the study groups to assure a successful position paper should also:

1. Describe and portray the anticipated environment within which the strategic objectives are to be accomplished and the potential influences on the objectives resulting from significant environmental changes.

2. Identify the organizational opportunities within an assumed environment for achieving the strategic objectives, coupled with an evaluation of the risks associated with the identified opportunities (e.g., danger of being tied to an obsolescent technology, lagging behind the state-of-the-art, competitive pressures, etc.).
3. Make a forecast of the resources (human and nonhuman) required to attain the objectives, considering such factors as: resource availability as a function of time, the point in cost and time beyond which a resource commitment cannot be stopped without a serious impact on the company, the effects of alternative resource commitment, the availability of resources, etc.

To accomplish the final product described above, the study groups should perform the following tasks:

1. Study the preliminary organizational objectives set forth to determine the suitability of those objectives to total organizational purposes.
2. Examine the sufficiency and realism of the inherent premises and forecast data used to establish group strategic objectives and prepare amended or additional premises and forecasts if required.
3. Define data base input requirements from other study groups.
4. Validate, revise, add to, and/or delete organizational subobjectives as appropriate.
5. Recommend planning responsibility assignments necessary to support total organizational objectives.
6. Develop supporting subobjectives required to accomplish organizational objectives and recommend planning responsibility for subobjectives.
7. Forecast the necessary total resource requirements needed to attain the recommended objectives.
8. Prepare a position paper supporting the study group's selection of organizational objectives.

For instance, one company outlined the required substance of the position papers to be prepared by new product study groups as follows:

SUMMARY FORM. EVALUATION OF BUSINESS AREAS.

I. *Description of Business and Function It Serves.*

(Keep it brief. Be sure to identify in terms of product, service, software, leasing, franchise, etc.)

II. *Establish the Need.*

1. Timing.
   When should activity begin? Indicate milestones. Assess each milestone on the basis of present best estimates, also indicate range of dates that could occur. What options are available to management based upon the range of dates?
2. Growth.
   a. Show value of *specific* market by year as a curve. Indicate risk or uncertainty by showing maximim growth that could *reasonably* happen and minimum. (Could business be replaced, superseded, never materialize?)
   b. Table enumerating *"facts"* and *"assumptions"* with respect to timing and growth. Consider market, economy, technological development, politics, and public policy. Assess the risk of the assumption changing.

III. *Market.*

Who is the customer(s)? How is the marketing done now? Is it the only way? Why? Is there an international market? Should it be included now or later? Why?

IV. *Competitive Situation.*

Who are the present competitors? What are their strengths and weaknesses. Comparisons on basis of research and development capability, engineering, marketing, contracting base, manufacturing capability, and image with customer; i.e., price, service, reliability, maintainability.

V. *Capability Analysis.*

Chart of capability required vs. capability available. Short recommendation of way to meet the capability requirements where there is a company lack. Estimate cost and time required. Highlight strengths and weaknesses!

VI. *Investment.*

Dollars invested/dollar sales, industry experience in time and amount invested before there is income. Front end load? Typical return on investment. Demand analysis; i.e., what is history of ups and downs in the particular business area? Range of swings, length of cycles, etc.

VII. *Political, Social, and Economic Considerations.*

Relate to company and its image. Image of the business.

VIII. *Strategy.*

Based upon the picture developed in the steps above, what actions and strategies are recommended?

NOTE: With respect to everything listed, in your report question whether it is fact or assumption and then with respect to each assumption evaluate the risk; i.e., what limits can be placed about the assumption? What alternatives are available based upon the swings that could occur to the assumption.

4. PLANNING CONFERENCE Once the study groups have developed a set of position papers supporting their choices, an off-site planning conference can be conducted to further refine the output of the planning system. The planning conference can best be described by reviewing the goals of such a conference:

1. A better understanding of the problems confronting the organization and the business environment under which that organization will have to operate for the next several years
2. An appreciation of the other fellow's problems and points of view and a recognition of the interactions between organizations within the company
3. Definitive statements of strategic objectives which provide a standard of what can be expected of the organization over the next several years, and of who will be responsible for planning these accomplishments
4. Policy statements from which each organization can develop short-term operating objectives in support of approved company objectives
5. The realization that the strategic objectives that are developed can be regarded as vital issues affecting the course of future actions in the company

The purpose of the planning conference is to seek agreement upon company objectives and subobjectives which define what business the organization wants to be in and what means it will employ to get into that business. Appointment of personnel to "attack and defense" teams for review of the study group position papers will depend on the incumbents' organizational position as well as the particular credentials each one can bring to the process.

5. ATTACK/DEFENSE TEAMS The purposes of the attack/defense teams should be to:

1. Evaluate and recommend to the planning study groups a set of organizational objectives which are to be attained during the period for which plans are to be developed.

2. Recommend to the senior executives those individuals who are to be assigned planning responsibility for the planning objectives.

Moreover, the attack/defense teams should:

1. Evaluate and review the subobjectives contained in the study group position paper and modify, add to, and/or delete those subobjectives as necessary to develop a set of appropriate and consistent long-range organizational objectives.
2. Determine unique resource requirements of the recommended company objectives and delineate and define critical areas needing further study.

The objectives and data contained within the position paper provide the impetus for the attack/defense team discussions. Consequently, each team member must thoroughly study the position papers prior to the planning conference. Members of the attack/defense team should remember that their primary duty is to prepare a set of organizational objectives which they, as members of management, can recommend, support, and implement. The objectives recommended by the study groups should be carefully reviewed and evaluated in the spirit of constructive criticism. Since the purpose is to seek a consensus of the team, the individual should be prepared to modify a minority view. In initially reviewing a particular position paper, each attack/defense team member is urged to consider:

1. Possible inconsistencies and/or conflicts in data and objectives within the subject position paper and with other position papers
2. The need for the delineation of truly significant organizational objectives, clearly stated and implementable without creating an overpowering of lower-echelon plans
3. The requirement for additional inputs from other attack/defense teams
4. The impact of stated or implied resource requirements versus resource availability
5. The need for joint discussion with other attack/defense teams

Attention should be devoted to the criteria for selecting people to participate in the attack-defense effort. One company used personnel from the study groups to comprise their defense teams and the following guidelines were used in selecting people for attack teams:

1. Include *both* "young Turks" and more experienced people.
2. Choose people known for candor and rigorous thinking.

3. Make team interdisciplinary.
4. Choose people who are not committed to a "new idea" which they themselves are trying to "sell" to the company.
5. Make attack team members of comparable rank to those on defense team.

6. CORPORATE REVIEW After the teams have critiqued and evaluated the position papers, these documents should be reviewed by the study groups and forwarded to senior company management. The result will be an integrated set of position papers with a stated rationale such that the strategies contained in the papers:

1. Can be related to higher-level objectives.
2. Are feasible, understandable, measurable, and verifiable.
3. Are consistent with budgetary allocations.
4. Embody a single result.

The recommendations for organizational strategy growing out of the process outlined in Figure 11-8 should do much to alleviate the effects of organizational parochialism and should provide an integrated set of organizational strategies for the planning time period. The process will also encourage a high degree of personal involvement and participation, thereby enhancing the acceptance of the top-level strategies within the total organization.

SUMMARY

An operational planning system consists of a system of plans, a planning process, and a management system for motivating and guiding the planning process.

A system of plans is an integrated set of documents which treat the various important areas of an organization's current and future activities. The system of plans includes a strategic plan—which specifies broad goals and strategies; a development plan—which focuses on new generations of products and services; and an operational plan—which focuses on current products, services, and markets. All of these high-level plans are supported by specific project plans in the system of plans. These project plans are the basic building blocks of the system.

A planning process is a formal phased set of activities which must be effected in order for planning to be accomplished and implemented. The process involves the establishment of general goals, the collection of information, the making of assumptions and forecasts, the establishment of specific objectives, and finally, the development of the products of the process—the plans.

However well designed are the system of plans and the planning process, little will happen if the planning is not "managed." This is particularly true in organizations where formal planning has not been the rule. Planning will occur and be implemented only if a favorable climate is developed. It must be performed in a framework which requires objective analysis and accountability for planning activities. Further, it will be effectively integrated as a continuing process in an organization only if it involves diverse interest groups within the organization and only if it has the continuing support of top management.

EXERCISES

1. What is a strategic plan? Describe its contents.
2. What is the organizational development plan? How is it related to: the strategic plan?; the divestment plan?; the diversification plan?; the R&D plan?
3. One of the things specified by an organizational development plan is the technique(s) which will be used to screen new investment opportunities. Why is it important that an organization have this specified: in advance?; at a high planning level?
4. What is the operations plan? How is it related to: project plans?; functional plans?
5. What are the two basic plans which directly support the strategic plan? Generally describe their salient differences. (Hint: See Figure 11-1.)
6. How are project plans related to: the operations plan?; the corporate development plan? Illustrate the sort of project plan which might be involved with each of these basic plans.
7. Which of the plans in the system of plans is the most specific? What details does it involve?
8. How do functional plans differ from project plans?
9. Relate strategic, development, operational, and project plans *(a)* schematically, *(b)* in illustrative relative-time horizons, *(c)* in terms of content.
10. What is the difference between a goal and an objective? How should they be related?
11. "To permit the explicit consideration of uncertainty in decision problems, forecasts should be made in terms of ranges of the environmental variables which have a reasonable likelihood of occurrence." Using the techniques and terminology of Chapter 10, describe in detail what this statement means.
12. "Our planning is done each September and October for the coming year." What are the possible difficulties with the process described by this statement of a business executive?
13. What is the relative role of line managers and staff planners in the planning process?

14. "Plans are very important. How they are developed is more or less irrelevant. The important thing is the product which planning produces." Critique this statement.
15. Why is the "planning climate" so important to good planning? What is the single best way to enhance the planning climate?
16. What can the manager do with regard to "change as a way of life" in order to enhance the climate for the planning process? Why is this so important?
17. "The best laid plans of mice and men. . . ." Complete this old adage and relate it to planning in organizations.
18. What is the advantage in having corporate-level analysts involved in the planning process? How should these analysts be rewarded? Why?
19. What is the role of past plans in a current planning process?
20. The military services have often assigned military project managers to weapons systems development projects for periods of three years. Then, they are reassigned elsewhere. What deficiencies can you see in such a procedure? What other policy might be followed? Can you see any defects in it?
21. Bob Jones has been project manager for an aerospace firm's Project X since its beginning nine years ago. The project has not been awarded to his firm; both his firm and another are performing studies on it. Recently, top management asked a consultant to review the project. When he discussed "strategy" with Jones, the latter responded by saying that his strategy was secret and would be kept so by him. Discuss this situation in terms of possible problems and difficulties which might exist. What would you do if you were the consultant?
22. The organization should have a hierarchy of goals and objectives. Explain.
23. What is the relationship between a planning process and the organization's information system?
24. One company began a formal planning process only to find that supporting information was lacking. As a result, formal planning was deferred and an information system development project was undertaken. Comment on this.
25. What is the most important reason for the use of study groups in a planning effort?
26. What kinds of areas can study groups focus on in a planning process?
27. What is the role of the attack-defense phase of the process described in Figure 11-8? What would you think would be a good way of defining the membership of the attack and defense teams? How might they be related to the study groups?
28. What are the three key elements of a successful operational planning system?
29. Distinguish between a "planning system" and a "system of plans."
30. Why is it necessary to "manage planning" as well as to define a system of plans and a formal planning process? What is meant by the "management of planning" in this context? How is the use of the term "management" in this context related to the "decide-do" model of management?

REFERENCES

Ackoff, R. L., *A Concept of Corporate Planning,* John Wiley & Sons, Inc., New York, 1970.

Bell, E. C., "Practical Long-range Planning: Case Histories Show How It's Actually Done," *Business Horizons,* December, 1968.

Brown, Harold, "Planning Our Military Forces," *Foreign Affairs,* January, 1967.

Cherikos, Thomas N., and A. C. R. Wheeler, "Concepts and Techniques of Educational Planning," *Review of Educational Research,* June, 1968.

Hartley, Harry J., *Educational Planning—Programming—Budgeting: A Systems Approach,* Prentice Hall, Inc., Englewood Cliffs, N.J., 1968.

Herrmann, Cyril C., "Systems Approach to City Planning," *Harvard Business Review,* September, 1966.

Novick, David, "Long-range Planning through Program Budgeting: A Better Way to Allocate Resources," *Business Horizons,* February, 1969.

Smalter, Donald J., "Influence of Department of Defense on Corporate Planning," *Management Technology,* December, 1964.

Steiner, G. A., *Top Management Planning,* The Macmillan Co. of Canada, Limited, Toronto, 1969.

12

planning methodologies

The various stages of the planning process require that decisions be made, forecasts of the future environment be developed, and the impact of these forecasts on the organization be assessed. In Chapter 10, the formal analysis of decision problems was considered. Although this process is an important part of planning, it is not the only methodology which the manager can draw on to aid in the planning process.

In this chapter, we shall introduce a number of other planning methodologies. In dealing with each of them, the reader should be aware that, like the formal framework for the analysis of decision problems, their utility is not limited to the planning function. Rather, they are widely applicable throughout the range of management activities.

DEVELOPING GOALS AND OBJECTIVES

We have already discussed the development of goals and objectives in Chapters 9 and 11. Here, we wish only to stress that *involvement* is particularly crucial to developing organizational goals—particularly if the organization is not business-oriented.

Goals are broad statements of direction. *Objectives* are specific desired accomplishments which can be measured in a specified time period. Objectives thereby support goals in that the achievement of broad goals will usually require the achievement of a number of consistent objectives.

For instance, a school district might establish the goal of "improving reading skills" in pupils. Among the associated objectives might be:

Sixth-grade pupils will be able to read and pronounce with 80 percent accuracy a list of words selected from the Stanford Achievement Test.
Sixty percent of eleventh-grade pupils will be able to score at least the county average on a standardized reading test.

But precisely how are such goals and objectives to be set? Clearly, goal development requires a variety of inputs from the organization's various clientele groups. For instance, the goals of a school district must be developed and agreed upon through a process of involvement of various community leaders and groups. Planning study groups can be set up to draft a statement of goals in the manner described in Chapter 11. These can then be reviewed and reused as necessary. The key point, however, is *involvement.*

Objectives are more specialized than goals, and therefore the process of developing them must require greater technical expertise. The basic questions to be answered are:

1. What *output indicators* are appropriate to what we wish to achieve?
2. What *success criteria* are to be applied?

For instance, are the measurements made by a standardized reading test appropriate indicators of attainment of the goal of improved reading? What does the community mean by "good reading"? How does it measure it? What constitutes success? If 60 percent of eleventh-graders can do it, will the community regard this as a step toward the achievement of broad reading goals?

Of course, the resolution of these questions involves discussion, negotiation, and consensus, as does the process of establishing goals. Here, however, the process is more parochial, since it will primarily involve educators who will first develop objectives and then report back to their clientele groups for their evaluations of how well the objectives suit the goals.

The combination of internal expertise and clientele evaluation is very critical from two standpoints:

1. It continues clientele involvement beyond the goal-setting stage.
2. It forces technicians (educators, in this case) to explain their objectives to nontechnicians.

In most fields, technicians have tended to develop jargon and "sacred cows" to which they sometimes adhere without realizing it. By forcing them to explain their objectives in English, the manager is nurturing a process which may be at least as helpful to the technician as it is to the nontechnician.

THE INVENTION OF ALTERNATIVES

Once objectives have been clearly stated, the question of "inventing" various ways of attaining those objectives becomes paramount. Again, this is a part of the planning process which is highly subjective and "people-dependent."

Brainstorming

One device which has been used successfully in one form or another in generating new alternatives is the technique called "brainstorming." In a brainstorming session, a group is called together and encouraged to discuss possible new alternatives—say, new products or fruitful areas for new products. "Out-of-the-blue" thinking aloud is encouraged. No critical analysis of ideas is permitted, although it is hoped that each "wild" idea may lead to another and perhaps eventually to a radical idea which has merit. The interaction of the individual members of the group is believed to have a stimulating effect. Often during later analysis, although most of the recorded ideas which have been put forth in the brainstorming session are discarded as impractical, a few ideas which merit study are uncovered. Some who have participated in apparently fruitless sessions have felt that, at the very least, the sessions served to open new avenues of thought in the minds of individuals who might later produce meritorious ideas.

Pseudosophisticates may consider the technique of brainstorming to be "old hat." Indeed, there is no scientific evidence that it really works. Nonetheless, the pragmatist will find that the basic idea is both useful and used (although the term "brainstorming" may not always be applied). A good illustration of its utility is the important role it plays in the operations of the Van Dyck Corporation, which specializes in devising new products for clients such as Olin Mathieson, J. C. Penney, and Textron. *Business Week* describes Van Dyck's brainstorming session as follows:

> *The staff bats around ideas, and scrawls them down on scraps of paper, which are tossed into a huge fishbowl. . . . Any idea goes in if it has aroused even a glimmer of response from the group. Later on, a two-man team—always one engineer and one industrial designer—cull out the most promising candidates.*[1]

FORECASTING

If an organization is concerned about the future environment in which it will operate—as it is when it devotes its attention to planning—it must assess the relative future influences of the various environmental factors

[1] *Business Week,* July 2, 1966, pp. 52–54.

upon the organization. To do this requires that *forecasts* (or predictions) of the future be made.

Explicit forecasts of the future are inherently uncertain, and yet so, too, are the guesses, conjectures, or hunches which so often are a part of any informal process which is used as an alternative to formal forecasting. Although there are many risks involved in forecasting, their implications, as far as long-range plans are concerned, can be minimized by (1) tempering the corporate long-term planning to the basic secular trends in the economy rather than to forecasts which reflect yearly fluctuations in the business cycle and (2) planning for fluctuations in the economy and long-term changes in the political, technological, social, and legislative climate.

The role of forecasting in long-range planning is to project future conditions and organizational performance as related to the future environment. Forecasting does not itself incorporate the function of altering policies and strategy to take advantage of or protect against future economic or social conditions. However, important management decisions are expected to result from the forecast. The function of forecasting is clear: given the present position of one organization, it seeks to determine what will be the effect of the environment on the long-term objectives of the firm.[2]

Specific forecast information includes a broad base of economic, political, and social phenomena derived from the environment in which the business is functioning and from records and individuals within the firm itself. Collectively, this intelligence provides the knowledge required to make assumptions and decisions about the organization's future posture. Application of these economic, political, and social data depends on the answer to the basic planning question: "What is it that we are trying to do?" The information used in the forecasting process may include, but is not necessarily limited to:

1. *External Information*
   Government statistics (employment, housing, etc.)
   Trade association data
   Information on social mores (fashion and styles)
   Legislative actions
   Political environment
   Market potential
   International affairs
   Capital availability

[2] Abraham Lincoln said, "If we could first know where we are and wither we are tending, we could better judge what to do and how to do it."

Government expenditures and controls
Gross national product changes
Demographic changes
Technological progress
Industry trends
Competitors' actions
Business-cycle behavior
Government fiscal policy

2. *Internal Information*
Organizational posture
Corporate objectives
Present competitive position
Availability of human and nonhuman resources
Organizational history
Executive expectations

Various kinds of forecasts are commonly derived from this sort of input elements: among the most important are time-series forecasts, forecasts using statistical economic indicators, and technological forecasts.

Time Series Forecasts

A "time series" is a sequence of values of some quantity corresponding to particular points or intervals of time. For example, production or price data arranged chronologically represent a time series.

Time-series forecasts seek to extend a historical data series into the future as a prediction of future values of the quantity in question. The methods used to do so range from simple extrapolations done "by eye" on a graph of the series to sophisticated statistical techniques which seek to assess long-term trends and seasonal and cyclical effects. As noted earlier, long-range planning must be based primarily on secular trends—the long-run growth or decline pattern of the series. However, it is often necessary to evaluate seasonal and cyclical factors in order to measure the trend.

The basic models used in statistical forecasting relate to these salient elements. For example, an additive model might be used:

$$O = T + S + C + R$$

where $T =$ trend
$S =$ seasonal effect
$C =$ cyclical effect
$R =$ random (or unexplained) variation
$O =$ basic time-series historical data

Thus, the assumption of the model is that each point in the series occurs as a result of four factors (*T,S,C,* and *R*) combining in additive fashion.[3] By using the known quantities in the series (the *O*'s), we can estimate the other quantities. Then, having estimated these quantities, we can in turn forecast future values of the *O*'s.

Forecasting Statistical Economic Indicators

Whereas the time-series forecasts assume that the past can be projected into the future, the use of economic indicators in forecasting assumes that combinations of things happening today can be used to predict things which will happen in the future. For example, a number of "leading indicators" are used by economists in the belief that they consistently precede changes in business activity. Thus, they serve as a basis for predicting such changes. For example, such indicators as constant-dollar GNP, an index of industrial production, and sales of retail stores may be viewed (in the context of their historical time series) as indicative of the level of *future* business activity.

The relationships used may be either qualitative or quantitative. Statistical techniques are often used to estimate the numerical relationships involved. The directional changes in the indicators, taken as a group, may be used to predict qualitative directional changes in business activity. Business periodicals often report such happenings in such terms as, "All but two of the leading indicators turned down last month."

Technological Forecasts

Technological forecasting has been described in Chapter 7 as one of the techniques which can be used for forecasting technological needs. Technological forecasts aim at establishing timetables for the meshing of the need for products and services and the technological feasibility of producing them. Such forecasts are of obvious importance to long-range planning since the development of new products is a result of a myriad of decisions involving capital investments, research expenditures, market analyses, etc., which must be made sequentially over a long period of time.

[3] Of course, there is nothing sacred about an additive model. Many other forms have been used to relate these four basic factors. See National Industrial Conference Board, *Forecasting in Industry,* Studies in Business Policy, no. 77; and National Industrial Conference Board, *Forecasting Sales,* Studies in Business Policy, no. 106. A basic reference is Julius Shiskin, *Electronic Computers and Business Indicators,* National Bureau of Economic Research Occasional Paper 57, 1957.

Forecasting Techniques

Two techniques which can be useful in forecasting are sufficiently important and pervasive to warrant attention in this nontechnical treatment of forecasting. The two approaches—least squares analysis and the Delphi technique—illustrate the wide range of possible approaches to forecasting, since one is a mathematical approach and the other is subjective in nature.

LEAST SQUARES ANALYSIS The technique of "least squares" is applicable to situations involving two or more quantities—one of which is to be predicted or forecast. Suppose, for example, that the manager or planner wishes to forecast a quantity symbolized as *"y"* based on his knowledge of one symbolized as *"x"*. Further, suppose that he has data for *x* and *y* from other time periods or other organizations. *One* such data point can be thought of as an *(x,y)* pair—a value of *x* which occurred in conjunction with a value of *y*.

Such data can be displayed in a *scatter diagram,* as in Figure 12-1. Obviously, from those data, the two quantities are related—however imperfectly—since high values of *x* tend to be associated with high values of *y* and low *x* with low *y*. Two quantities having this property are said to have a degree of *correlation.*

The term "correlation" is a part of almost everyone's vocabulary. Like many other terms used in management analysis, its precise meaning is somewhat different from its more nebulous meaning in everyday usage. "Correlation," as used here, is the degree of *linear dependence* existing

FIGURE 12-1 *Scatter diagram.*

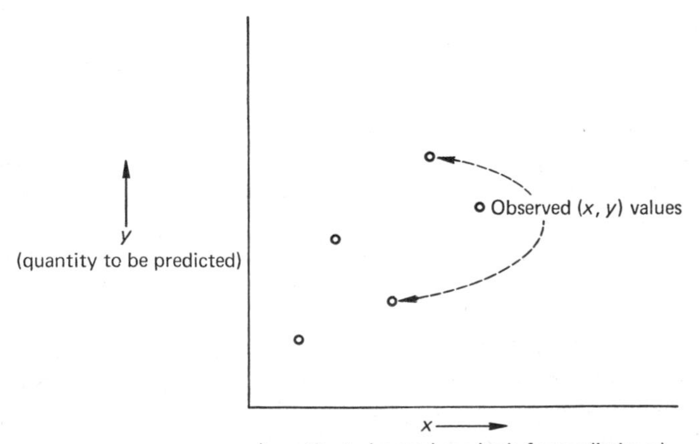

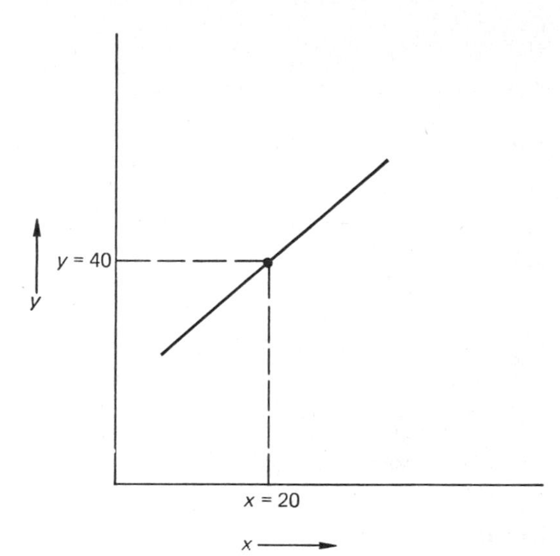

FIGURE 12-2 *A predictive line.*

between the variables in question, i.e., the degree to which the points on the scatter diagram tend to lie in a straight line.

In this situation our objective is to predict y if given a knowledge of x. In other words, if we know x to be 20, we would like to have a straight-line relationship estimated on the basis of past experience to use to predict y. Such a relationship is shown in Figure 12-2, where the prediction of y is 40.

The question is, How do we get such a straight line? The answer is given by statisticians in the principle of "least squares." This principle simply says, "Try all of the possible straight lines which could be fit through the data points of the scatter diagram. Choose as a basis for prediction that line which has the least sum of squared errors."

Two of the many possible straight lines are shown in Figure 12-3. The errors associated with using line 1 for predictive purposes are shown as solid vertical distances. The one marked "error," for instance, is the error that would be made if the line were used as a means of getting a prediction of y from the value of x which is associated with that (x,y) data point. This value of x is indicated to be $x = 20$ in the figure. So, too, are each of the other vertical lines the errors associated with using the line as a means for obtaining a prediction of y from x.

The errors associated with line 2 are shown in dotted vertical lines. One can intuitively see that line 1 is better than line 2. This would be formally determined by taking all of the errors for a given line, squaring each so that positive and negative errors do not cancel one another, and

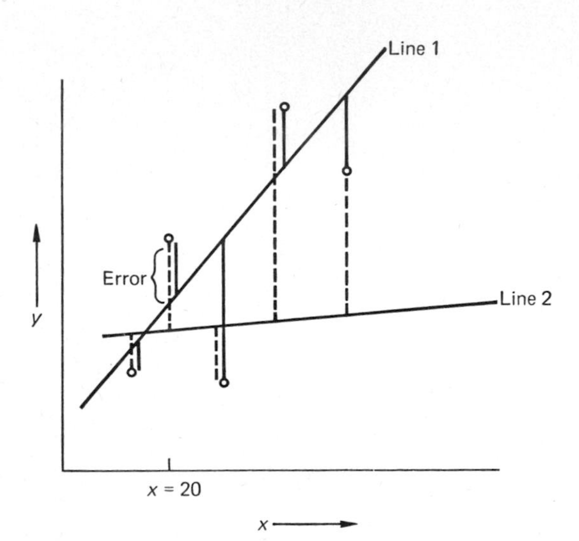

FIGURE 12-3 *Two possible predictive lines on scatter diagram.*

finding that line which has the *smallest sum of squared errors.* Hence the name "least squares."

Fortunately, all of this has been done for us, and the results are summarized in a few simple formulas. If y^* is the quantity to be predicted from a knowledge of x, and $\bar{x}$ and $\bar{y}$ are the average values of x and y in the scatter diagram, the equation shown in Figure 12-4 is the equation of the best-prediction line—the "least squares line."

FIGURE 12-4 *Least squares predictive line.*

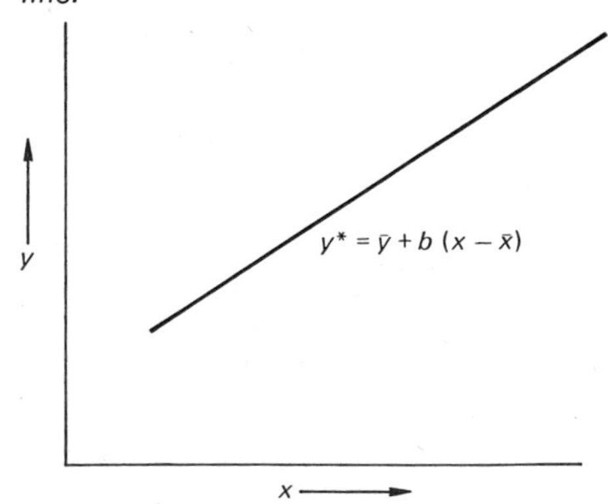

y^* – Value to be predicted
$\bar{y}$ – Average of y's in scatter diagram
x – Known value of x
$\bar{x}$ – Average of x's in scatter diagram
b – Parameter of prediction line

To determine b, we use the equation[4]

$$b = \frac{\text{Sum}(xy) - \dfrac{(\text{Sum } x \text{ Sum } y)}{\text{\# of observations}}}{\text{Sum}(x^2) - \dfrac{(\text{Sum } x)^2}{\text{\# of observations}}}$$

A numerical illustration of this is shown in Table 12-1. Only four data points are available for analysis. Table 12-2 shows the partial calculations which are necessary for the determination of b using the above formula. The result is a b equal to 0.004.

The resulting predictive equation is

$$y^* = 10.25 + 0.004\,(x - 1{,}725)$$

where the 10.25 and 1,725 are the averages taken from Table 12-2.

If the manager wished to use this equation to predict the value of y which would result in a situation where x was 2,000, he would calculate:

$$y^* = 10.25 + 0.004\,(2{,}000 - 1{,}725) = 11.35$$

DELPHI FORECASTS Opinion-based forecasting is often valuable to the planner and manager. For instance, McGraw-Hill surveys of businessmen's plans for making expenditures on plant and equipment are published regularly in *Business Week* and are widely used as forecasting bases. Other forecasts are made by obtaining the collective judgment of groups. However, forecasts based on composites of group opinions have often been found to represent *compromises* rather than *consensuses* since such things as the prestige or personality of individuals can inordinately influence the judgment of a group.

Delphi is a technique which enables a group of experts to contribute to one another's understanding and to refine their opinions as a result of

[4] For further details on the development of this result, see any basic statistics text—e.g., W. S. Peters and G. W. Summers, *Statistical Analysis for Business Decisions*, Prentice-Hall, Inc., Englewood Cliffs, N.J., 1968.

TABLE 12-1

Predictor x	*To be predicted* y
2,000	10
3,000	16
1,000	8
900	7

TABLE 12-2

$$\text{Sum } x = 2{,}000 + 3{,}000 + 1{,}000 + 900 = 6{,}900$$
$$\text{Sum } y = 10 + 16 + 8 + 7 = 41$$

$$\bar{x} = \text{Average } x = \frac{\text{Sum } X}{\#\text{ of observations}} = \frac{6{,}900}{4} = 1{,}725$$

$$\bar{y} = \text{Average } y = \frac{\text{Sum } Y}{4} = \frac{41}{4} = 10.25$$

$$\text{Sum }(x)\text{ Sum }(y) = 6{,}900(41) - 282{,}900$$
$$\text{Sum }(xy) = 200(10) + 3{,}000(16) + 1{,}000(8) + 900(7) = 32{,}300$$
$$\text{Sum }(x^2) = (2{,}000)^2 + (3{,}000)^2 + (1{,}000)^2 + (900)^2 = 14{,}810{,}000$$
$$(\text{Sum } x)^2 = (6{,}900)^2 = 47{,}610{,}000$$

$$B = \frac{82{,}300 - (282{,}900/4)}{14{,}810{,}000 - (47{,}610{,}000/4)} = 0.004$$

interaction with other experts.[5] Delphi physically separates the experts, however, so that some individuals and their rationales do not become submerged in the overt activities of a group.

Delphi involves a series of steps:

1. Predictions by each expert
2. Clarification by a "neutral" investigator
3. Requestioning of experts combined with feedback from other experts

The process of requestioning is designed to eliminate misinterpretation and to bring to the attention of each expert elements of knowledge which may not be known to all.

A typical Delphi session has been described as follows:

Round 1: In the first round of questioning each expert is asked to list developments that he thinks will occur in his field, say, within the next century. An expert in eugenics might predict that man will be able to control heredity to a large extent within the next 100 years. Another expert might predict that automated highways will be common in use long before then. The investigators then edit these predictions and prepare a questionnaire for the second round.

Round 2: Each expert is given a questionnaire which asks him to estimate certain numerical quantities—say, the date by which we shall have the chemical control over hereditary defects through

[5] *Business Week* reports on Delphi and derivations of it used by TRW, Inc., in "New Products: Setting a Timetable," May 27, 1967, pp. 52–56.

molecular engineering, or the probability that all highways in the United States will be automated by the year 1990. By asking for such simple numerical estimates, the investigators can easily summarize the responses in statistical fashion.

In one actual experiment, the median of the responses to the question about molecular biology mentioned above was the year 1993. Half the answers were between 1982 and 2033; a quarter of the experts saw hope of achieving control before 1982, and the remaining quarter of responses indicated that we would not achieve such control until sometime after 2033, or possibly never. In other words, the range of responses could be conveniently divided into a lower quartile range of 25% of the answers, an upper quartile range of 25% of the answers, and an inter-quartile range of 50% of the answers. The inter-quartile range is taken as indicating the consensus to date. Of course, these percentages are chosen arbitrarily, one could just as well consider the lower 10%, the middle 80%, and the upper 10%. But, the percentages suggested here have worked out well in many experiments.

Round 3: The digested information from the second round of questioning is then fed back to each of the experts and they are all asked to scrutinize their first responses. In addition if an expert's first response fell outside the middle (inter-quartile) range of responses, it has been customary to ask him to express his reasons for his "extreme" opinion.

Experience has shown that obtaining reasons from respondents whose opinions fall outside the inter-quartile range achieves two things:

1. *It permits the investigator to refine his question. If he takes into consideration some of the reasons given for the more extreme responses, he can reword or modify the question so that the range of response will narrow itself down.*
2. *It permits him to state the minority opinions, and to feed this information back to the whole group on a fourth round of questioning.*
   Almost invariably, a number of the experts change their estimates after rethinking the question, and the consensus narrows as a result.

Round 4: Once the investigator has digested the new estimates and calculated the new consensus on the year of achievement, he can approach his experts with a questionnaire for the fourth round.

If it seems desirable to do so, the questioner can manipulate the results of this round and go on to a fifth or even a sixth round.[6]

PROJECT SELECTION

After alternatives have been invented and forecasts have been made, some alternatives must be selected for further study. This area of *project selection* has a wide literature.[7]

The basic problem can be handled using a decision theoretic approach such as that of Chapter 10. However, *the involvement of people in the evaluation is of as great an importance as is the criterion to be applied.*

Here we present an approach used by one company for project selection by its new product task forces. This simple, yet attractive, scheme was designed to promote objective evaluation and to achieve a degree of comparability across the various task forces which were evaluating new business opportunities.

PROJECT EVALUATION PROCEDURE

1. *Task force leaders* meet and develop a list of criteria[8] such as:
   a. Current size of market
   b. Potential for growth of market
   c. Competitive position of company in the market
   d. Investment required
   e. Compatibility with company expertise
2. *Task force leaders* agree on allocation of 100 "importance points" to these various criteria.
3. *Individual task forces* evaluate each of their potential business opportunities by rating them on a 3 or 5 point scale for each criterion. For instance, the criterion "potential for growth of market" might have a scale:
   a. Certain to grow dramatically
   b. Likely to grow substantially
   c. Likely to grow modestly
   d. Stable
   e. Declining market

[6] From "The Basic Delphi Method," *Harvard Business Review,* May–June, 1969, pp. 80–82. Used by permission.

[7] For instance, see B. V. Dean and M. J. Nishry, "Scoring and Profitability Models for Evaluating and Selecting Engineering Projects," *Operations Research,* vol. 13, no. 4, July–August, 1965, pp. 550–569.

[8] Note the similarity of these criteria to the required contents of the task force position papers given on pages 286–288.

Each business opportunity would be rated at *one* of these scale positions for each criterion.

4. Overall, numerical ratings can then be developed by multiplying importance weighting of criterion by scale factor (10, 8, 6, 4, 2) for scale position and summing.

Illustrative of this procedure is the business opportunity evaluated in Table 12-3.

TABLE 12-3

Criterion (importance weight)	*(10) Very Good*	*(8) Good*	*(6) Neutral*	*(4) Bad*	*(2) Very Bad*	*Score*
Current size of market (30)			X			180
Potential for market growth (20)	X					200
Competitive position (20)		X				160
Investment required (10)	X					100
Compatibility with company expertise (20)				X		80
	Total score					720

The "total score" represents an index which can be compared across product opportunities. So long as too great a significance is not read into these numerical scores,[9] they can be useful devices for project selection.

PROJECT PLANNING

One of the planning techniques most frequently used in the development of project plans is *network analysis.* Although this approach is widely useful beyond project planning, it is introduced here in that context. Subsequently, we shall focus more specifically on its use in the control function of management.

The Project Network

Network plans are developed by first studying the project to determine the approach, methods, and technology to be used and then breaking the

[9]See William R. King, *Quantitative Analysis for Marketing Management,* McGraw-Hill Book Company, New York, 1967, pp. 122–132, for a discussion of the assumptions and limitation of such techniques.

project down into elements for planning and scheduling purposes. The elements of a project can be classified as follows:

Project objectives. These are the goals to be accomplished during the course of the project. In most cases, the project objectives are specified before the network plan is prepared; the plan merely prescribes the course to be followed in achieving the objectives.

Activities, tasks, jobs, or work phases. These elements identify and describe the work to be performed in accomplishing the project objectives. They normally utilize time and other resources.

Events or milestones. These are points of significant accomplishment—the start or completion of tasks and jobs, the attainment of objectives, the completion of management reviews and approvals, etc. They are convenient points at which to report status or measure and evaluate progress.

After the elements of the project have been determined, they are arranged in the sequence preferred for their accomplishment. This is a process of synthesis that must consider the technological aspects of the activities and tasks, their relationships to one another and to the objectives, and the environment in which they will be performed.

A network is used to reflect these factors in portraying the sequence in which the project elements will be accomplished and the various interrelationships of the elements. Networks are composed of events, which are represented by points interconnected by directed lines (lines with arrows) which represent activities. Technical constraints are also represented as directed lines. Elements of the network correspond to elements of the project as follows: Points in the network represent the accomplishment of specific project objectives, called "events" and "milestones"; the lines between the points represent project "activities," with the direction of the line indicating a precedence or sequential relationship; and other directed lines (broken) describe constraints.

Activities are the jobs and tasks, including administrative tasks, that must be performed to accomplish the project objectives; activities require time and utilize resources. The length of the line representing an activity has no significance. The direction of the line, however, indicates the flow of time in performing the activity.

Events are usually represented by circles or squares. Numbers or letters are inserted in these circles and squares to identify the events and the activity that connects two events. Events represent particular points or instances in time, so they do not consume resources; the resources to accomplish an event are used by the activities leading up to it.

Figure 12-5 shows events #1 and #2 and a single activity which begins with event #1 and ends with event #2. Event #2 also represents

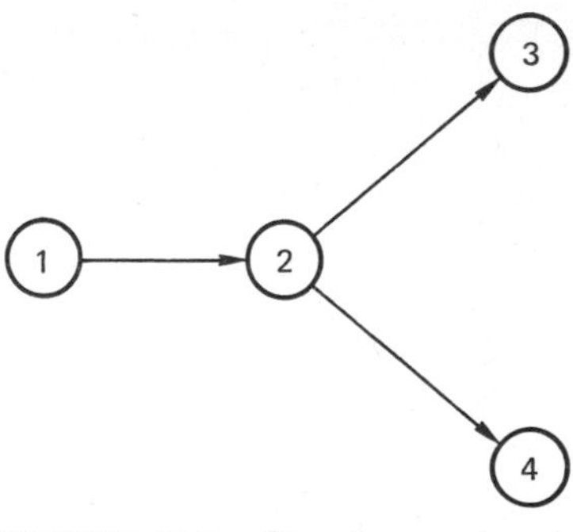

FIGURE 12-5 *Simple network configuration.*

the beginning of two other activities which culminate with events #3 and #4, respectively. The activities in such a network are conveniently labeled by the numbers of the events. For instance, the leftmost activity would be designated activity 1-2, etc.

Constraints in network plans represent precedence relationships resulting from natural or physical restrictions, administrative policies and procedures, or management prerogatives, and they serve to identify activities and events uniquely. Constraints, like activities, are represented in a network plan by directed lines. However, constraints indicate precedence only; they do not require resources and normally do not require time. Those constraints which require neither time nor resources are represented by broken directed lines, which are often referred to as "dummy" activities.

The network plan is constructed by drawing directed lines and circles in the sequence in which the activities and events are to be accomplished. The network begins with an event called the *origin,* which usually represents the start of the projcct and from which lines are drawn to represent activities. These lines terminate with an arrow and a circle representing an event, which may be the completion of a project element or an activity. All activities that are to be performed next are then added to the network plan by drawing directed lines from their respective predecessor event. For example, if activities #2 and #3 are to be simultaneously performed on completion of activity #1, Figure 12-5 describes their precedence relationships.

To progress from one event to another in a network model requires that an activity be completed. As each activity is added to a developing network, its relationship to other activities is determined by posing the following questions:

What activities must be completed before this activity can start? Activities that must be completed first are *predecessor* activities.

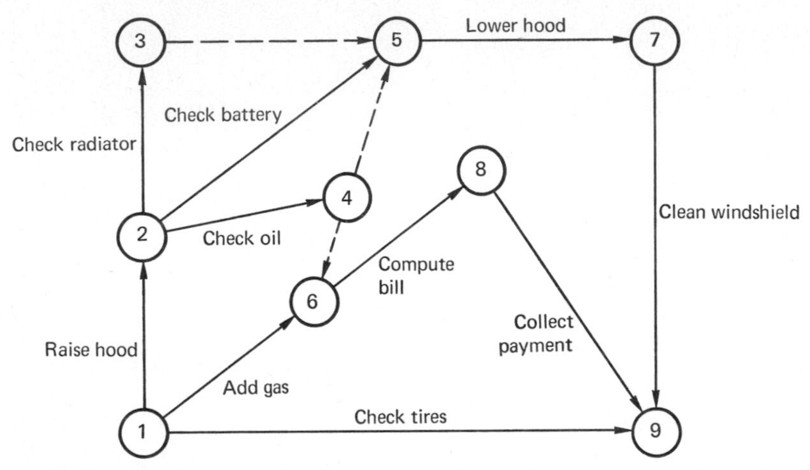

FIGURE 12-6 *Network plan for auto service project.*

What activities can start after this activity is completed? Activities that can start after are *successor* activities.
What activities can be performed at the same time as this activity? These activities are *concurrent,* or parallel, activities.

To illustrate the preparation of a network plan, let us consider as a project the servicing of an automobile at a service station.[10] This example will be slightly exaggerated in order to emphasize the interrelationships between project activities that must be considered. The project situation is described as follows:

> *Automobiles arrive at a service station for gasoline. Services provided by the station include cleaning the windshield and checking the tires, battery, oil, and radiator. Sufficient personnel are available to perform all service simultaneously. The windshield cannot be cleaned while the hood is raised. Customers are charged only for gasoline and oils.*

Figure 12-6 shows the network plan. Events 1 and 9 are the origin and terminal events, respectively representing the start and completion of service. Three constraints, or dummy activities, are used to sequence the activities properly.

The constraint between events 3 and 5, denoted as activity 3-5, is used so that the activities "check radiator" and "check battery" will not

[10] This illustration is adapted from David I. Cleland and William R. King, *Systems Analysis and Project Management,* McGraw-Hill Book Company, New York, 1968.

have common predecessor and successor events. The dummy activity 4-5 is used for the same reason. The constraining 4-6 is used to indicate that the activity of computing the bill cannot start until the activities "check oil" and "add gas" have been completed.

Thus, the network vividly displays the project's relevant activities, events, and constraints for the planner. The planning value of such a simple model is clear. For instance, the same work group would probably not be assigned to the "raise hood," "check tires," and "add gas" activities in this auto service illustration, since these activities *can* be performed simultaneously. The most efficient way to perform them is in parallel with a different work group assigned to each. On the other hand, it would be reasonable to assign the "compute bill" and "collect payment" activities to the same work group, since they occur in sequence and since the "learning" which has taken place in the former might well lead to more efficient accomplishment of the latter.

Activity Time Assessment

For purposes of detailed planning and control, the manager should have estimates of the *time required to complete each of the activities.* Such estimates can be obtained from the people who are charged with responsibility for the various activities. Although there will be a large degree of uncertainty associated with most of them, such estimates are generally infinitely superior to no time estimates at all.

The manager wishes to have time estimates for the various activities so that he can assess those activities in which delays will be critical to the total project. If the activities are defined so that they represent short, controllable tasks, and if one of these critical activities encounters difficulties, some control action may be taken to hasten its completion. These activities which should be carefully monitored are formally termed "critical activities"—i.e., *those in which a delay will cause a delay in the completion of the entire project.* Identification of these activities is accomplished by determining the *critical path—the path through the network which has the longest duration.*

Figure 12-7 is a hypothetical project network in which a time estimate has been made for each activity.

The longest path (the critical path) through the network is the lower one (through events 1, 2, 4, 5, 6, and 7), which requires an estimated total of 17 weeks to proceed from event 1 to event 7. The middle path (through events 1, 2, 6, and 7) requires 16 weeks, and the upper path (through events 1, 2, 3, 6, and 7) requires an estimated total of 15 weeks to proceed from event 1 to event 7. In this simple network there are only these three

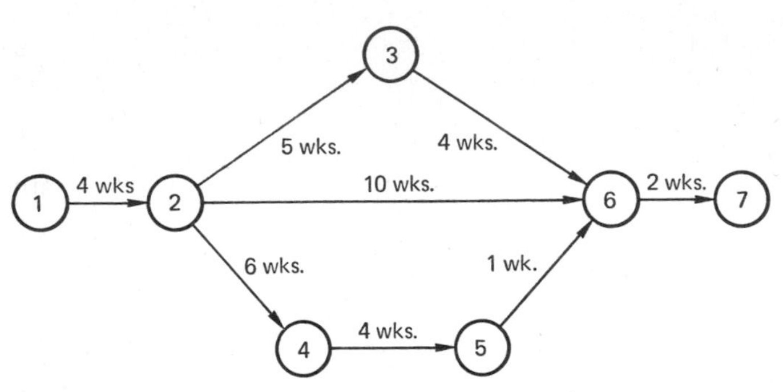

FIGURE 12-7 *Project network with time estimates.*

possible paths from event 1 to event 7. The lower path is the most time-consuming one, so it is the critical path. *Critical activities are those which are on the critical path.*

It is easy to see that a delay on any activity on the critical path must necessarily delay the completion of the project. If all activities except the critical activity between events 4 and 5 actually took exactly the estimated time and this activity actually required 5 rather than 4 weeks to complete, event 7 would not be accomplished until 18 weeks after event 1. However, if an activity which is not on the critical path is delayed, the completion event (7) may not be delayed. If all activities go according to schedule except the one between events 3 and 6, for instance, and that one slips by one week, the final event (7) will still be completed 17 weeks after event 1 (even though the upper path requires 16 weeks instead of 15). This is because the *final event cannot be reached until all preceding activities are completed, and this cannot take place any sooner than the time required for the longest path through the network*—the critical path.

A second important feature of the network (in addition to the critical path) is the positive *slack* associated with the noncritical paths. "Slack" is the extra time available along the path, i.e., the extra time the path could require before its duration would become equal to that of the critical path and, hence, before it would become an alternate critical path. The upper path in Figure 12-7 has two weeks of slack and the middle path has one week. (These periods are determined by subtracting the estimated time required for the path from the time required for the critical path.)

The manager may use the network and critical path for a number of purposes, primary among which are planning, scheduling, and control. We have already noted how certain initial planning may be accomplished by using only the basic network. Additional planning may involve the reallocation of resources from activities which are not on the critical path to critical activities. Of course, this presumes that the time estimates are valid (or at least of the same degree of inaccuracy) and that it is possible to hasten the completion of activities on the critical path by expending greater resources on them. Clearly, if this were so, it would be advantageous to lengthen the duration of noncritical paths by reducing the resources being expended on activities on these paths and to reassign those resources to critical activities. The resulting shortening of the critical path would produce a gain (in terms of a quicker project completion date), and no loss would be incurred (in terms of the completion date) as long as the increased duration of any noncritical path was not greater than its slack.

After such initial reallocations have been made, scheduling simply involves the establishment of beginning and completion dates for the various activities as indicated by the revised activity duration estimates (revised after the initial resource allocations). In subsequent chapters, we shall delve more deeply into this and other "control" uses of network planning and some of the more sophisticated network techniques which have been developed.

SUMMARY

A variety of methodologies—formal and informal, subjective and objective—may be applied to the planning process. Among those which warrant attention are methods for the development of goals and objectives, methods for the invention of alternatives, methods for forecasting, and methods for project planning.

The key element for ensuring success in the development of goals and objectives is involvement. Clientele must be involved in both phases. In the development of goals—broad statements of direction—clientele play a central role. In the development of objectives—specific accomplishments desired in a specified time frame—internal technicians play a primary role with clientele representatives serving as reviewers.

Brainstorming is one useful approach to the much-neglected area of the invention of alternatives. Too often, planning processes go on under the implicit assumption that alternatives are specified and well understood by all. This often leads to the omission from consideration of potentially good alternatives. A process such as brainstorming reduces the likelihood that good alternatives will be overlooked by planners.

Forecasts of the future environment are an essential part of planning. Here, time-series forecasts, statistical economic indicator forecasts, and technological forecasts are discussed in terms of their potential utility for the planner. Least squares analysis and the Delphi technique are illustrated as methods for operationally obtaining forecasts using, respectively, objective numerical data and subjective judgmental data.

Project planning using critical path analysis can provide the basis for the project plan. These approaches rely on network models which depict projects as a complex of relationships among those activities and events which are necessary for the fulfillment of project objectives. Their utility extends beyond planning to include the control of the project once it has been planned.

EXERCISES

1. Relate the accomplishment of objectives to the achievement of goals. Illustrate.
2. How should an organization go about establishing goals and objectives? How should the process for the establishment of goals differ from that for the establishment of objectives? Why?
3. One of our firm's goals is to "improve its image." Suggest a rational process for establishing objectives related to this goal.
4. What is brainstorming?
5. Why do alternatives have to be invented using an approach such as brainstorming?
6. Forecasts of the future are inherently uncertain and inaccurate. Why should we then use them in planning for the future?
7. Distinguish between planning and forecasting.
8. Distinguish between time-series forecasts and forecasts using statistical indicators.
9. "Leading" economic indicators are those in which changes are believed to lead (precede) changes in the economic activity to be predicted. For instance, the overall state of the economy might be predicted from a number of indicators such as "wholesale prices" and "carloadings," which have historically turned up prior to an upturn in the economy and turned down prior to a downturn. How might the "least squares" approach be used in this regard?
10. Return to Figure 7-5 and try to determine the decisions which might be involved in using the technological forecasting chart depicted there. What additional information might be required? How might it be obtained?
11. What is a scatter diagram?
12. Explain what is meant by the principle of "least squares."
13. Suppose that least squares techniques are used to develop the following predictive equation:

    $$y^* = 106.5 + 0.6\,x + 1.4\,Z$$

    where x and Z are quantities to be measured and used in predicting values of y.
    a. If x is 100 and Z is 52, what prediction of y would result?
    b. How does this use of the results of least squares analysis differ from the illustrations in the chapter?

c. If y were a measure of overall economic activity and both x and Z were leading indicators (see exercise 9), what would be the relationship among the time periods in which the measurements of x and Z and the forecast of y were made?

14. How might least squares analysis be used to analyze time series data?

15. Describe how you might try to evaluate the worth of brainstorming and the Delphi technique by objective scientific experimentation.

16. What is the nature of the questions and forecasts that you believe the Delphi technique to be most suited to?

17. How might network plans for a project fit into the system of plans discussed in Chapter 11?

18. Relate the camping trip work breakdown structure of Figure 7-3 to a network plan.

19. What role is played by "dummy" activities in a network plan? How do they differ from "real" activities? Give illustrations of some management edicts which might be transcribed into dummy activities in a project plan.

20. What is the value of a project network in and of itself (i.e., without activity time estimates, etc.)?

21. What are "critical activities"? In what sense are they "critical"?

22. What is "slack"? What is its importance to the project manager?

23. Summarize the management uses of a project network.

24. Using the project selection model of Figure 12-3, compare an opportunity which is rated "good" on every criterion except the first, on which it is rated "bad," with the project which is rated in that table. Which of the two projects would you select? Why?

25. *a.* Determine the critical path in the network shown below (time estimates in weeks).

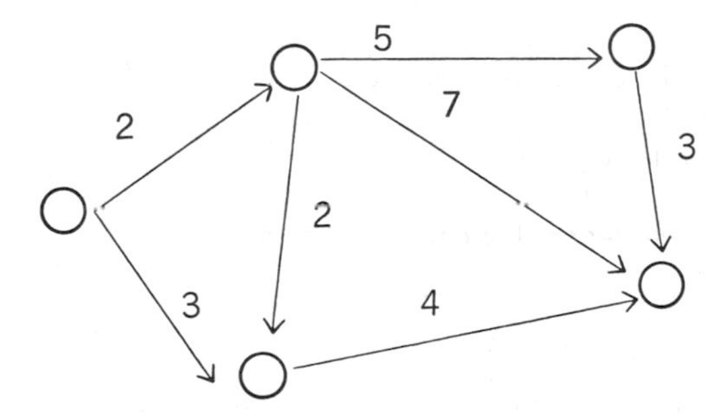

b. Suppose 8 weeks have elapsed and the activity anticipated to take 4 weeks is now expected to take another 3 weeks to complete. All other activities are on schedule. What management action, if any, might be indicated?

REFERENCES

Ackoff, R. L., *A Concept of Corporate Planning,* John Wiley & Sons, Inc., New York, 1970.

Brett, E. C., *Business Forecasting,* McGraw-Hill Book Company, New York, 1958.

Cleland, D. I., and W. R. King, *Systems Analysis and Project Management,* McGraw-Hill Book Company, New York, 1968.

Handy, H. W., and K. M. Hussain, *Network Analysis for Educational Management,* Prentice-Hall, Inc., Englewood Cliffs, N.J., 1969.

Kiershaw, J. A., and R. N. McKean, *Systems Analysis and Education,* RAND Corporation, Santa Monica, Calif., RM-2473-FF, 1959.

King, W. R., *Quantitative Analysis for Marketing Management,* McGraw-Hill Book Company, New York, 1967.

Steiner, G. A., *Top Management Planning,* The Macmillan Co. of Canada, Limited, Toronto, 1969.

Young, M. (ed.), *Forecasting and the Social Sciences,* Heinemann Educational Books, Ltd., London, 1968.

"PERT"

te = expected time ea activity

T_E = earliest expected time

T_L = latest allowable time

$$te = \frac{a + 4b + c}{6}$$

a = most optimistic time

b = most likely time

c = most pessimistic time

section five

EXECUTION

13

organizing—authority and responsibility

The simple dichotomous model of management which is used in various places in this book sees the manager as performing two essential functions—*planning* and *execution.* In the previous section, we have treated the planning function of management. Equally important, and in some ways more complex, is that function which has to do with the execution of plans and the carrying out of decisions.

The first step in the execution process is the establishment of organizational patterns which can contribute to the effective and efficient accomplishment of the organization's objectives. Intrinsic to this "organizing" function are the concepts of "responsibility" and "authority." Authority and responsibility constitute the framework of organizational management—the basis on which any "organization" is founded.

AUTHORITY

According to a dictionary definition, "authority" is the ". . . legal or rightful power; a right to command or to act."[1] Responsibility, a corollary of authority, is the obligation one has to act in response to an order issued by higher authority.

A Simple View of Authority

Authority is essential to any group effort. The authority exercised by a manager springs from his organizational position through the process of delegation from the next higher level. The ultimate source of authority can be traced back to the right of private property in our society. Thus, if one

[1] *Webster's New Collegiate Dictionary,* G. & C. Merriam Company, Publishers, Springfield, Mass., 1961, p. 60.

is employed by a large corporation that is operating under a democratic form of society, the consummation of the employment contract creates the agreement whereby an employee agrees to accept the authority of a superior as a condition of his employment. By accepting the contract, the employee agrees to offer some part of his physical and/or mental abilities —his private property—to aid in the attainment of organizational objectives.

A simple view of authority rests on several basic assumptions:

1. The organization chart is a realistic descriptive model of an organization.
2. Legal (or line) authority is delegated down through the "chain of command." Therefore, if one has legal authority, he can demand the obedience of others.
3. Given sufficient authority, an individual can accomplish organizational objectives regardless of the complexity of the forces that are involved.[2]

Under such assumptions, authority would appear to be a straightforward concept. To accomplish anything in an organization operating under these assumptions, one need only follow the chain of command running from the chief executive to the lowest-ranking operative. Orders are promulgated from above, and those members in the chain of command see that they are carried out.

Total Authority—An Evolving Concept

But it just does not work that way in the real world! For instance, FBI Director J. Edgar Hoover had direct informal links with seven different United States presidents. He conveyed information to them and undertook special assignments for them even though his formal authority did not encompass such a role and even though his formal superior was the attorney general of the United States—one of the President's cabinet officers.

This *legal authority* (the right to *demand* obedience) and *influence* (the ability to *obtain* obedience) both operate in defining the *total authority* of an individual. Most modern organizational managers operate through a combination of legal and informal authority since ". . . simply being in

[2]Point 3 is paraphrased from John C. Ries, *The Management of Defense,* Johns Hopkins Press, Baltimore, 1964, p. 38.

an executive hierarchy does not mean that one can direct freely those below him."[3]

Even though authority has often been viewed in the one-dimensional vertical context by management theorists, the dual nature of total authority has long been recognized. For instance, Fayol defined authority as ". . . the right to give orders and the power to exact obedience."[4] However, he complemented this by saying:

> *Distinction must be made between a manager's official authority deriving from office and personal authority, compounded of intelligence, experience, moral worth, ability to lead, past services, etc. . . . personal authority is the indispensable complement of official authority.*[5]

Thus, the concept of a "total authority," which is at least two-dimensional, represents an evolving view of authority. This changing concept of authority has severe implications for those who function in organizations. If one's ability to get things done depends on more than his legal authority, he must develop relationships and ways of doing things which go beyond the issuing of orders. Indeed, managers have discovered this at all levels. For instance, every president of the United States within the remembrance of the authors has at one time or another publicly expressed his surprise at the small degree to which his formal authority actually enables him to influence things. All have established informal staffs and contacts to supplement their formal authority.

There are also severe implications of the changing nature of authority to those in nonexecutive positions in organizations. Max Ways has summed this up well:

> *A bright and well-educated young man, whether he enters a large business or a small one, can reasonably expect to achieve a fulcrum of influence and responsibility, a position where his weight counts. The limits upon how much he counts will be determined not by any rigid social structure or superior power center. The weight he swings will depend mainly upon how he fares in competition and cooperation with other bright and well-educated young men of his own generation. For the typical power conflicts in contemporary American*

[3] Herbert A. Simon, Donald W. Smithburg, and Victor A. Thompson, *Public Administration,* Alfred A. Knopf, Inc., New York, 1950, p. 404.

[4] Henri Fayol, *General and Industrial Management,* Sir Isaac Pitman & Sons, Ltd., London, 1949, p. 21.

[5] *Ibid.*

economic life are much more "horizontal" (i.e., among peers) than "vertical" (i.e., between inferiors and superiors).[6]

Operationalizing Authority

To deal with the authority concept, one must dissect it into its various constituent parts and view it in various operational contexts. To accomplish this, we first analyze the concept of authority in its two major dimensions—legal and de facto—and then consider some "kinds" of operational authority—e.g., functional and staff authority.

LEGAL AUTHORITY An operational concept of total authority really involves the ability to influence others to do what one wishes them to do. It is exercised *both* by virtue of an individual occupying an organizational position *and* by the influence an individual is able to wield in his environment. Thus, there are really two kinds of authority:

1. *Legal (de jure) authority,* which rests in an organizational position and which may be exercised by an individual occupying the position.
2. *Informal (de facto) authority,* which comes about because of the individual's personal characteristics and capabilities. This type of authority may be exercised regardless of the presence or absence of legal authority.

The first type of authority, legal authority, is associated with the idea of being able to order others to do something or to cease doing something already under way. For example, the threat of physical punishment, the threat of depriving someone of his livelihood, or the threat of depriving an individual of his daily sustenance can be very effective in conditioning behavior, at least on a short-run basis.

Legal authority resides in an organizational position and is delegated through the media of position descriptions, functional statements, organizational rank and title, standard operating procedures, policy letters, policy manuals, organizational charters, organizational charts, and other such documentation. An example of legal authority is found in a company policy in the Aerospace Division of Martin-Marietta Corporation. This particular policy states, relative to the delegation of authority to the vice president of the Denver Division:

The primary responsibility of executing assigned programs or tasks and the local management of facilities, personnel, and other Martin

[6] Max Ways, "More Power to Everybody," *Fortune,* May, 1970, p. 292.

> *Company resources contained within or allocated to the Denver Division is hereby vested in and there is hereby created the Vice President of the Denver Division.*
>
> *To fulfill the responsibilities defined herein concerning the Denver Division, the Vice President of the Denver Division is hereby granted the necessary authority to execute proposals and contracts and amendments thereto on behalf of and in the name of the corporation for parts, materials and services; and to take such other action as may be necessary. The Vice President of the Denver Division is accountable to, and reports to the President of the Martin Company.*
>
> *To the extent necessary the authority herein contained may be delegated. All further delegation of authority granted shall be in writing and maintained in file by the Secretary of the Corporation.*[7]

This delegation of authority comes from a policy document signed by the president of the Division. This policy, written to guide the exercise of legal authority by the vice president of the Denver Division of the Martin-Marietta Corporation may be contrasted with the functional statement established for the finance department of the same company. This functional statement is reproduced in its entirety in Figure 13-1. As can be determined by comparing the two policy documents which delegate authority in the company, the functional statement for the finance department has considerably more specificity. This is to be expected, since the functional statement is written for a lower echelon in the organization than the policy for the vice president's authority. As authority is delegated downward in organization, one can expect to find the limitations and descriptions of the media stated in more explicit, narrow terms. An ultimate recipient of authority (such as an individual running a machine tool in a machine shop) may have no legal authority over others. His authority does not include the right to order or command the activities of the others, although he may have the right to plan, organize, and control his own job within the constraints laid down by his supervisor. However, he may exercise influence (and hence authority) over his peers and associates in an informal fashion.

Thus, there is indeed a broader range of formal authority at the higher levels of organizations. This is illustrated in Figure 13-2, which shows the foreman and worker with narrower scopes of authority than the president.

LIMITATIONS ON LEGAL AUTHORITY The exercise of legal authority is severely constrained in any organization by laws, societal morals, poli-

[7]Martin Company Policy Number AD-4, Revision 1, dated April 15, 1963, subject: Delegation of Authority to the Vice President of the Denver Division, Martin-Marietta Corporation. Used by permission.

FINANCE DEPARTMENT

The Director Finance is responsible to the Executive Director—Management Operations and responsive to the Vice President—Finance for developing and operating an effective system of financial management and control to safeguard the interests of the Company and the Customer.

This responsibility will be discharged as follows:

a. Support Programs in conformance with Functional Statement No. "PS".

b. Develop and operate an integrated accounting system to service the present and anticipated needs of the Denver Division.

c. Develop and implement a comprehensive system to provide cost and other data to be used by program estimating personnel. Develop information pertaining to the state-of-the-art in the estimating field and disseminate information to guide estimators in the use of preferred estimating techniques. Train and provide qualified price estimators for assignment to programs. Review and approve major program estimates and participate with Program Finance and Program Contracts in contract price determination.

d. Develop and operate a standard financial and manpower budgeting system for all organizations and report to management the Division's performance against the budgets.

e. Negotiate overhead rates and direct labor rates with the customer.

f. Conduct internal financial audits of the Denver Division's operations in accordance with sound business practices and report thereon to management.

g. Administer the Division's insurance programs as delegated by the Corporation.

h. Develop a budget training program to effectively improve capabilities of personnel working in budgetary activities.

i. Administer the Division's tax obligations as delegated by the Martin Company General Office Staff.

j. Prepare, or assure the preparation of, all required consolidated financial and manpower reports for the Denver Division.

k. Maintain, or assure the maintenance of, a comprehensive statistical bank of accurate financial data.

l. Develop and manage a comprehensive cost reduction program for the Denver Division in order to elicit from the Division and its suppliers the lowest priced, highest quality product.

FIGURE 13-1 *Functional statement.*

cies, procedures, contracts, etc. Thus, the broad scope of authority at the top level of organizations may be illusory. For instance, the legal articles of incorporation limit the authority of the directors of the corporation. The officers are further limited by general corporate objectives and policies. Laws also limit each manager—e.g., laws prescribing standards for collective bargaining and recognizing unions as agents for workers.

The temporal values of a society represent no less important a constraint on managerial authority. For example, many Christian churches have found the charge of heresy to be a formidable tool for maintaining eccelesiastical authority. In recent history, the charge of heresy has become less in concert with prevailing values, and it has therefore become much less used than it was during earlier eras. Indeed, some churches have proceeded to reject the concept.[8]

Other examples of the limitations of legal authority have been occurring in that most hierarchical of organizations—the Catholic Church. Pope Paul's encyclical on birth control was not received supportively by many

[8]"Episcopalians: An End to Heresy," *Time,* Aug. 25, 1967, p. 39.

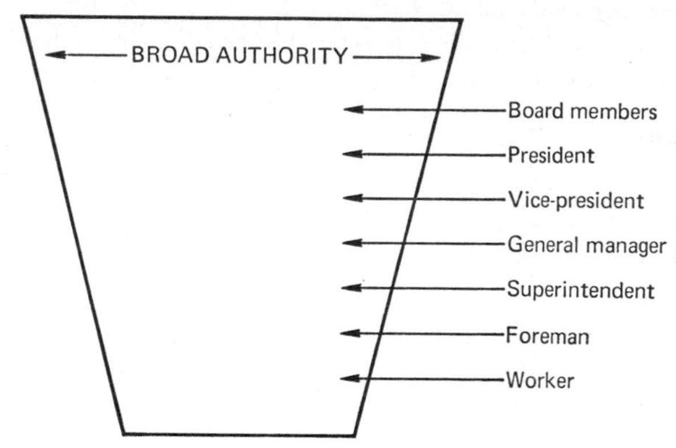

FIGURE 13-2 *Scope of authority.*

clergy. This reflected less their opposition to the topic than their opposition to the unilateral authority of the Pope. According to Athos:

> *The young priests, and their supporting bishops and even cardinals were saying in effect that the tradition of the past was not sufficient to justify the authority of their leader, that recent custom was just as insufficient, that long acceptance of the role of the leader was no longer solely determining. They were saying that the locus of authority now rests in the man in the role, and that their own perception of what was right or true was relevant in judging their superior's view. And they were insisting that the judgments of their superiors be related to the values which they held.*[9]

In brief, the clerics were expressing doubt as to the unilateral authority emanating from an organizational position and flowing downward depending on how much each successive level wanted to delegate.

Another informal form of limitation on the authority of managers comes about because of the physical or mental limitations of people. An individual cannot be "ordered" to develop a computer program if he does not have the appropriate experience or training; nor can someone be commanded to "think objectively" about a problem, particularly if he is emotionally involved.

Thus, legal authority is not an absolute power to order things into being. Many rulers have found this to be so when they became the object of a revolution. To be meaningful, authority must be accepted by those individuals who are the objects of authority, since in our society the individual has a great freedom of choice.

[9]Anthony G. Athos, "Is the Corporation Next to Fall," *Harvard Business Review*, January–February, 1970, p. 56.

ACCEPTING AUTHORITY The basis of the acceptance theory of management is that authority has no force or influence unless it is voluntarily accepted. One author's definition of authority highlights this idea:

> *Authority is the relationship that exists between individuals when one accepts the directive of another as authoritative. That is, when the individual receiving the directive weighs the consequences of accepting it against the consequences of rejecting it, and decides in favor of acceptance. The authoritative nature of the directive is confirmed when the person accepting the directive acts in accordance with it, within the confines of his understanding and ability.*[10]

Such a definition leaves little to the imagination—the manager has no meaningful authority unless the subordinate (or the peer, associate, etc.) confers it upon him. The sanctions that can be imposed by a superior may make the price of refusing to accept the order too high. But if the subordinate accepts the order with some reservations, the effect can be the same as if he had chosen not to accept the order at all.

Of course, there are always "excuses" which can be claimed by an individual as a way of refusing to accept an order without incurring too severe a penalty. For example:

1. Not understanding the order
2. Rationalizing the order as being inconsistent with existing policy and value systems
3. Professing ignorance of the order
4. Claiming physical or mental inability to carry out the order
5. Accepting the letter of the order while failing to accept its spirit and intent

Indeed, even though an individual who refuses to obey an order—either overtly or covertly—risks being discharged from his position and possibly incurring the social disapproval of his peer group, there are many instances in which the real risk is minimal, especially if an order is not blatantly refused. Most individuals in an organization possess an expertise about their own job which is sufficiently superior to that of others to permit them to fail to accept orders while appearing to accept them.

There are clear examples of situations where the basis of knowledge may provide the rationale for refusing to accept an order. A penetrating example is offered by the military establishment, where refusal to accept an order in earlier days would have perpetrated a court-martial.

[10] Daniel J. Duffy, "Authority Considered from an Operational Point of View," *Journal of the Academy of Management*, vol. 2, no. 3, December, 1959, p. 167.

> *Soldiers at all levels are expected to use their heads, especially in those numerous situations where their knowledge is superior to that of their commanders. Recently, CBS News showed a Vietnam incident where a company of men, suspecting an ambush, had refused to go down a road. Mutiny? A few days later the deputy commander of the brigade of which the unit was a part told CBS News that there would be no punishment. "Thank God, we've got young men who question," he said. The communications that are still called "orders" are frequently initiated from below—where the knowledge is.*[11]

The acceptance idea explains the way in which recent United States presidents have set up staffs with personal allegiance to them. These staffs may at times come into conflict with the executive departments of the government, but they function well because their members accept the authority of the President and attempt to carry out his directives in both fact and spirit.

DELEGATION OF AUTHORITY Delegation is the process of having one or more persons represent another person or persons. A *delegate* is an individual sent and empowered to act on behalf of another person or persons. Most of us are familiar with a political delegate, such as a member of the U.S. House of Representatives, who is empowered to act on behalf of his constituents in making laws and carrying out other legislative responsibilities. He therefore has the *authority* to act for another.

Often delegation is the process of empowering an individual to exercise legal authority given from a higher organization level. More specifically, delegation:

1. Involves the assigning of tasks to certain individuals;
2. Involves the granting of specific authority to act in performing the tasks assigned;
3. Exacts responsibility;
4. Creates obligations that cannot in themselves be delegated.

In the delegation process the superior typically provides authority to a subordinate through job descriptions, policy documents, procedures, and related media. Delegation can be given through a documented contract, or it may be verbal in nature. Whatever the form, the delegation process creates both the authority and the obligation to act or not to act. One of the best examples of delegation is found in the "agency power" that one general partner has in committing a partnership to a course of action.

[11] Ways, *op. cit.*, p. 174.

Delegation is a means of spreading the executive's attention over a greater quantity of affairs than would otherwise be possible. It is essential to the management function, since it is the basic process involved in the creation of the superior-subordinate relationship. Often, the delegation process provides the legal basis for dealing with other organizations—e.g., when the industrial relations manager is assigned authority to represent the company in negotiations with a union.

Thus, delegation is a means to cope with the problem of increasing organizational size and complexity. More intrinsically, however, it is basic to the manager's role as a *decider* rather than a *doer.* The only way in which he can avoid actually carrying out all of the details required to implement each of his decisions is to delegate authority to others.

DE FACTO AUTHORITY By "de facto authority" we mean the informal influence that an individual has over the people with whom he interacts. This kind of authority transcends legal authority and operates with subordinates, peers, associates, superiors, and friends—in fact, with everyone in any social situation. This kind of authority has to be earned and to be granted by those who will be affected by it. It comes into being through an individual possessing talents such as:

1. Superior knowledge
2. An ability to persuade people to his way of thinking
3. A suitable personality and the ability to establish rapport with his fellow man
4. A favorable reputation with his peers and associates
5. A record of accomplishments which lends credence to his experience and reputation
6. An ability to build confidence in his peers and associates
7. Patience to listen to the problems of his subordinates and his peers and a willingness to help out when asked or when the need to help is sensed
8. An ability to resolve conflict between his peers, subordinates, and associates

De facto authority is earned or granted by the social group in which the individual finds himself; it does not follow the chain of command, but often cuts across formal organizational lines into other departments and areas.

Informal authority is often created by an "informal organization" which exists within the formal one. The informal organization comes into being because of the interaction of the people, and it is based on social relationships rather than the formal relationships that are prescribed by organizational documentation. Since authority in the informal organization is earned by the individual and granted by the members of the group, it is more unstable than formal authority. It can change as frequently as members change their attitude toward their fellow members. An informal leader, such as ". . . an old-timer who is looked upon as the expert on job problems, a listener who serves as counselor, and a spokesman who is depended upon to convey key problems to the manager,"[12] may enjoy considerable authority and deference from his associates. In such a case authority is obviously granted and not imposed.

STAFF AUTHORITY According to classical management theory, the authority of an individual acting in a staff capacity comes from his right to advise, assist, and counsel the line official to whom he reports. Often the staff officer is expected to act in a staff capacity to many other line officials in the organization. For instance, a personnel officer is generally thought of as a staff official; he can prescribe certain personnel policies for the organization within the jurisdiction of his job description and the latitude given to him by his superior. His line authority is limited to just those individuals related to him in a superior-subordinate sense. He does not issue orders to line officials except through the authority of the senior executive of the organization. Thus, much of the authority of the staff official comes from de facto sources which depend on his expertise and willingness to assist the line officials.

FUNCTIONAL AUTHORITY This type of authority is the legal right to act with respect to specific activities, processes, practices, and policies in the organization. Either line or staff officials can exercise functional authority; it is a small element of the authority of the senior executive which is delegated through the organization's "chain of command." The need for this delegation emanates from the extreme degree of division of labor—the assignment of persons or groups to specialized tasks—which is found in most modern organizations.

The delegation of functional authority often results in certain line managers being deprived of some of their authority. For instance, a quality control man who is responsible for determining the quality of the product being produced can have sufficient authority to stop the production line if the products are not coming up to the necessary standard. His authority usually cannot be overruled by the production manager, except if the senior line official concurs in such action.

[12]Keith Davis, *Human Relations at Work.* McGraw-Hill Book Company, New York, 1967, p. 214.

Total Authority

Authority is an essential element of the management of an organization, since it establishes the basic "web of relationships" which permit the organizational elements to interact and work toward a goal. However, when authority is viewed as a force to use to require compliance with orders, it must be tempered by the personal freedom that people require in their daily work activities.

A manager will find that his ultimate success depends on total authority—a combination of legal authority and de facto authority—particularly as the latter authority comes about as a result of the acceptance and respect that his subordinates and peers give him. In today's complex organization simply complying with directives is not enough. More than simple physical presence and blind obedience of the organizational member is required if the organization is to be successful. Loyalty, enthusiasm, initiative, and creativity as well are required of the subordinate. Particularly, the support of many peers and associates is required in coordinating and accomplishing ends which involve elements outside of the chain of command. While legal authority is a necessity, it can never replace the additional influence which one can have by being able to develop de facto authority. Often, the potential support that his subordinates, peers, and associates can give the manager make the difference between a mediocre and an outstanding job.

In some cases, de facto authority may be the only source of influence that an individual may have. Since legal authority is attached to an organizational position, the individual who has a job that requires the extensive crossing of organizational lines to discharge his responsibility may have very little in the way of formal authority. For example, a machine operator may frequently find it necessary to coordinate his affairs with those of many of his peers—e.g., in the providing of maintenance and repair for his machine, in arranging for supplies and material, and in providing information on when and what he is to produce.

In the planning which must be done for the development of a school district, many groups outside the formal organization must be dealt with and included in the decision-making procedure. For instance, parents, students, teacher unions, state and local governments, and many other interest groups usually wish to be involved. If they are not, they can take actions to ensure the failure of any plan which is adopted without their concurrence. Yet, the educational planner has no formal authority with respect to these groups. He must develop de facto authority, using all of the formal and informal relationships which exist or may arise.

One of the most important instances where de facto authority is of paramount importance is the case of project management. The project manager's authority and responsibility cut across the traditional functional areas of the organization, such as production, marketing, finance, and quality control. The extensive diagonal and horizontal relationships found in such an environment usually create conditions where the legal grant of authority is insufficient to meet the need for favorably influencing the many managers with whom the project manager must deal. Since the project manager cannot command the regular functional line officials, he can gain considerable influence by creating alliances with them. In this way, his de facto authority is established and he may use it in combination with his legal authority to achieve his project objectives.

RESPONSIBILITY

Responsibility is an inescapable corollary of authority. The oft-quoted management principle of "parity of authority and responsibility" indicates that an individual should be given the necessary authority to discharge his responsibility. Responsibility indicates the individual's assumption of an obligation, i.e., an agreement, promise, contract, or recognition of duty. Responsibility also means what one is expected to do or not to do as a responsible person. Thus, the accepted obligation implies some constraint on behavior, as the result of either a specific grant of authority or what is expected as the result of holding a particular office.

For instance, a parent is expected to provide the means of support for his children. The court systems of our society will bring sanctions against an individual who fails to accept and discharge this responsibility toward his dependents. A trustee appointed by a company or an investment banker to represent the interests of bondholders is expected to act in a responsible manner toward the bondholders and the corporation. His duties would include the collection and disbursement of interest and principal, protection of the security, protection of the bondholders, and such related matters. The society in which the trustee functions—i.e., the financial community—further expects the trustee to assume these basic obligations and duties. The specific contract between the trustee, the bondholder, and the issuing corporation will spell out the detailed relationships.

Thus, each individual on assuming a given organizational position takes on many general implicit responsibilities of the position as well as more definitive formal responsibilities associated with the position.

Delegation of Responsibility

The delegation of authority is a clear necessity if the manager is to perform his basic role as a decider. However, what about the delegation of responsibility?

This point is one of the most controversial in all the field of management. One point of view is displayed by Koontz and O'Donnell:

> *Responsibility cannot be delegated. While a manager may delegate to a subordinate authority to accomplish a service and the subordinate in turn may delegate a portion of the authority received, neither delegates any of his responsibility. Responsibility, being an obligation to perform, is owed to one's superior, and no subordinate reduces his responsibility by delegating to another the authority to perform his duty.*[13]

Realistically, however, there are some organizational and geographical limitations on the fixation of responsibility at high levels in the organization. Can the senior executive be held responsible for all of his subordinates, particularly if he has taken *reasonable* care to see that they are acting in a responsible manner? One would reasonably expect that a supervisor could be held responsible for the immediate subordinate in the next echelon in the chain of command. Beyond the first level of subordination, the issue can become clouded by intervening levels and policy, and procedural documentation can often serve to insulate the level at which responsibility can be fixed.

Thus, the process of delegation, while providing the legal basis for the subordinate to act, cannot really relieve the responsible executive of his responsibility. Indeed, in complex organizations, it may be impossible to delegate responsibility even if it were desired. For instance,

> *General John P. McConnell, recently retired Air Force Chief of Staff, once made the point to Congress that when he operated a flying unit and a squadron commander goofed, he fired the commander. But in procurement, he said, it was virtually impossible to find the right man to fire because too many people at too many levels had too much to say about major programs.*[14]

Often, the pattern of delegation ensures that the superior is "included into" important actions taken by the subordinate using authority which has been delegated to him. For instance, the patterns may be along one of the following lines:

1. Subordinates have complete freedom to make decisions in allocating and using human and nonhuman resources.

[13]Harold Koontz and Cyril O'Donnell, *Principles of Management,* McGraw-Hill Book Company, New York, 1964, pp. 55–56.

[14]"Defense: The Dogfight Over the F-15," *Business Week,* Dec. 20, 1969, pp. 96–98.

2. Subordinates may make decisions provided these decisions are reported to higher-level executives.
3. Subordinates may make decisions only after the counsel and guidance of the superior have been solicited.
4. Subordinates may use judgment in electing whether to check with superiors or not before a decision is made.

Often, the right of the subordinate to delegate may be restricted. There is good reason for such restrictions. The superior may wish to ensure that a given level of experience or executive rank is brought to bear on certain matters. By restricting the level to which authority is delegated, this can be accomplished. Also, there may be instances in which the authority to negotiate and administer is delegated but where the final approving power remains in the hands of the executive. This is often the case in collective bargaining, where the authority to conduct preliminary negotiations may be given to the director of industrial relations but final negotiations may be conducted by the chief executive of the firm. On the labor side, negotiation authority is given to the negotiating team but final approval authority may rest with the membership of the union.

COMBINING AUTHORITY AND RESPONSIBILITY

Responsibility, at least in the popular sense, is a continuing obligation which arises from a superior-subordinate relationship. The interpretation of responsibility to mean the obligation to do that which is expected extends the concept beyond the superior-subordinate relationship to others in a group situation—e.g., peers and associates. While this de facto responsibility is not identified with a specific obligation such as one which is established in an organizational policy or procedure, it still exists because of the *expectation and standards* of the peer group. Of course, there is a very good pragmatic reason for one to assume the responsibility arising out of a relationship with a peer, since that same peer may be in a position to aid and assist in the future. Much of the authority that one may have over peers and associates comes into being simply because he has demonstrated a willingness to reciprocate in accepting responsibility and to counsel and assist when the peer requires.

SUMMARY

Authority is the cement of organizations. A manager's total authority is made up of two parts: his legal authority, as defined by the legal basis of the position which he holds; and his influence, or de facto authority—

that which he possesses because of personal qualities such as technical competence, persuasive power, personality, personal alliances, etc.

Responsibility is the handmaiden of authority. While authority can clearly be delegated, it is not clear that responsibility can be delegated.

A number of conditions which should be sought in the establishment of authority and responsibility relationships succinctly summarize their role in management:

1. Authority and responsibility should be given and expected in equal amounts.
2. *Legal* authority is granted through organizational documentation and is complemented by *de facto* authority assumed by the individual, regardless of legal grant.
3. There are some practical operating limitations to the fixation of responsibility because of organizational and geographical expanses.
4. Social, biological, and legal constraints limit the nature and extent of authority that can be exercised.
5. Subordinates, peers, associates, and other people are both the object of, and a tempering influence on, the extent of authority that is given, expected, and received.
6. Authority patterns do not necessarily follow the chain of command; rather these patterns constitute a "web of relationships" between the participants involved in management situations.
7. Authority should be delegated and responsibility exacted to the extent necessary to accomplish the results that are expected.
8. Functional authority should be exercised in an impersonal manner through written orders, schedules, and other documentation.
9. People do have a degree of choice in accepting or rejecting the imposition of authority.
10. Authority and responsibility patterns in the organization may be modified by the informal organization.
11. The creation of legal authority must of necessity include the development of the sanctions to enforce the authority.
12. The successful delegation of authority depends on the superior's willingness to let go of decision-making power, and his willingness to trust subordinates.

EXERCISES

1. Distinguish between authority and responsibility.
2. What is the ultimate source of organizational authority?
3. Relate a simplistic view of authority to the bureaucratic model of organizations.

4. What is total authority? How is it related to legal authority and to influence?
5. How is the legal authority of an organizational manager made known to the members of the organization?
6. "Authority is a two-way street. It is not truly possessed by one until it is accepted by another." Is this a valid statement?
7. Comment on the concept of authority as it relates to expertise.
8. What is meant by "delegation of authority"?
9. What is the "informal organization"?
10. "The only real authority is the informal kind. Because of this, it is more important and more stable than legal authority." Comment on this.
11. Contrast and compare staff authority and functional authority.
12. Relate the concept of authority to the management of the planning process.
13. Discuss authority as it relates to the development of goals for a school district and a law enforcement agency.
14. Describe the authority status of the project manager.
15. Discuss the concept of the "delegation of responsibility."
16. "Authority may be divided and subdivided among various people until no one person possesses much. Responsibility, on the other hand, can be given to others, but it remains to the same degree in the hands of its original holder." Comment on this.
17. During the period of the various trials which grew out of the My Lai massacre incidents of the Vietnam war in the early 1970s, many public figures variously suggested that: *(a)* blame was being placed at too low a level in the Army chain of command; *(b)* the President of the United States should be tried for murder; *(c)* our execution of a Japanese general after World War II for war crimes committed by his troops without his knowledge or approval dictated that the commanding general of American forces in Vietnam should be brought to trial. Comment on these suggestions in the light of the concepts of authority and responsibility.
18. How might authority at all levels in an organization be limited by restrictions emanating from a given society? Give specific examples.
19. What type of authority does a teacher have in a classroom?
20. A clever individual can always find ways of avoiding the authority of somebody placed in a supervisory position over him. Is this so? Demonstrate by example that you understand the foregoing statement.
21. What examples might you give of situations where the student could refuse to accept the orders of a teacher?
22. Is it possible to delegate *de facto* authority? Justify your answer.
23. Differentiate between *line* authority, *functional* authority, *project* authority, and *staff* authority.
24. Why is specificity of authority in a project/functional interface essential?

REFERENCES

Athos, Anthony G., "Is the Corporation Next to Fall?" *Harvard Business Review,* January–February, 1970.

Barnard, Chester I., *The Functions of the Executive,* Harvard University Press, Cambridge, Mass., 1938.

Cleland, David I., "(The) Deliberate Conflict," *Business Horizons,* February, 1968.

______, "Project Management—An Innovation in Managerial Thought and Theory," *Air University Review,* January–February, 1965.

______, "Understanding Project Authority," *Business Horizons,* Spring, 1967.

______, and Wallace Munsey, "Who Works With Whom?" *Harvard Business Review,* September–October, 1967.

Duffy, Daniel J., "Authority Considered from an Operational Point of View," *Journal of the Academy of Management,* December, 1959.

Gaddis, Paul O., "Winning Over Indifferent Youth," *Harvard Business Review,* July–August, 1969.

Golembiewski, Robert P., "Authority as a Problem in Overlays: A Concept for Action and Analysis," *Administrative Science Quarterly,* June, 1964.

House, Robert J., "Role Conflict and Multiple Authority in Complex Organizations," *California Management Review,* Summer, 1970.

Mandeville, Merton J., "The Nature of Authority," *Academy of Management Journal,* August, 1960.

Ries, John C., *The Management of Defense,* The Johns Hopkins Press, Baltimore, 1964.

Steiner, George A., and William G. Ryan, *Industrial Project Management,* The Macmillan Company, New York, 1968.

Stieglitz, Harold, "What's Not on the Organization Chart," *The Conference Board Record,* September, 1964.

Thompson, James D., "Authority and Power in 'Identical' Organizations," *The American Journal of Sociology,* November, 1956.

Ways, Max, "More Power to Everybody," *Fortune,* May, 1970.

______, "Tomorrow's Management: A More Adventurous Life in a Free-Formed Corporation," *Fortune,* July, 1966.

14

organizing—the systems view

The systems model of the organization views the organization in terms of flows and processes. The elemental parts of the organization are people, and people must perform the processes and create and sustain the flow. They must therefore be organized in a way which facilitates the efficient and effective execution of the organization's task.

THE MATRIX ORGANIZATION

The two-dimensional nature of authority, as disccussed in the previous chapter, leads one to naturally question the traditional one-dimensional, top-to-bottom model of the organization.

Consider, for example, a company made up of two divisions. Division A is an operating entity which produces a standardized product in high volume. Within Division A are the functional departments through which the standardized work can flow.

Functional departmentalization is the traditional way of organizing in such an instance. A *major functional* department such as finance would normally be comprised of a number of *minor functional* departments such as credit, disbursements, fund control, and accounting. Division A may be thought of in terms of four major functional departments—production, engineering, personnel, and finance. The managerial emphasis in these departments would, because of the nature of the activities which they perform, be primarily on improving operating efficiency.

Division B of the company, on the other hand, is a "job shop" operation which performs contract work for the government. Their work load is composed of various projects, each with rather specific objectives and a well-defined point of completion. *The managerial emphasis here needs to be on the timely completion of these projects.*

The level of participation by various functional entities in each of

Division B's projects is highly variable as each project progresses toward completion. In the early stages of a project, the emphasis may be on the choice of materials and components, while at a later stage this facet will be deemphasized. This is the case because of the natural "life cycle" of a project. Thus, the combination of people and resources that can best jointly pursue project goals changes from time to time in each of Division B's projects. Here, then, the primary management emphasis must be on organizing and controlling the projects, rather than the seeking of greater efficiency.

The *matrix organization* of the hypothetical two-division company is illustrated in Figure 14-1. Division A needs no matrix concept because of its continuing standardized work load. Division B has all of the same functional departments as does Division A, but each of them provides facilities and functional support to the two major activities of the division —Projects C and D.

Each of the project organizations depicted in the lower portion of Figure 14-1 is comprised of a project manager and work groups from the various functional departments. The project manager is given the author-

FIGURE 14-1 *Matrix organization.*

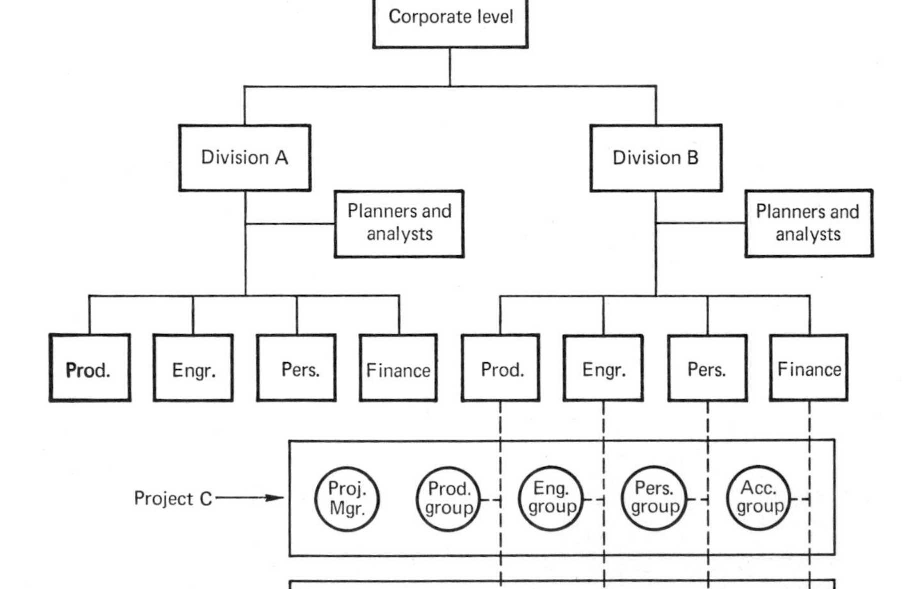

ity and responsibility for the achievement of project goals. The line organization is left to function by providing support for the various projects. Typically, the work groups for each project are assigned on a full-time basis from their various functional units to the project. When the project is completed, or when their services are no longer essential to it, they are assigned back to their functional unit.

The matrix organization is the realization of a two-dimensional organization which emanates directly from the two dimensions of authority. Two complementary organizations—the *project organization* and the *functional organization*—are merged to create the matrix organization.

Systems Relationships

An organization may be thought of as facing a "stream of projects"—a flow of activities which demand its attention. These projects are horizontal rather than vertical—they cut across the traditional top-to-bottom functional units and chains of command.

Each project in the "stream" which provides work for the organization is in a different stage of completion—one may be an idea undergoing feasibility study, another may be in development, some may be in production, and others may be being terminated in favor of new projects. The various projects also range widely in size and complexity—e.g., from developing a major new product line to working a single order being processed through a job shop.

The classical bureaucracy, with its strong superior-subordinate relationships and its vertical patterns of authority and responsibility, is not an adequate organizational pattern for implementing the systems view. Flows and processes are horizontal in nature; so, too, are projects horizontal. On the other hand, the bureaucracy—as exemplified by the classical organizational chart—is vertical.

A partial resolution of this horizontal-vertical dilemma is found in the establishment of *project teams.* These teams bring together the varied resources and human skills which are necessary to do such things as solve a problem, develop a product, and implement a new computer system. Thus, the teams are charged with accomplishing *project* goals.

An entire business or governmental organization might be viewed as a collection of teams acting in a coordinated way. Each team has a leader and followers—each having specialized work to perform. Each individual also has to relate and coordinate his efforts with those of the other members of his team and with those of other individuals in other teams. Managers have to relate individual effort and team effort. The result is a web of relationships existing between the participating individuals and teams in the organizational system.

The project concept clearly requires the crossing of many different chains of command. Figure 14-1 shows that these people assigned to the two projects have dual responsibilities—to the project and to their functional organization.

The effective management of projects such as C and D in Figure 14-1 creates the need to formalize the crossing of organizational lines to accomplish project goals. Authority in the management situation can be assigned to an individual managing a special organizational unit built around the task rather than to the more traditional functional departments. However, this naturally creates conflicts and managerial problems which transcend those of traditional hierarchical organizations. To illustrate, let us examine the relationships that appear between the project manager (charged with the responsibility for managing a specific project) and the functional manager who is charged with managing a functional entity in support of the stream of projects that are flowing through the organization.

The Functional Manager's Role

The functional manager is responsible for one of the principal elements of the organization, such as finance, production, marketing, R&D, or a subelement thereof. He is responsible for providing information and services on the "state-of-the-art" in his discipline and for supporting all the projects in the organization. In most instances we think of the functional manager as exercising line authority.

The functional manager is responsible through the chain of command for his department. Since he is held responsible and accountable for his element, it is a natural thing for him to become parochial in his viewpoint. Functional managers concentrate most of their efforts within their own function; hence, they tend to become known as marketing, financial, or production "people," and, in truth, their views tend to be detailed by their function.

The performance of activities through the use of organized groups based on specific disciplines protects the integrity of the existing specialized competence. Yet, the functional manager can easily become overwhelmed by this provincialism and come to consider his function as the only important one—he may be prone to embellish it at the expense of the other organizational elements. In the very human desire to do a good job, the functional manager tends to develop "tunnel vision," which allows him to see things only within the narrow scope of his function and to conveniently ignore the "bigger picture."

An often-heard gripe in a variety of different organizations is that the organization seems to be run for the benefit of the accounting department or the elevator operators rather than to enhance the opportunity to achieve overall objectives. This is a natural outgrowth of overzealous functional management. Since the responsibilities of the functional manager are limited to his area, he seeks to make that area as efficient and effective as possible—often without regard to the effect of his actions on other functions or, more importantly, on the basic tasks which the overall organization must perform.

The Project Manager's Role

One important context in which the salient aspects of the project manager's role are easily seen is in the interorganizational context. For instance, in developing a major weapons system, the Department of Defense appoints a project manager, as does the contractor who will perform the development. The "project" is the development of the weapons system. The relationship of the two organizations—Defense Department and contractor—revolves about the relationship of these two project managers. This is illustrated in Figure 14-2.

In the relationship depicted in this figure, each respective project manager acts as a focal point to permit the channeling of major project considerations, such as the determination of basic policy and the making

FIGURE 14-2 *Interorganizational project manager relationships. Critical decisions involving policy and managerial prerogatives are directed through the central focal point. Decisions involve cost and cost estimating, schedules, product performance (quality, reliability, maintainability), resource commitment, project tasking, trade-offs, contract performance, and total system integration.* SOURCE: *David I. Cleland, "Project Management—An Innovation in Managerial Thought and Theory,"* Air University Review, *January–February, 1965, p. 19.*

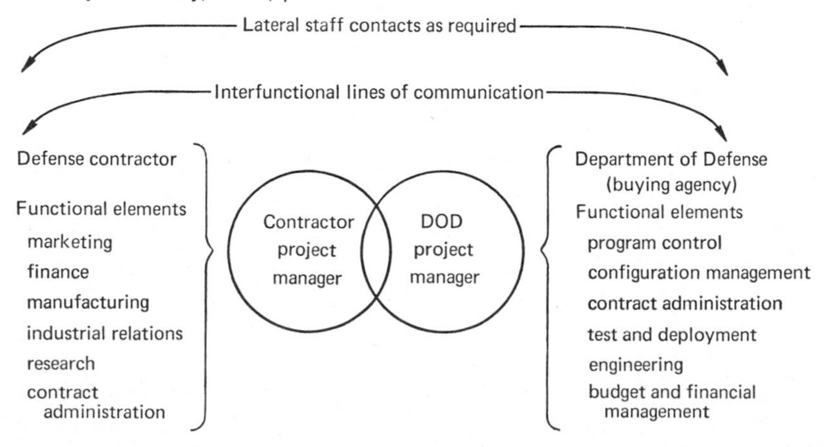

of strategic decisions. Since each project manager has a global view of the project, he has the perspective necessary to consider cost, performance, and schedule aspects of the project.

Of course, while major policy issues and decisions are channeled through the project managers, this does not preclude extensive lateral contacts between project participants—both government and contractor—as they go about their work.

The two project managers are acting as "general managers" of their respective project as far as the parent organization is concerned. By being in this focal position, they are better able to ensure that overall managerial prerogatives relative to the project are reserved for themselves and not diffused in the functional chain of command. In this face-to-face relationship the two project managers can better control and resolve both interfunctional and interorganizational problems arising during the course of the project. This organizational relationship precludes any one functional manager from overemphasizing his area of interest in the project to the neglect of other considerations. The overlapping of the fields of the two project managers implies an area of inescapable interdependence in matters involving the project.

The Project Manager as Integrator

In reviewing the focal position of the project manager one is struck by the emphasis that is placed upon the integrative functions in his operations. The requirement of the blending of the actions of many diverse groups and many parts into a systematic whole is always present. In this sense, the project manager is truly a "general manager," even though he may operate at a relatively low level in the overall organization. Since he is not normally at the presidential or vice-presidential level, he is not a "top manager," *but he must perform the same general management functions as do top managers*—he must integrate the efforts of a variety of functional managers to accomplish the goals of the project and the organization.

The Project-Functional "Conflict"

An important part of the task of both the project and functional managers in a project situation is to achieve harmony of activity among organizational elements with conflicting objectives. *The project and functional managers thereby are involved in a deliberate and purposeful conflict within the organization.* Figure 14-3 describes the relative authority and responsibility of project and functional managers in one aerospace company.

Project Manager

Project Planning and Direction

Determines *what* and *how much* effort will be accomplished and *when* it will be performed through the development, issuance, and maintenance of master project plans to include the project work breakdown structure and work statements with accompanying budgets and schedules.

Assures the accomplishment of the technical objectives, schedule requirements, and effective cost management of the project.

Project Control

Monitors total project cost, schedule, and technical results against project plans, initiates any necessary corrective action to assure accomplishment of project objectives, and monitors contractual reporting.

Configuration Management

Integrates the technical and administrative action of identifying functional and physical characteristics of an item, controls changes proposed to these characteristics, and provides configuration accountability information affecting the project.

Customer Coordination

Provides the prime contact with the customer and higher company management of project activities.

Administration

Approves the assignment and provides periodic reports on effectiveness of key functional personnel assigned to the project.

Functional Manager

Operation Planning and Direction

Determines *who* will perform specific tasks, *how* they are to be accomplished, and *how well* the work is performed.

Provides a stable base for the development of talent and skills to assure the maintenance and enhancement of technical capability.

Provides necessary facilities and services to support project requirements.

Operational Control

Assures that the technical excellence and quality requirements of assigned tasks are met and that the tasks are accomplished on schedule, and within budget.

Administration

Performs administrative services in support of personnel assigned to a project.

Initiates merit increases for all personnel within their organization.

Manages overhead effort in support of productive work applied to projects.

FIGURE 14-3 *Project and functional management authority and responsibility. Adapted from David I. Cleland, "The Deliberate Conflict," Business Horizons, February, 1968, pp. 78–79.*

In such a conflict, where does the decision-making authority rest? Legal authority to resolve the project-functional conflict ultimately rests with the common supervisor of the project and the functional individuals. But this conflict, which arises many times in the workaday world, cannot be resolved simply by dictation from above. Rather, the project or functional individuals must maintain open minds and be ready to negotiate the relative matters important to the issue, such as cost, schedule, and technical performance of the project. Also, there will arise questions, such as those involving resource allocations, personnel tasking, priorities, and the assignment and promotion of personnel who serve both the functional manager and the project manager. In such a continuing and purposeful conflict, the willingness to negotiate is essential. Human differences in role specialization and personality exist; provincial loyalities are inescapable and must be accepted as a normal course of affairs in bringing about the coordinated effort.

Under these circumstances the managers must build an environment for open and free expression in which the project-functional feedback is meaningful and real and in which people seek the opportunity to "have their day in court" to negotiate their differences and relate the action that each is taking in his job.

This continuing integration of project-functional interests presents a different authority-responsibility relationship from that envisioned under the line-staff concept. Little commanding occurs between the participants since it is extremely difficult to "command" those over whom one has no legal authority. De facto authority takes on considerable importance in this environment; attention must be given to identifying and integrating the respective roles, functions, interests, and contributions of the participants.

In the management of large projects, the authority patterns are different from those found in the purely functional organizations. Project authority manifests itself within the legitimacy of the project; it extends horizontally, diagonally, and vertically within the parent organization and to other participating organizations. Traditional line-staff relationships are modified in such an environment, since in this situation the functional managers (such as a production manager) give advice, counsel, and specialized support to the project manager.

The interplay between functional and project teams includes a mix of line and staff people. In a project that is highly technical in nature, the administrative executive must follow staff advice—the legal decision may be his, but the technical judgment is the prerogative of a staff official. In such circumstances the staff official exercises a type of authority by providing value judgments for the executives. Line functional people play a role in the project which is different from their role in their function. Each of the project participants plays a mix of roles depending on the na-

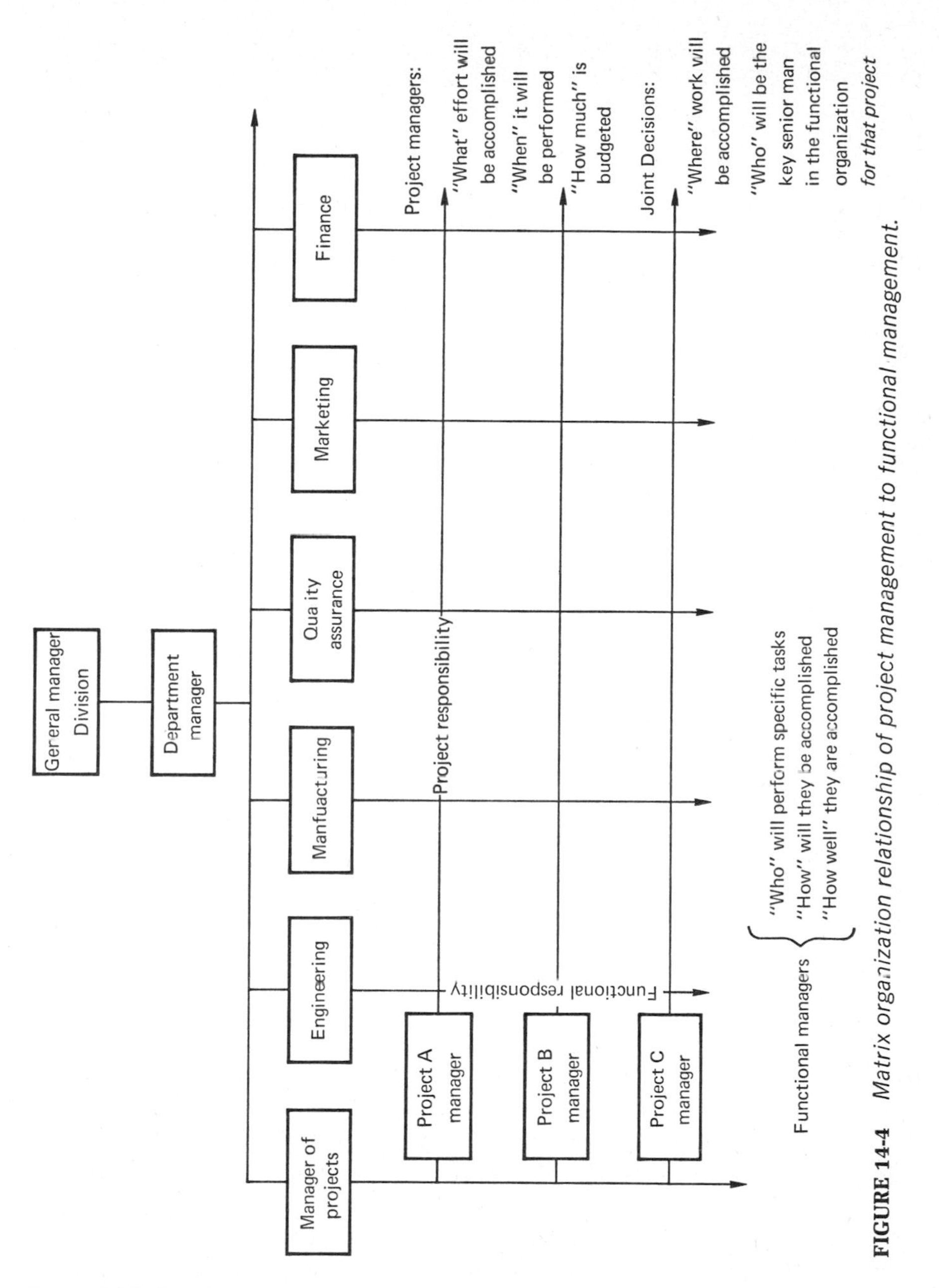

FIGURE 14-4 *Matrix organization relationship of project management to functional management.*

ture of his involvement in the project. The line-staff relationship gives way to other relationships, such as peer-to-peer and specialist-to-generalist.

Another way of defining the reciprocal authority and responsibility between the project and functional managers is shown in Figure 14-4. There, the emphasis of the definition of the matrix organization is on:

Who will perform specific tasks?
How will the tasks be performed?
How well are the tasks accomplished?
What effort will be accomplished?
When will it be performed?
What quantity of resources (money, time, and material) are available?
Where will the work be accomplished?
Who will be the senior participants?

The complementary nature and merging of the interests and concerns of project management and functional management are typified by the matrix representation of Figure 14-4. There, a "manager of projects" is envisioned as someone who is responsible for directing and evaluating the activities of a number of individual project managers. Such a position would probably be warranted only in a large organization which is predominantly project-oriented. However, the existence of the position illustrates the need for some coordination when many project managers are involved, and it illustrates the need for a project management policy to guide the various managers.

Figure 14-5 focuses primarily on the detailed responsibilities of project and functional managers. Often, "work package managers" will be delegated some of the project responsibility, as shown in the lower right corner. However, the key to making the matrix organization operate is *the sharing of authority and responsibility between project and functional managers.* They are interdependent entities, and each must therefore depend on the other to cooperate in accomplishing organizational objectives.

The Impact of Conflict

Any conflict in an organization—be it deliberate or not—will inevitably create tensions and frustrations in those involved. "This is particularly true in business where the individual often finds himself forced to choose among personal values and ultimate loyalties that may be in sharp contrast with each other, with the values held by others, . . . or with urgent organizational considerations."[1] The conflicts arise naturally out of the conflict between personal and organizational values and between different sets of organizational values. One's values are based on such things as

[1] Edmund P. Learned, Arch R. Dooley, and Robert L. Katz, "Personal Values and Business Decisions," *Harvard Business Review,* March–April, 1959, p. 118. The ensuing discussion of conflicts in value systems is adapted from this reference.

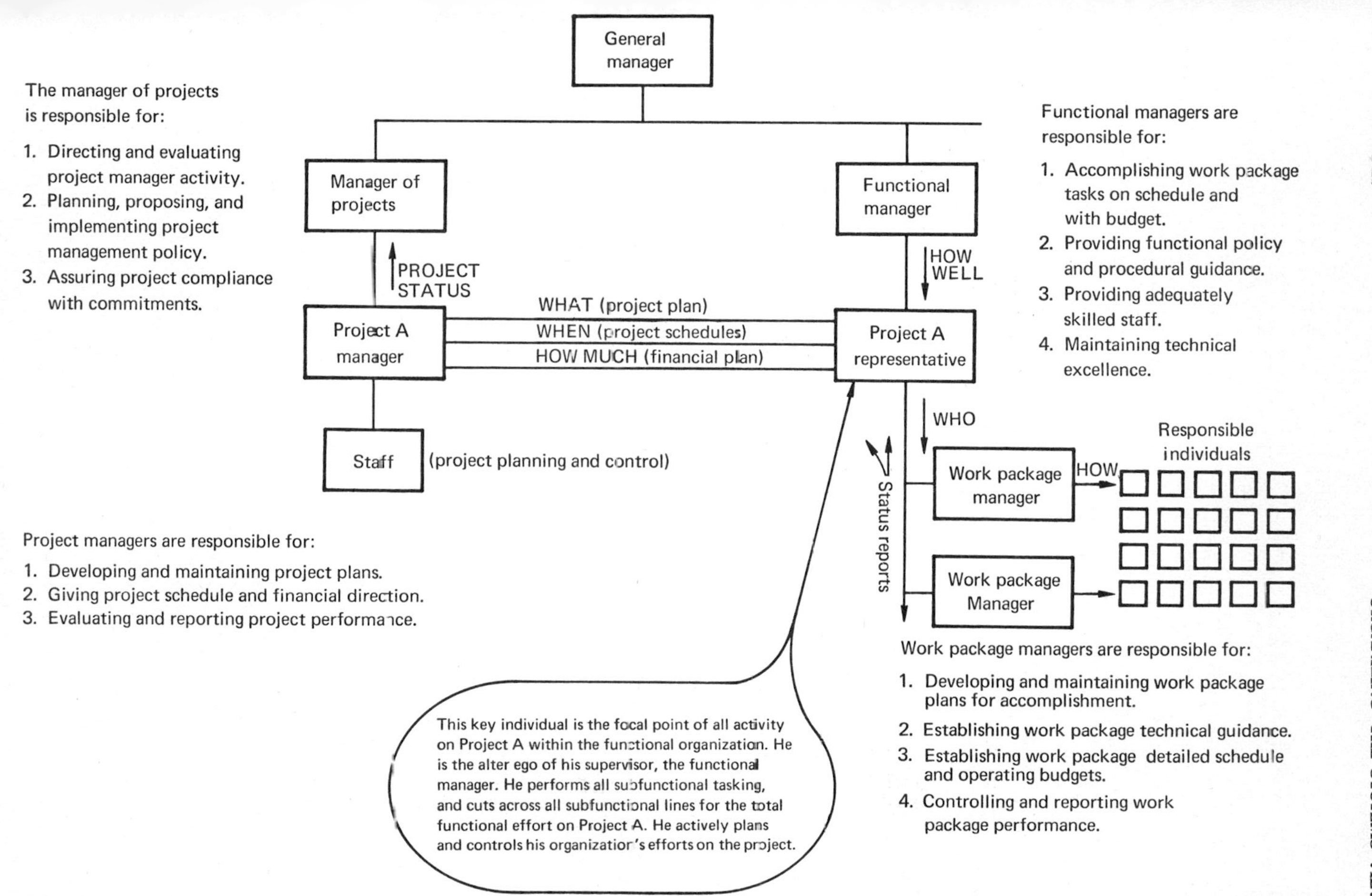

FIGURE 14-5 *Matrix organization relationship of project management to functional management. Adapted from: Weapons Department Management Guide, Astronuclear/Underseas Divisions, Westinghouse Electric Corporation, Baltimore, Md., p. 1–25. Used by permission.*

his individual personality, his perceptions of a particular situation, and his formal and informal roles in the organization.

The formal relationships which are specified in describing the role and position of each manager are reflected in a set of documents such as job descriptions. This documentation is more likely to describe *what the incumbent does* rather than how he *relates* to other groups. Position responsibilities and activities are directed toward improving a particular function's performance rather than toward the overall organizational effort. The basic conflict in the organization arises out of the inherent tendencies of the two managers to surrender to the pressures of suboptimization, that is, to emphasizing their function at the expense of the overall effort.

Conflict over issues and approaches in an organization is to be expected, not as a detriment to organizational efficiency, but rather as an expected matter resulting because people view problems and issues from the perspective of their loyalties. Deliberate conflict is not necessarily always bad or good—the conflict can become detrimental if it deteriorates to personal attacks and petty jealousies. "Deliberate conflict" is created by the organizational structure and can be consciously managed by the perceptive manager.

Of course, there are many conflicts which are in addition to those conflicts which come about through the specialization of skills. Ewing has cited a few of the more evident ones: ". . . the profit motive versus the community service motive, gentleness versus harshness, self-interest versus the organization's interest, and others. One of the most elusive ones is competitiveness versus cooperation."[2]

Deliberate conflict is not all blessing for the manager—conflict which fixes on personalities rather than issues can be weakening to the organization and frustrating to the members. But the tension that arises out of conflicting values can lead to ". . . better, more imaginative performance"[3] and to a stimulation of organizational esprit de corps.

"Needs" of Functional and Project Managers

To summarize the matrix organization, it is interesting to focus on the "different needs" of project and functional managers as they work to make the matrix organization effective.

THE FUNCTIONAL MANAGER'S NEEDS The functional department is typically organized to facilitate a continuous, homogeneous flow of work

[2]David W. Ewing, "Tension Can Be an Asset," *Harvard Business Review*, September, 1964, p. 26.

[3]*Ibid.*, p. 78.

in supporting the product or service being created by the organization. In such a setting the functional manager's needs are the same as the needs of any manager. Table 14-1 describes these needs, together with rationales for them.[4]

THE PROJECT MANAGER'S NEEDS In contrast, the needs of the project manager center around his focal position in the organization and his responsibility for the integration of project and functional activities.

[4]This table is based on material in P. W. Timmermans, "The Program Management Philosophy of the Martin Company," Air Force Institute of Technology, Wright-Patterson Air Force Base, Ohio, August, 1966.

TABLE 14-1 "Needs" of the effective functional manager

Need	*Rationale*
To . . .	. . . in order to . . .
. . . be placed in chain of command	. . . permit delegation of authority from superior
. . . have total authority to hold his subordinates responsible	. . . facilitate exacting of responsibility from subordinates
. . . have his authority equal his responsibility	. . . maintain unity of command
. . . have a job description which clearly states his authority and responsibility	. . . avoid divided loyalties and to achieve better working relationships for his subordinates
. . . accept his responsibility	. . . avoid overlap of authority and responsibility with other managers
. . . have a fair standard to evaluate and reward his performance	. . . respect the principle of the absoluteness of responsibility*
. . . have a high degree of technical competence in his field of specialization	. . . guide the actions of technical people under his functional jurisdiction
. . . maintain an adequate span of control	. . . be able to anticipate the direction which the technology of his function will take
. . . have a flexible organization alignment	. . . effectively manage the work of his subordinates
. . . be placed in a position to effectively facilitate his leadership	. . . reflect a realistic "layering" of organizational levels
	. . . meet the changing demands of his discipline
	. . . assure that he is recognized as the leader in his discipline and his responsibility for the managerial function within that discipline

*This principle means that "no superior can escape, through delegation, responsibility for the activities of subordinates, for it is he who has delegated authority and assigned duties." Harold Koontz and Cyril O'Donnell, *Principles of Management,* McGraw-Hill Book Company, New York, 1964, p. 65.

TABLE 14-2 "Needs" of the effective project manager

Need	*Rationale*
To . . .	. . . in order to . . .
. . . have well-defined and documented authority and responsibility	. . . gain acceptance by all participants
. . . have an adequate information system	. . . permit the reporting and evaluation of progress
. . . understand general management problems	. . . enhance the "authority of knowledge"
. . . have sufficient executive rank and organizational level	. . . have a management philosophy that does not have organizational or functional constraints
	. . . permit direct contact with the general manager
. . . be organized outside of the functional area in a separate support section	. . . provide general administrative leverage in dealing with supporting executives
. . . have a maximum of co-location of the project participants	. . . maintain a close proximity of working relationships
. . . maintain a flexible organization relationship	. . . maintain an unbiased position to better integrate across all functional elements
. . . be the project proposal manager	. . . be involved in the early vital planning that determines how the participating organizations will carry out their respective roles
	. . . have an appreciation of the complexity and diversity of a project
. . . have a strong voice in establishing a basic team throughout the life of the project	. . . facilitate qualified people from each functional area to work together as a team and be measured on team results
. . . have continuity of project understanding	. . . provide the necessary retention of specialized project knowledge
. . . have a definite contractual status with the rest of the project participants	. . . improve coordination
	. . . permit smooth transition during the passage from conceptual, definition, and other phases of the project life cycle
	. . . be accepted as the legal and de facto agent of the organization
	. . . strengthen the ability of the top executives to hold the project manager responsible for project results
	. . . avoid friction in the many cross-functional and cross-organizational relationships

TABLE 14-2 (Continued)

Need	*Rationale*
	. . . provide administrative leverage in planning, organizing, and controlling the project affairs . . . provide tenure of responsibility during the life cycle of the project . . . preclude an abdication of responsibility on the part of the participating organizations and/or members . . . motivate the project participants to accomplish adequate planning

Table 14-2 illustrates the project manager's specialized needs and some rationales for them in the fashion of Table 14-1.

THE DECENTRALIZED ORGANIZATION

As any organization grows, managers are forced to delegate authority and responsibility. The formal recognition of this in terms of the organization's form and structure is usually referred to as "being decentralized." Since, of course, no organization is ever completely centralized or decentralized, the question of decentralization is a matter of degree.

General Motors Corporation is usually thought of as a "model" of decentralization. Its division managers operate their organizations like independent businesses, subject to the framework of broad corporate policy. The advantages of this approach are seen as:

1. *Speed and lack of confusion in decision-making.*
2. *Absence of conflict between the top management and the divisions.*
3. *A sense of fairness in dealing with executives, confidence that a job well done would be appreciated, and a lack of politics in the organization.*
4. *Informality and democracy in management.*
5. *Absence of a gap between the few top managers and the many subordinate managers in the organization.*
6. *The availability of a large reservoir of promotable managerial manpower.*
7. *Ready visibility of weak managements through results of semi-independent and often competitive divisions.*
8. *An absence of "edict management" and the presence of thorough*

> *information and consideration of central management decisions.*[5]

At GM, each division (Chevrolet, Buick, Frigidaire, etc.) designs, develops, manufactures, and distributes its own products. As a result, the great bulk of management decisions are made and executed within the decentralized divisional structure. Only a small proportion of the total decision-making load remains for executives at the corporate level. Of course, these relatively few decisions made at the corporate level are of extraordinary significance.

In essence, the major operating entities of a decentralized organization are in competition with each other for resources and with outside firms for business. Usually, one division is not required to use the products of another; it may instead purchase a competitor's products if they are deemed more desirable. While this leads to many apparent paradoxes in organizations—e.g., one's own valve plant operating at less than peak capacity while a sister division purchases valves from an outside source—it imposes high standards of performance on subunits and their managers.

Such apparent paradoxes of the decentralized organization are often revealed to the layman in terms of the organization's apparent size. When an outsider visits a decentralized corporation's headquarters, he is usually amazed at the small size of the corporate-level activity. Some giant corporations are so decentralized that they literally operate with a hundred or so people at the corporate headquarters. One who is aware of the firm's reputation as an employer of tens of thousands cannot but be amazed by it.

Of course, decentralization is usually accompanied by centralization in a number of areas—policy formulation, financial controls, and the organization of some functional and service areas.

Policy Formulation in the Decentralized Organization

In the decentralized organization policy is formulated by corporate-level executives with the advice and counsel of divisional managers. At GM, policy decisions are recommended to the board of directors by various top management committees—executive, finance, etc.—whose work is performed by various subcommittees (termed "policy groups"). These committees and subcommittees are aided by corporate staffs and divisional managers, but the primary authority and responsibility rest with top management.

[5] P. F. Drucker, *Concept of the Corporation,* The John Day Company, Inc., New York, 1946, pp. 47–48.

Overall planning must also be accomplished at the top level of decentralized organizations to ensure that the goals of the various subsystems are consistent and that production and marketing policies fit into the organization's "big picture." This can take place in the manner described extensively in Chapter 11.

Control in the Decentralized Organization

If centralized control were not exercised over the various "independent" divisions of a decentralized organization, each division might proceed to seek its own objectives regardless of their compatibility or incompatibility with overall organizational objectives. Moreover, if financial controls were not imposed, each division might decide to expend a disproportionate share of overall organizational resources. Thus, control procedures are extremely important in a decentralized organization. (We shall take up the specifics of such procedures in Chapter 16.)

Centralized Functions in the Decentralized Organization

The advantages of size cannot be experienced if the organization is completely fragmented into independent operating entities. Therefore, in certain functions of the decentralized organization, centralization may be appropriate.

Computer facilities and operations represent one area of activity which has come to be more centralized in American business in recent years. Schoderbeck and Babcock have researched this and conclude that EDP's (electronic data processing's) level ". . . has been moving up in the managerial hierarchy. At the present time almost three-fourths of the EDP departments are located at the upper management level. . . ."[6]

Other functions and services which may well be centralized in a decentralized organization are finance, public relations, and engineering. Depending on the size of the organization and the costs involved, a subsystem manager can often have much greater expertise available to him if such services are centralized than he could ever hope to afford by himself.

Motivations other than economic efficiency also can often justify the centralization of certain functions in the decentralized organization. For instance, audit and accounting controls can be made more effective by having divisional financial officers report to corporate-level as well as to

[6] P. P. Schoderbeck and J. D. Babcock, "The Proper Placement of Computers," *Business Horizons,* October, 1969, p. 42.

divisional managers. This ensures objective comparisons of divisional performance in terms of both costs and rates of return.

Decentralization of the Executive

The simple demands of time have led some large organizations to decentralize the top executive position in the organization. An "executive office" or "president's office," in which several executives operate almost as "associate presidents," is becoming increasingly prevalent.[7]

Such a decentralized executive function does not usually operate as a committee. Each member has special talents and is encouraged to act as president in his assigned field—thus giving the executive more eyes and ears. However, the chief executive is still identified as such, and he still retains ultimate responsibility.

The Systems View of Decentralization

The similarities between the decentralized organization and the matrix organization are great. Both are output- or product-oriented. Divisions of GM are defined by their product lines, just as a project is defined by the "product" to be produced.

The management implications of decentralization are also similar to those of the matrix organization. Much must be accomplished informally, even though the organization may have many well-defined procedures. Managers must be able to work for more than one boss. Delegation of authority is a way of life in both kinds of organizations. The project manager acts as organizational executive for his assigned project, as the associate president acts as president for his assigned area in the "executive office" concept.

In each case, the need is to provide more managers with the broad scope of vision, authority, and responsibility which has become necessary to get things done in a complex world. The systems concept has supplied a way of doing this, and although it is sometimes not thought of in systems terminology, the systems concept continues to pervade modern organizations in their decision making, operations, and organization.

MODELING THE ORGANIZATION'S RELATIONSHIPS

Systems-oriented organizations may be depicted in terms of traditional organizational models such as the organization chart. However, when these

[7] "Management Outlook," *Business Week*, Dec. 23, 1967, p. 40.

models are constructed to be useful rather than to simply serve as decorations for the walls of managers' offices, they often become unwieldy and cumbersome.

For instance, the chart in Figure 14-6 depicts the general matrix concept as applied at TRW Systems. However, it does not adequately describe specific relationships, and if it were extended to greater detail, it would be difficult to understand. Even if such a chart is complemented by organizational relationships being defined by some form of documentation such as a policy letter, job description, organizational charter, or

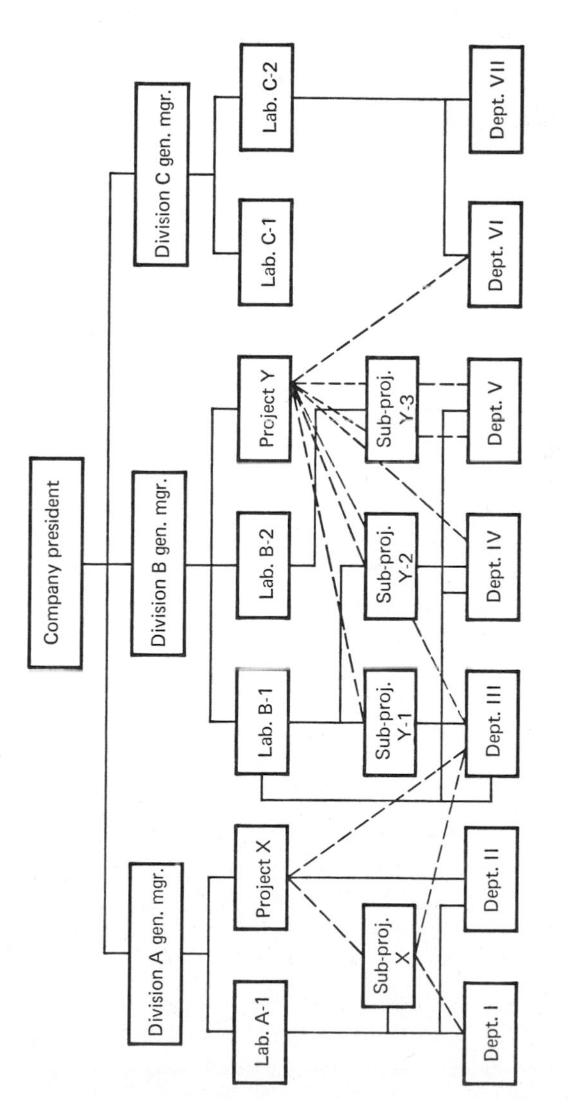

FIGURE 14-6 *Matrix organizational chart. Reprinted from the March 20, 1971, issue of* Business Week *by special permission. Copyrighted © by McGraw-Hill, Inc.*

related media, the total package often lacks specific definition as to how organizational positions (and, consequently, people) relate to each other.

Multiple-project organizations, the integration of de jure and de facto authority, activity specialization, and informal working arrangements all combine to make the conventional pyramidal organizational chart a relatively inadequate analytical tool to use for understanding inter- and intraorganizational relationships. Even the use of policy manuals, job descriptions, and other media to delegate authority and exact responsibility does not operationalize the way people relate to each other in their organizational roles, if only because often policy manuals are so voluminous that people simply do not take time to read them.

The classical organizational chart depicts the vertical interconnections of formal authority; a technique is needed which portrays the relationships existing among peers, superiors, subordinates, colleagues, and others in the organization. For example, Figure 14-7, an organization chart of an equipment test division of an electronics company, shows position titles, vertical flows of authority and responsibility, and levels of organizational alignment. Contrast this with the chart reflected in Figure 14-8, which depicts a total organizational system in the form of a systems *linear responsibility chart.* The latter chart shows the interrelationships between several individual jobs and a given organization task; it displays in matrix format the essential information published in policy manuals, job descriptions, and associated media.[8]

The advantage of the systems linear responsibility chart is that it enables the user to see his own role in the organization as well as the reciprocal roles of the individuals with whom he must work. Within any organization there are a myriad of authority-responsibility relationships; to identify and chart all of these relationships would be difficult if not impossible. Certain recurring relationships do emerge and serve as a focal point for the organizational deliberations among job positions and tasks. These relations center around the following:

Work is done. This relationship designates the individual who actually performs the task.
Direct supervision. The individual who exercises supervisory jurisdiction of the "work is done" incumbent.
General supervision. This relationship is assigned to the administrative supervisor of the person performing "direct supervision."
Intertask integration. A person who has functional authority over how a particular task is to be accomplished.

[8]Adapted from David I. Cleland and William R. King, *Systems Analysis and Project Management,* McGraw-Hill Book Company, New York, 1968, pp. 196–224.

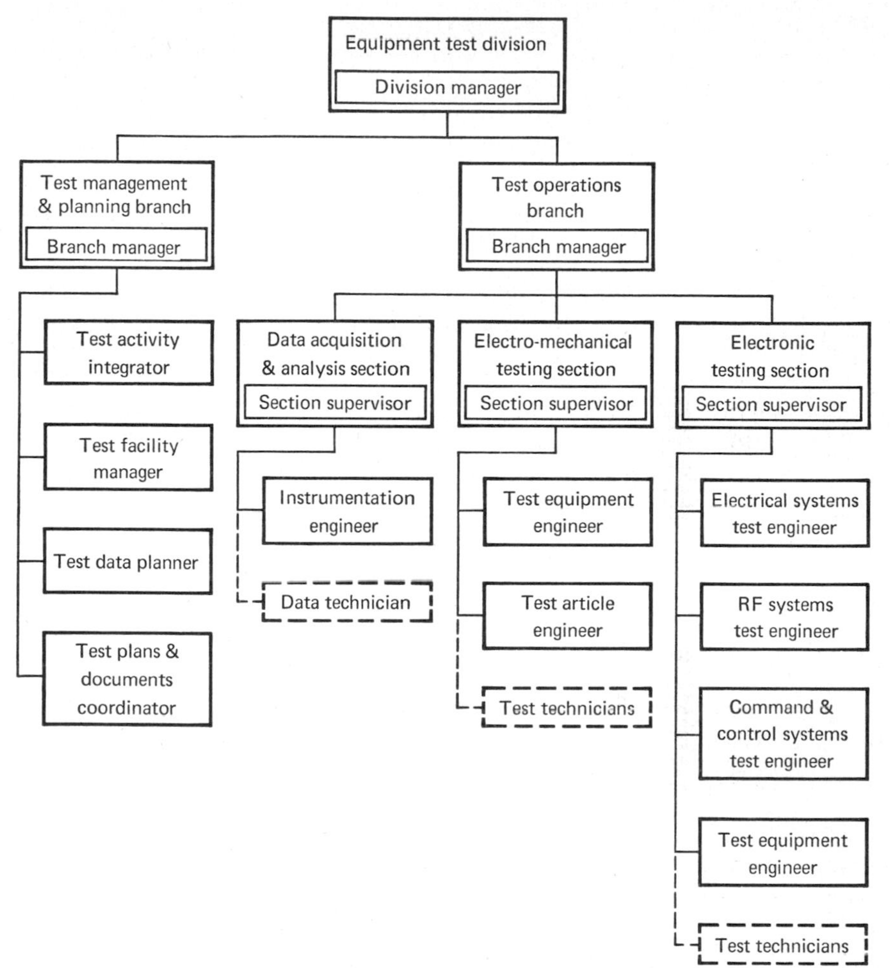

FIGURE 14-7 *Pyramid organization chart for equipment test division. Adapted from David I. Cleland and William R. King, Systems Analysis and Project Management, McGraw-Hill Book Company, New York, 1968, p. 216.*

Occasional intertask integration. This relationship is similar to intertask integration except that the latter deals with general or routine tasks whereas "occasional intertask integration" deals with tasks of a specialty nature.
Intertask coordination. This relationship is one of routine providing of information or notification of actions about to be taken. It includes the "what" and "when" nature of such things as intertask sequencing, schedule consistency, quantities, qualities, etc., but not control over the "how" of task accomplishment.
Occasional intertask coordination. This is basically the same relationship

Key to symbol titles

- ■ Work is done
- ◩ Direct supervision
- □ General supervision
- ▲ Intertask integration
- △ Occasional intertask integration
- ● Intertask coordination
- ○ Occasional intertask coordination
- ⬢ Output notification mandatory

		Test management and planning branch					Test operations branch	Data acquisition and analysis section		Electro-mechanical testing section			Electronic testing section				
Task/job relationships	Division manager	Branch manager	Test activity integrator	Test facility manager	Test data planner	Test plans & documents coordinator	Branch manager	Section supervisor	Instrumentation engineer	Section supervisor	Test equipment engineer	Test article engineer	Section supervisor	Electrical systems test engineer	RF systems test engineer	Command & control systems test engineer	Test equipment engineer
Major functional area: test program activities																	
Approve test program changes	■	●					●	○									
Define test objectives	◩	■					▲	●		●			●				
Determine test requirements	□	◩	■	●	●	●	▲	○		○			○				
Evaluate test program progress	◩	■	●	○			●										
Make test program policy decisions	■	●					●										
Write test program responsibility documents	□	◩	■				▲			○			○				
Major functional area: integration of test support act																	
Chair test working group	□	◩	■				▲			○			○				
Prepare milestone test schedules		◩	■				▲	⬢		⬢			⬢				
Write test directive		◩	■	●	●	●	△			○			○				
Write detailed test procedures			○	○	⬢	⬢	□	▲	△	◩	▲	■	◩	■	■	■	▲
Coordinate test preparations			△	▲	○		◩	■	●	■	●	●	■	●	●	●	●
Verify test article configuration			⬢			●			▲			■		■	■	■	
Major functional area: all systems tests																	
Certify test readiness	◩	▲	△	△			■	●		●			●				
Perform test director function	■	●	●	○			▲										
Perform test conductor function	◩						■	▲		▲			▲				
Analyze test data					●	○		▲	▲	◩	■	■	◩	■	■	■	■
Resolve test anomalies	□				○		◩	●	●	■	●	▲	■	▲	▲	▲	●
Prepare test report	□	▲	○		○	⬢	◩	■		■			■				

FIGURE 14-8 *Systems linear responsibility chart for equipment test division. Adapted from David I. Cleland and William R. King, Systems Analysis and Project Management, McGraw-Hill Book Company, New York, 1968, pp. 214–215.*

as the preceding one except that only specialized instances of notification are required.

Output notification mandatory. Used to denote a job position that must receive an input factor of information from the "work is done" job relationship. This relationship involves the transfer of information and not any coordinative responsibility.

The completed chart is a linear responsibility model of a managing subsystem which helps to clarify how organization positions relate to each other. The chart contains the information of the pyramid chart reflected in a different form; it also reflects the flow of work among the various subsystems of the organization. The chart can serve both as a planning tool in determining the best arrangement of authority and responsibility and as a control technique in ensuring that tasks are carried out as originally planned. Managers who use the chart will find that it forces "thinking through" the desired interrelationships, thereby giving users a better understanding of their reciprocal roles and obligations.

SUMMARY

The matrix organization represents a combination of traditional vertical functional departmental organization with project organization. In a matrix organization, projects cut across functional lines. The project teams are composed of people from the functional departments who are assigned to the project for a specific period or for the duration of the project. When their assignment is complete, they return to their function.

In a matrix organization, the project and functional managers have different roles; yet each is dependent on the other. The project manager exerts a general management viewpoint with regard to his project. The functional manager is responsible for maintaining the integrity of his function. Yet neither can operate without the support of the other.

The purposeful conflict established in such an organization between the project and functional managers serves to create an atmosphere in which negotiation is the only basis for action and progress. However, the conflict can be counterproductive for the organization if it degenerates into personalities and activities aimed at "hurting the other fellow" more than they are at helping. Hence, the matrix environment is a dynamic and exciting one that can create great accomplishments as well as great disorder, depending on the skill with which it is managed.

Like the matrix organization, the decentralized organization is created to take advantage of systems characteristics. Although most decision making in such an organization is performed at low levels, the most significant decisions are treated by top organizational management. Most frequently policy formulation, control, and some other functions remain centralized even in the most decentralized organization.

To model the relationships inherent in such complex organizations requires more sophisticated approaches than the traditional organization chart model. The system linear responsibility chart—which entails a matrix of relationships—permits the clear description of the relative roles of individuals in the organization. As such, it is an organizational model which focuses on the systems characteristics of organizations.

EXERCISES

1. What is a "matrix organization"? How is it compatible with the systems view of an organization?
2. What is meant by "the project manager's need to cross organizational lines in order to achieve project goals"?
3. In a matrix organization, which manager has the responsibility for keeping up with the "state of the art" in a particular discipline?
4. What is the value of operating a matrix organization in which disciplinary elements remain intact? Why not just have projects and do away with the disciplinary organizations?
5. A manager wishes to consummate a special business deal for his firm under a set of unusual conditions prescribed by the customer. He believes that the deal will be very profitable, but he is told that the accounting system is not set up to handle such a transaction. Discuss this situation in terms of project and functional management.
6. Discuss the project manager's role in interorganizational project management.
7. "The face-to-face contact between the project managers in two interfacing organizations requires that other project personnel not be in direct contact with their counterparts in the other organization." Discuss.
8. Why is the project manager referred to as performing a general management role?
9. Someone has said that the project manager acts as the organization's chief executive with regard to his project. Discuss.
10. What is meant by a "purposeful conflict" between the project and functional manager? How is this conflict to be resolved?
11. Contrast project-functional authority with line-staff authority.
12. What is the role of a "manager of projects"?
13. A project manager resorts to an appeal to the common supervisor in order to settle a dispute with a functional manager. Would you say that this action reflects a failure of the system or an indication that the system is working smoothly?
14. Discuss the project and functional managers' respective roles in terms of suboptimization.

15. Discuss the salient differences between the "needs" of the project manager and the functional manager.
16. "In a decentralized organization, relatively few decisions are made at the highest organizational level. Therefore, even though top management in such an organization is obviously important, it is relatively less significant to the organization than is the top management of a centralized firm." Comment on this statement.
17. When a decentralized organization is operated according to a "profit center" concept, each profit center manager is compensated on the basis of the profits produced within his domain. Sometimes, this leads to difficulties in developing joint activities involving a number of profit centers. Explain this situation using your knowledge of human behavior. What might be done to resolve such a problem?
18. Discuss *(a)* policy formulation, *(b)* financial control, *(c)* management control, *(d)* centralized functions, in decentralized organizations.
19. Discuss the similarity of the concept of a "president's office" to that of project management.
20. Contrast organization charts and linear responsibility charts. For what purposes might each be best?
21. One of the advantages of the linear responsibility organizational model is that it forces the modeler to "think through" the relationships which exist, or should exist, in the organization. Of course, this is true of any modeling process. However, this point of view can become a rationalization for an ineffective modeling process. How would you evaluate whether the development of a linear responsibility chart has had this value in your organization?
22. Develop a linear responsibility chart for an organization with which you are familiar (for instance, a university).
23. Distinguish between the planning and execution responsibilities of the project and functional manager—i.e., how are their responsibilities different in the areas of planning and execution?
24. Why is the project manager referred to as an integrator generalist? Would it be proper to refer to the functional manager in like terms?

REFERENCES

Cleland, David I., "Organizational Dynamics of Project Management," *IEEE Transactions on Engineering Management,* December, 1966.

Delbecq, Andre, et al., *Alternative Strategies of Organization Design: A Toxonomy of Micro-Organizational Variation,* prepublication draft, Matrix Management Project Group, Graduate School of Business, The University of Wisconsin, Madison, 1968.

Dickson, G. W., "Management Information-Decision Systems: A New Era Ahead," *Business Horizons,* December, 1968.

Flaks, Marvin, and Russell D. Archibald, "The EE's Guide to Project Management, III—Network Systems—A Project Management Tool," *Electronic Engineer,* June, 1968.

______, "The EE's Guide to Project Management, IV—The Computer's Role in Network Management Systems," *Electronic Engineer,* June, 1968.

French, Wendell, "Processes Vis-à-Vis Systems: Toward a Model of the Enterprise and Administration," *Journal of the Academy of Management,* March, 1968.

Fricks, Robert E., "Structural Problems in Organization Theory," Systems Research Center, Case Western Reserve University, Cleveland, 1967.

Goodman, Richard A., "A System Diagram of the Functions of a Manager," *California Management Review,* Summer, 1968.

Higgans, Carter C., "The Organization Chart: Its Theory and Practice," *Management Review,* October, 1956.

Hodgetts, Richard M., "Leadership Techniques in the Project Organization," *Academy of Management Journal,* June, 1968.

Holmes, Roger, "Power and Consent, A Social-Psychological Analysis of Organizations," London School of Economics, London, September, 1966.

Lazar, R. G., and A. D. Kellner, "Personnel and Organization Development in an R&D Matrix-overlay Operation," *IEEE Transactions on Engineering Management,* June, 1964.

Macko, Donald, "Hierarchial and Multilevel Systems," Systems Research Center, Case Western Reserve University, Cleveland, 1967.

Mee, John F., "Ideational Items: Matrix Organization," *Business Horizons,* Summer, 1964.

Mockler, Robert J., "The Systems Approach to Business Organization and Decision Making," *California Management Review,* Winter, 1968.

Pondy, L. R., "Systems Theory of Organizational Conflict," *Academy of Management Journal,* Summer, 1966.

Sengupta, S. S., and R. L. Ackoff, "Systems Theory from an Operations Research Point of View," *IEEE Transactions on Systems Science and Cybernetics,* November, 1965.

Shull, Fremont A., *Matrix Structure and Project Authority for Optimizing Organizational Capacity,* Business Research Bureau, Southern Illinois University, Carbondale, Ill., October, 1965.

Stewart, John M., "Making Project Management Work," *Business Horizons,* Fall, 1965.

15

motivation—the human subsystem

The human element is the breath of the organization's life. However, the human subsystem of an organization is more than just the people who devote their working lives to the achievement of organizational goals; it also includes the myriad of formal and informal interpersonal relationships which pervade an organization.

The changes which have taken place in organizations in recent years are more apparent in other subsystems—e.g., the technological subsystem. After all, one might say, "People are people, and they are today much the same as they were yesterday." In fact, this is not the case. The changes which have occurred in the attitudes and motivation of workers and in the way in which they are "managed" are even more striking than are the advances which have taken place in technology.

THE MANAGER AND INDIVIDUAL BEHAVIOR

The manager is a person who accomplishes results through working with others in the group situation; he makes things happen *through the efforts of other people.* His concern is to accomplish results by attaining objectives. While he may deal with his subordinates and attain his objectives through directing their efforts, he must also work closely with (and for) others in the group situations. Thus, just as an individual plays many roles—husband, father, division manager, PTA member, etc.—during a given day, so, too, does a manager play different roles in his interpersonal relationships within an organization. He must, for example, work:

1. With *superiors* who direct his efforts and who depend on him to accomplish their ends,
2. With *peers* toward common organizational goals,
3. With *associates* in organizations that are reciprocally supporting him in his environment,

4. With his *subordinates* by doing the things necessary to provide the proper economic, social, and psychological environment for them to work together with personal satisfaction.

To successfully operate in the organization's human subsystem, the manager must have an understanding of individual and group behavior and the things which motivate people. He must be both knowledgeable and reflective about such questions as:

1. Why do people behave in the manner they do?
2. Why don't people agree with my views?
3. Why can't people understand things that are perfectly clear to me?
4. Why won't people do the things I want them to do?
5. Why do people resist change when the change is for the better?
6. Why can't I predict how people are going to react to a situation?

To aid in developing a better understanding of individual behavior and the way in which the manager can exercise some control in directing people toward the organization's objective, let us consider the problems and opportunities involved in "managing individual behavior."

A Job Performance Model

A simple model of job performance, such as that shown in Figure 15-1, can provide the basis for the manager's understanding of the ways in which he can influence individual behavior. The model shows that job performance is a function of two primary elements—the individual's *ability* and his *motivation*. Ability, in turn, depends on mental capacity, experience,

FIGURE 15-1 *A simple job performance model.*

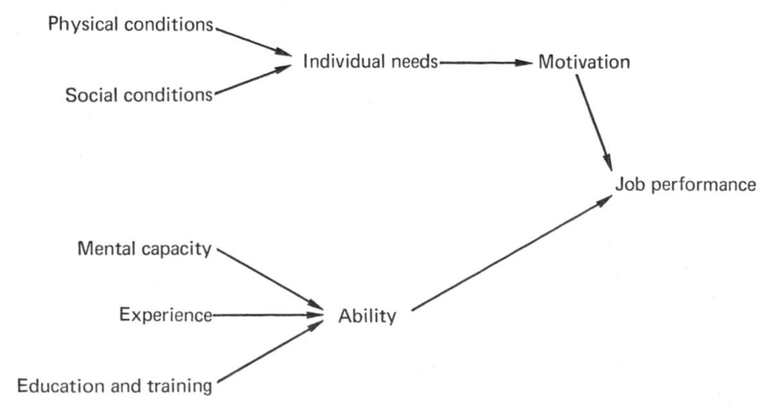

education, and training. An individual's motivation depends on his "personal needs" and the way in which his physical and social surroundings satisfy him.

The elements which the manager can affect are at the left of the diagram in Figure 15-1. Although he cannot affect the basic abilities or needs of the people who function in his organization, he can select people on the basis of their needs, mental capacity, experience, education, and training, and he can provide physical and social conditions which contribute to the fulfillment of individual needs.

Individual Needs

To better understand the operational methods which the manager may use to affect job performance through personal needs, let us begin by concentrating on the concept of individual needs. Two alternative views of human behavior, as put forth by Douglas McGregor in his book, *The Human Side of Enterprise,* show a basic starting point for the analysis of individual needs. McGregor calls his alternative views Theory X and Theory Y. Theory X states:

> *The average human being has an inherent dislike of work and will avoid it if he can. . . . Because of this human characteristic of dislike of work, most people must be controlled, directed, threatened with punishment to get them to put forth adequate effort toward the achievement of organizational objectives. . . . The average human being prefers to be directed, wishes to avoid responsibility, has little ambition, wants security above all.*[1]

Many of the beliefs associated with Theory X are implicit in much of the traditional literature of organizations and management; they are operationalized in the stress that management places on "productivity," or in the concept of the "full day's work" and of the evils of featherbedding and the restriction of output. In other words, management's view of human behavior which is reflected in these practices presumes that man has an inherent dislike of work. However, according to McGregor, Theory X neglects some basic assumptions about the motivation of people. These neglected assumptions are that:

The human being is a wanting individual—as soon as one need is satisfied, another takes its place. Human needs follow some pattern of importance.

[1]Douglas McGregor, *The Human Side of Enterprise,* McGraw-Hill Book Company, New York, 1960, pp. 33–34.

Once a need has been satisfied, that need is no longer a motivator of behavior. Thus, if a thirsty man has satisfied his thirst, he will not be motivated to find something to drink. When basic or primary needs such as satisfaction of hunger and thirst are satisfied, needs of a higher order assume considerable importance. Social needs such as acceptance, belonging, association, and forgiving and receiving friendship and love take on importance as the individual enters into the many social units found in modern society. The highest order of needs—even above the social needs—are egotistic needs and relate to one's self-esteem, self-respect, and self-confidence, his need for achievement, independence, knowledge, and needs for status, and those needs that relate to one's reputation. Highest of all man's needs are the needs for self-fulfillment, to become that which one is capable of becoming in the broadest sense of the term.[2]

McGregor's Theory Y concept has the following elements:

1. *The expenditure of physical and mental effort in work is as natural as play or rest. The average human being does not inherently dislike work. Depending upon controllable conditions, work may be a source of satisfaction (and will be voluntarily performed) or a source of punishment (and will be avoided if possible).*
2. *External control and the threat of punishment are not the only means for bringing about effort toward organizational objectives. Man will exercise self-direction and self-control in the service of objectives to which he is committed.*
3. *Commitment to objectives is a function of the rewards associated with their achievement. The most significant of such rewards, e.g., the satisfaction of ego and self-actualization needs, can be direct products of effort directed toward organizational objectives.*
4. *The average human being learns, under proper conditions, not only to accept but to seek responsibility. Avoidance of responsibility, lack of ambition, and emphasis on security are generally consequences of experience, not inherent human characteristics.*
5. *The capacity to exercise a relatively high degree of imagination, ingenuity, and creativity in the solution of organizational problems is widely, not narrowly, distributed in the population.*
6. *Under the conditions of modern industrial life, the intellectual potentialities of the average human being are only partially utilized.*[3]

[2] Adapted from *Ibid.*, pp. 33–43.
[3] *Ibid.*, pp. 47–48.

There is a vivid contrast between the Theory Y assumptions about human behavior. Theory Y points up the fact that the limits of human cooperation in organized activity are set by how people are managed and motivated, not by inherent characteristics of individual personality. Of course, Theory X provides an easy out—if people do not cooperate with you, it can be attributed to their inherent traits, not to your lack of management ability. Theory Y places the responsibility for motivation and attitudes of workers squarely on the manager. If people are indifferent, lazy, unwilling to assume responsibility, uncooperative, and uncreative, then the cause lies in the manager's method of handling his people.

In developing his Theory Y concept McGregor speaks of the needs of people. A. H. Maslow has devised a "hierarchy of needs" concept which places the various needs of individuals in relationship to one another.[4] His concept of a hierarchy with certain "higher-order" needs and "lower-order" needs identifies five basic levels:

1. The basic sociological needs, such as hunger, thirst, the activity-sleep cycle, sex, and evacuation
2. The safety and security needs for protection
3. The love needs for satisfactory associations with others
4. The esteem needs for self-respect and for the respect of others
5. The self-actualization needs of achieving one's potential for maximum self-development

These basic needs are illustrated in hierarchical form in Figure 15-2. The low-level needs—those related to the body rather than the mind—and the higher-level needs are thought of as a hierarchy in the sense that lower-priority needs will tend to monopolize the individual's conscious thoughts and actions until they are fulfilled. Then emphasis can be given to the fulfillment of higher-order needs.

Maslow's model does not assure that the lowest "box" of needs must be completely filled before the second one can be begun; rather, one can be fulfilling all levels of needs at once. But if certain minimum levels are not attained for the basic needs, little attention will be devoted to the higher ones. Of course, each person's "satisficing value"[5]—that degree of fulfillment of a given need which will permit him to go on to devote conscious attention to the next higher-order need—is different for various need levels. A mountain-climbing adventurer may require the fulfillment of only a small percentage of "normal" human needs from level 1 for a

[4] A. H. Maslow, "A Theory of Human Motivation," *Psychological Review*, vol. 50, 1943, pp. 370–396, and A. H. Maslow, *Motivation and Personality*, Harper & Row, Publishers, Inc., New York, 1954.

[5] A term borrowed from March and Simon and applied here. See J. G. March and H. A. Simon, *Organizations*, John Wiley & Sons, Inc., New York, 1967, pp. 140–141.

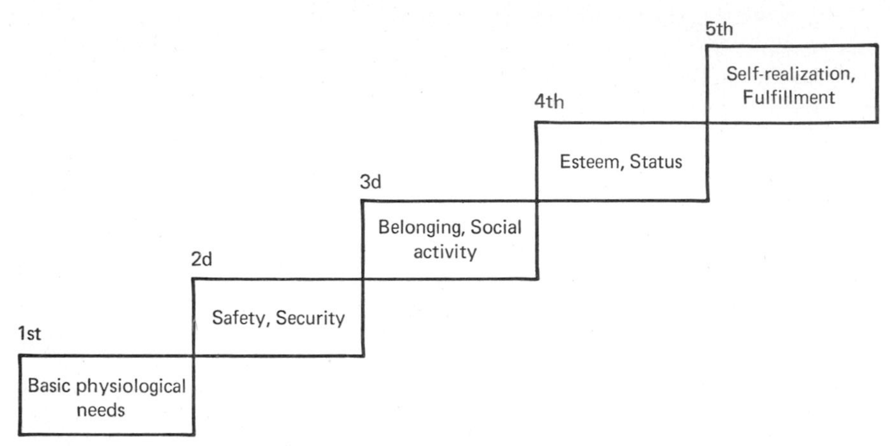

FIGURE 15-2 *Order of priority of human needs. Adapted from Keith Davis, Human Relations at Work, McGraw-Hill Book Company, New York, 1967, p. 26. Used by permission.*

period of time while he seeks to fulfill needs at a higher level. The classic "harried housewife" with four children to care for may have such high satisficing values for the basic needs that she never gets to seek out the higher needs. However, the mountain climber's satisficing value probably returns to normal after he has conquered the peak, and the housewife may get the opportunity to fulfill herself after her children are in school and she has some free time to devote to higher-order needs.

The Manager's Role in Satisfying Needs

Modern managerial thought assumes the "hierarchy of needs" model and the Theory Y concept to be essentially valid—i.e., they are believed to be reasonably good descriptions of reality. How, then, does the manager influence the individual to work toward the organization's goals?

The answer to this question is not simple and direct for a variety of reasons. First, the higher-level needs, which have to do with the mind and the spirit rather than the body, are not readily assessable. According to Davis, since they are nebulous in nature,

> *Dissatisfied workers usually attribute their dissatisfaction to something more tangible, such as wages. Many wage disputes are consequently not really wage disputes, and meeting the wage request does not remove the basic dissatisfaction which existed. . . .* [such] *needs have the following characteristics:*
>
> **1.** *They are strongly conditioned by experience.*
>
> **2.** *They vary in type and intensity among people.*

3. *They change within any individual.*
4. *They work in groups, rather than alone.*
5. *They are often hidden from conscious recognition.*
6. *They are nebulous feelings instead of tangible physical needs.*
7. *They influence behavior. It is said that "we are logical only to the extent our feelings let us be."*[6]

We can often recognize the effects of physical illness—and we would encourage an individual who has, or thinks he has, a physical ailment to seek medical attention. A person's mental attitude may or may not have overt symptoms—yet the deprivation of his higher-order needs may have real behavioral consequences. The individual whose drive for recognition or status is frustrated may well feel physically ill, or he may exhibit hostility or passivity. In such a case, his illness or other manifestation may be as real as one which can be attributed to a physical cause. However, it is much more difficult to assess.

Another difficulty in assessing higher-level needs and in utilizing this assessment to the organization's advantage revolves about the question, "How much is enough?" Should an individual be provided with conditions which are conducive to the *total* fulfillment of his needs? If this is done, these fulfilled needs no longer act as motivations. Or, alternatively, should each individual be provided with conditions which are conducive to fulfilling basic needs to the satisficing value so that higher needs, which are presumably more important to the organization, can be given attention?

The third difficulty has to do with the question, Which needs are important for whom? For instance, is the manager to be concerned about higher-level needs only with regard to executives, or should he show such concern for all white-collar workers or for everyone in the organization? Can he reasonably concentrate solely on low-level need fulfillment for "workers" and solely on high-level need fulfillment for managers, for instance? Or must he do everything at once?

These difficulties are both real and apparent, and while no universally correct answers are available which obviate them, some empirical results suggest the answers and, therefore, the ways in which the manager may go about operationalizing his knowledge of human behavior. First, however, let us consider the controllable variables with which he has to work.

THE CONTROLLABLES OF HUMAN BEHAVIOR Theory Y suggests that "controllable conditions" may make work a source of either satisfaction or punishment.

[6] Keith Davis, *Human Relations at Work,* McGraw-Hill Book Company, New York, 1967, p. 24.

Consider some of the motivating influences related to controllable conditions which might be relevant to an individual who performs as a member of an organizational team:

Steady employment
Respect for me as a person
Adequate rest period or coffee breaks
Good pay
Good physical working conditions
Chance to turn out quality work
Getting along well with others on the job
Chance for promotion
Opportunity to do interesting work
Pensions and other security benefits
Having employee services such as office, recreational and social activities
Not having to work too hard
Knowing what is going on in the organization
Feeling my job is important
Having a voice in decisions
Having a written description of the duties in my job
Being told by my boss when I do a good job
Getting a performance rating, so I know how I stand
Attending staff meetings
Agreement with team's objectives
Large amount of freedom on the job
Opportunity for self-development and improvement
Chance to work without direct or close supervision
Having an efficient supervisor
Fair vacation arrangements[7]

This list suggests that there are many different kinds of reasons, working in different combinations, to motivate one to do his best work. Therefore, there are many different controllables for the manager—from vacation arrangements to the interest level of the work. One may think of these in terms of some basic categories—perhaps those on the left side of Figure 15-1—physical conditions, social conditions, mental capacity, experience, and education and training.

This list suggests that there are many different reasons working in different combinations to motivate us to do our best work. After the in-

[7] This list is based on one which was suggested by Gordon L. Lippitt of Leadership Resources, Inc., Washington, D.C. during a lecture at Wright-Patterson AFB, Dayton, Ohio.

dividual feels that his pay is adequate for the type of work he is doing, other factors come into play. For the creative scientist, a "large amount of freedom on the job" may be the most important factor. On the other hand, an hourly rated employee might place the most importance on "good physical working conditions." This suggests that the individual is truly unique, both in terms of himself and in terms of how he relates to the team or group with whom he affiliates. Why is this so? Is it because each individual has had a singular background of experience, or are there more subtle reasons? Certainly, the experience we have had in our lifetime conditions our behavior and how we react to various social stimuli. When an individual is brought into an organization, he brings about some influence on the existing members of the group. When an individual is hired for a job, his employer seeks out his mental and physical potential, for which the individual is paid a wage or salary and extended a variety of fringe benefits such as hospitalization and life insurance, paid vacations, etc. But the individual who is hired brings many things into the employment situation besides just his physical or mental capacities. Certainly, these other things are not the capabilities for which he is to be paid. These intangible factors are, of course, difficult to measure, but they do have some influence on the people in the work element; these factors affect the work of the individual and the work of his associates. From the standpoint of the individual, these intangible factors include such things as:

1. His psychological attitudes
2. His physiological condition
3. His political beliefs
4. His moral standards
5. His professional standards
6. His prejudices and habits[8]

Unfortunately, we are not able to measure these factors too well during the process of hiring the individual, although the psychological testing that can be conducted can point out some of the obvious things of one's value system and expected behavior patterns. Since this is true, why is it important that we recognize the existence of these factors? By doing so we can better understand why the individual acts as he does. His actions and attitudes come about not only because of the beliefs and prejudices that he has, but also because these beliefs and prejudices harmonize and conflict with those of his peers and associates.

The manager can do little to affect mental capacity except to hire

[8]Adapted from Michael J. Jucius, *Personnel Management,* Richard D. Irwin, Inc., Homewood, Ill., 1963, p. 86.

those who demonstrate it. He can, however, "create experience" through the jobs and tasks which he assigns. The common practice of rotating people through a variety of jobs is an attempt at "managing experience" rather than just letting it happen, for instance.

The manager can do a great deal in the way of education and training. He can institute formal courses at all levels—from those which seek to remedially upgrade basic reading and writing skills to advanced seminars on new techniques of management. Most large organizations have recognized this controllable and have developed extensive training programs at all levels. One need only read through the extensive management seminar offerings of the American Management Association and of individual industrial firms and government agencies[9] to recognize the scope of such activities at the higher managerial levels.

The provision of adequate physical conditions is another area in which management has control of the degree to which employees' needs will be fulfilled. The physical conditions surrounding a worker may influence his job performance—although the precise payoff is difficult to measure. Certainly, physical conditions affect the worker's status and role. Linton[10] defines *status* as relating to the positions in a social system occupied by designated individuals. *Role* has to do with the expected behavior pattern for a particular position. Thus, while status and role have nothing to do with physical conditions per se, physical symbols are often used to designate status. The badge of office worn by the military commander is only slightly less subtle than the three secretaries, one administrative assistant, outer office, private bath, carpet, and preferred parking space which help define the status of the bank president. So, too, are such appurtenances as a corner office, a good view, a private secretary, and other subtle differences in physical conditions used to designate status in such a "classless society" as a university.

Other physical conditions which are pervasively controlled, but not usually granted selectively, are job and plant layouts, lighting, heat, air conditioning, rest periods, and perhaps "music to work by." The decision to expend funds on these elements is dictated by a desire to fulfill the basic low-level needs which will "make employees happy."[11]

The most significant class of controllables is also the most difficult to manage—social conditions. Social conditions have to do with the makeup of both the formal and the informal organization and the leadership

[9] For instance, Westinghouse Electric Corporation offered such a wide variety of successful seminars that it spun off the Westinghouse Learning Corporation to develop more and to market them to non-Westinghouse organizations.

[10] Ralph Linton, *The Study of Man,* Appleton-Century-Crofts, Inc., New York, 1936.

[11] We shall later question the worth of such elements as positive motivators. See page 373.

relationships which are exercised therein. Such tangibles as personnel policies, wage levels, performance ratings, and the size of the organizational units are a part of "social conditions," but so, too, are such intangibles as leadership styles, team structure, group cohesiveness, etc. These controllable elements are so important that we shall devote considerable attention to them later. Now it is best to turn to some theory and empirical evidence which more clearly define the positive and negative actuators of behavior and which provide the manager with an approach to motivation.

AN OPERATIONAL APPROACH The various categories of Figure 15-1—such as social conditions, physical conditions, experience, and training and education—provide a general guide to the "buttons" which the manager can push to influence job performance. How, though, does he decide which button to push, and when, and how hard?

One general answer to these questions is suggested by the work of Herzberg[12] and others. They suggest five *determiners of job satisfaction:*

Achievement
Recognition
Work itself
Responsibility
Advancement

and a number of major *dissatisfiers:*

Company policy and administration
Supervision
Salary
Interpersonal relations
Working conditions

The satisfiers—labeled *motivators* to suggest their importance—are internal to the job: content, achievement, recognition, responsibility, and growth; while the dissatisfiers—labeled *hygiene factors*—are external environmental aspects. The motivators are positive in the sense that they can lead to increased job satisfaction and better performance. The hygiene factors are negative in the sense that when they are not satisfactory, they may lead to dissatisfaction and poorer performance, while their being satisfactory will not bring complete contentment with the job.

[12] F. Herzberg, B. Mausner, and B. Synderman, *The Motivation to Work,* 2d ed., John Wiley & Sons, Inc., New York, 1959, and F. Herzberg, *Work and the Nature of Man,* The World Publishing Co., Cleveland, 1966.

Examples of the successful application of a Theory Y approach through Herzberg's model have been well publicized. American Telephone and Telegraph is reported to have reduced turnover by 27 percent, nearly doubled productivity, and eliminated jobs for a net saving of over $½ million in 18 months by allowing girls in their stockholders relations department to research, compose, and sign their own responses to stockholders' queries without being checked by superiors.[13] Other organizations have had similar success: ". . . Texas Instruments pulls production people off the line for troubleshooting conferences with . . . engineers. Aerojet-General . . . presents its line workers with . . . production problems and asks what they would do to solve them."[14]

Herzberg has applied the phrase "job enrichment" to operational programs which seek to enhance job satisfaction through controlling the motivators. *Fortune* has described these programs as having

> *. . . certain elements that appear characteristically wherever job enrichment is going on. Central, of course, is the basic idea of giving the worker more of a say about what he or she is doing, including more responsibility for deciding how to proceed, more responsibility for setting goals, and more responsibility for the excellence of the completed product. It can also mean, in appropriate kinds of plants, allowing the worker to carry assembly through several stages, sometimes even to completion and preliminary testing, rather than doing just one small operation endlessly. Automobile assembly jobs offer dreary, repetitive, relentless work, and it is this uninspiring monotony that managers seeking to enrich jobs strive to avoid.*[15]

THE MANAGER AND GROUP BEHAVIOR

The human subsystem is indeed a *system,* with all of the interactions and interdependencies which the term presumes. While the individual can truly achieve much in large organizations, he can do so only through other people and with the support of other people.

Even the production worker who operates a machine tool finds that he is dependent on many people in his environment *other than his supervisor.* If he requires the services of an electrician or maintenance technician, the responsiveness of these people to his needs can influence his production rate. Other individuals in the production system who provide

[13]"Management: Making a Job More Than a Job," *Business Week,* Apr. 19, 1969, pp. 88–89.

[14]"Job Motivation: Money Isn't Everything," *Newsweek,* Apr. 22, 1968.

[15]J. Gooding, "It Pays to Wake Up the Blue-Collar Worker," *Fortune,* September, 1970, p. 133.

him with a flow of materials to be processed can also influence his effectiveness and the satisfaction he finds on the job. Furthermore, the production worker supports other individuals who depend on him in much the same manner. How successful each individual is on his job depends not only on the particular mental and physical skills he brings to the job, but also on how well he sees his job in relation to the other people and jobs in the interfacing systems.

The systems view of the organization is not an easy one for most individuals to live with. In effect, it says that each of us must share in the responsibility for our subordinates, peers, associates, and superiors; none of us can be an island unto himself.

Teamwork

The manager is simultaneously a member of many teams or groups. Since he may be a leader in one group (e.g., his department at the office) and a follower in another (e.g., the PTA), he plays many roles and is subjected to a wide variety of different pressures. In his role as a PTA member, he may be under pressure to take time from his busy schedule for regular attendance at meetings or to head a membership drive. The pressures at the office may be very different; for instance, a pressure for attendance would be very unusual in that organizational environment.

THE INTERNAL TEAM ENVIRONMENT The individual gives up some of his individuality when he is integrated into the human subsystem of the organization. He is subject to common motivations, goals, and attitudes; there are social forces which encourage the individual to conform with what the team as a whole desires. The team (a family, section, department, company, etc.) has its own structure, processes, definite norms of acceptable behavior, and specific set of values. The members of the team tend to do things in certain ways, and the individual who departs from the expected way of doing things may risk incurring the disfavor of the group.

A "team" comes into being when individuals are placed in the position of depending on each other in accomplishing a common objective. Churchman has referred to such groups as a ". . . manifold of persons, identifiable over a period of time, and sufficiently integrated so that its actions and objectives are identifiable."[16] All of us are familiar with the cohesiveness of a winning football team—it seems to operate as an individual as the various individual members subjugate their personal whims to integrate their efforts toward the pursuit of a common goal.

[16]C. W. Churchman, *Prediction and Optimal Decision,* Prentice-Hall, Inc., Englewood Cliffs, N.J., 1961, p. 299.

In any organizational context, the manager's task is to facilitate the integration of individual efforts toward the achievement of a team goal, whether the team be the entire organization, the department, or some team which is established on an ad hoc basis. The football coach does this when he insists that the star running back block for a teammate when the teammate is to carry the ball. So, too, does the executive do this when he directs some personnel to prepare the speeches which others will present before an audience. In such cases, the good of the whole is being put above the good of the individuals who comprise the team. Each is not left to optimize for himself. Thus, since no statistics are kept on blocking, the running back is doing little for himself by blocking for his teammate. However, he does it in part because the team (and coach) expect it of him, because he subscribes to their common goal (to win), and because he knows that he will be supported by his teammates only if he supports them also. Functioning as an integrated entity, the team can thereby produce a synergy—a whole greater than the sum of its parts.

The greater accomplishments that teams can produce are only in part due to synergistic effects, however. In truth, each individual brings more to a team effort than is apparent from a superficial assessment of his skills and abilities. He adds other intangibles that can contribute to or detract from overall organizational effectiveness. Often, these other elements are beyond the awareness of the individual members, yet they have a significant effect on both the organization in its totality and on other members of the organization. Most people want to participate meaningfully in the activities of their teams. Their motivations transcend the usual monetary ones which we have traditionally associated with organizational activities. There is no better example of this than the many nonprofit, educational, and charitable organizations which are aided on a nonremunerative basis by business and government executives whose professional services could never be afforded. The $50,000-per-year executive who spends many hours working as a member of the board of directors of a nonprofit nursery school is obviously motivated by something other than money; he is a member of a team, and his principal rewards come via that membership.

THE EXTERNAL TEAM ENVIRONMENT Often, teams may be so large that they must operate as subteams. In large and complex organizations, the manager is faced *both* with integrating individual performances toward the common goal *and* with integrating his team's effort with the efforts of other teams. The nuances of inter- and intragroup decision making make this no simple task.

A lack of individual contact with the external team environment can affect the individual who operates as a part of a team. Unfortunately, today's large and complex organization *can* tend to frustrate the individual who is trying to find himself and his role. It must be emphasized that *this need not occur* and that illustrations of the successful avoidance of the "small cog in a large machine" syndrome are common even in the largest of organizations.

The worker's problem of identity and loyalty may be greater than the manager's, since it is he who usually has limited contact with the world outside his team. Many managers may be able to identify with "the company" because of their activities outside of "the office." But most must deal with many of their team members who do not have much contact with the environment which is external to their own office, section, or team. These workers cannot be expected to readily identify with a broad entity—the organization—much of which is alien to them. Their loyalties are to themselves and their fellow team members rather than to the overall organization.

Committees

Much of the day-to-day "work" and decision making of management is done through formal and informal teams referred to as "committees." Committees and the "committee system" are much maligned in the folklore of business and governmental management. Comments such as, "If you want to insure that nothing gets done, refer the matter to a committee!" are commonly heard in the offices and meetings rooms of large organizations.

However, managers must not really believe that committees are ineffective, or they would not make such great use of them. Tillman reports that 94 percent of large companies (of over 10,000 employees) employ committees in their day-to-day operations.[17] Dartnell Corporation conducted a study which shows that sales and marketing executives attend an average of 20 meetings per year, conduct 10 more themselves, and spend $2,000 in addition to attend outside conferences—the total time spent in meetings amounting to about two months.[18] Since more higher-level managers than lower-level ones tend to serve on committees,[19] it is apparent that organizations behave as if committees have great value. Why is this so? What is it about committees which make them such an

[17] R. Tillman, Jr., "Problems in Review: Committees on Trial," *Harvard Business Review,* May–June, 1960, p. 8.

[18] Results quoted in *4A Newsletter,* American Association of Advertising Agencies, vol. 9, no. 1, Jan. 4, 1971, p. 6.

[19] Tillman, *op. cit.*

important part of the human subsystem? To answer this, we need concern ourselves with the decision-making processes of groups and the characteristics of group decisions.

Group Decision Making

The basic premise of group decision making is that a group can arrive at "better" decisions than can an individual because of the varied points of view and sources of information which through the group can be brought to bear on a problem. The cynic will respond to this by saying that it ensures that action will never be taken, since unanimity is unlikely to be achieved.

Indeed, there is a basic logical problem involved in group decisions. In theory, a group decision involves the development of a *social preference device* through which a set of known individual preferences are translated into group preferences. Thus, each member of the group has his own preferences for the alternatives which are available, each has different preferences, and the group must develop an overall group preference from these differing individual preferences.

We have already alluded to this problem in a slightly different context in Chapter 9. Here we repeat the "voting paradox" illustration to refresh the reader's memory and to emphasize the larger problem that it illustrates—that of developing group preference functions.

Arrow's[20] voting paradox displays the difficulty in justifying even a simple "majority rule" approach to group decision making. Consider Table 15-1, in which the preference order for three individuals for three alternatives—*A, B,* and *C*—is displayed. It is clear that *A* will be preferred to *B* by the group, since both individuals 1 and 2 prefer it to *B,* and "the majority rules." Similarly, *B* will be chosen in preference to *C* by the group by a vote which goes against the preferences of the middle individual. As a result, since *A* is preferred to *B* and *B* is preferred to *C* by the group, one might infer logically that alternative *A* should be preferred to *C.* After

[20] K. Arrow, *Social Choice and Individual Value,* John Wiley & Sons, Inc., New York, 1951.

TABLE 15-1 Voting paradox

	Preference order		
Individual	*Most preferred*	*Second most preferred*	*Least preferred*
1	A	B	C
2	C	A	B
3	B	C	A

all, if one likes chocolate ice cream better than vanilla, and vanilla better than strawberry, the logical conclusion is that chocolate is preferred to strawberry. A glance at the table of individual preferences indicates that this apparently logical deduction is not valid for this group. In fact, two of the three individuals prefer *C* to *A,* and hence the group, by virtue of majority rule, prefers *C* to *A.*

Thus, this simple situation displays a basic logical problem with group decision making. Indeed, Arrow has shown that the voting paradox phenomenon is pervasive.[21] Others have argued that his conclusions are based on too restrictive a set of assumptions.[22] However, the fact remains that one cannot "prove" the value of group decisions in logical terms.

Since groups do indeed "work" and produce results, and since the logical *process* is open to argument, one might consider it better to focus on the *nature of the decisions* which groups produce.

THE NATURE OF GROUP DECISIONS Studies have been made of group decision making in various environments and sets of circumstances. These studies seem to demonstrate a phenomenon which has become known as a *"risky shift."* Basically, the idea of a risky shift means that groups produce riskier decisions than do individuals.[23]

One can hypothesize various possible reasons why this is so. In a group, one can "spread the blame"; hence, he might be willing to go for a riskier choice in a group than he would alone. Or perhaps individuals do not wish to appear to be "conservative" in an era when youth, risk-taking, and liberalism are popular, so one is less conservative in group situations when others can view his behavior than he is when he is alone and not being viewed by others.

The Manager's Role in Utilizing Groups

Because he has little contact with the environment outside his team, the worker may have primary loyalties to the team rather than to the organization. Of course, it is the manager's task to cope with this problem and to

[21] *Ibid.*

[22] G. Tullock, "The General Irrelevance of the General Impossibility Theorem," *Quarterly Journal of Economics,* May, 1967, pp. 256–270, and J. Coleman, "The Possibility of a Social Welfare Function," *American Economic Review,* December, 1966, pp. 1105–1122.

[23] A wide literature exists on this topic. For example, see D. Bem, M. Wallach, and N. Kogan, "Group Decision-Making under Risk of Aversive Consequences," *Journal of Personality and Social Psychology,* May, 1965, and J. Bower, "Group Decision Making: A Report of an Experimental Study," *Behavioral Science,* July, 1965, pp. 277–290.

turn it into an opportunity to enhance productivity. One way of doing this is to establish formal teams rather than simply allowing informal ones to exist. The "team spirit" concept which is so commonly applied to sports teams can be developed in such a team, and once a worker becomes a part of it, he realizes that his co-workers count on him and will help him if he has difficulties and falls behind.

There is a general trend in industry toward establishing responsible autonomy for teams so that each team has a reasonably complete and satisfying "product" to produce with its efforts. This has been understood for some time. For instance, a major shift in mining techniques in the English coal industry involved a breakup of cohesive work groups and the establishment of a "production line" operation. As a result, productivity decreased and absenteeism, illness, and turnover among miners increased.[24] The result has been a gradual return to the "old way."

Observers who see the spirit and enthusiasm which can pervade a production operation which involves teamwork are particularly aware of the contrast to humdrum, assembly line plants where the atmosphere is often heavy and burdensome. There, the workers perform the same operation repetitively without seeing the finished products which they help to produce. And, although there is a close technological interdependence among the various production line functions, there is no *human* interdependence. Perhaps the difference is best summed up by *Fortune's* quote from a production worker who is part of a team: "There is this team feeling, people seem to help each other. I feel as if, when I am out, they miss me."[25]

The spirit which is fostered by the immersion of an individual into a team environment can indeed be used to the advantage of both the organization and the individual. He is more productive in a team, and he is rewarded and stimulated by his membership in it.

Indeed, it is increasingly difficult for management to avoid making extensive use of teams and committees. Galbraith argues that committees can be effective and efficient.[26] His argument for committees is based on the fact that important decisions must involve a greater range of knowledge and information than that possessed by any individual. In Japan, group effort has become a way of life. For example, Van Zandt notes: "The Japanese prefer to work as members of groups rather than as individuals. This characteristic is often cited as one of the most important in explaining Japan's economic success."[27]

[24] E. Trist and K. Bamforth, "Some Social and Psychological Consequences of the Longwall Method of Goal-Getting," *Human Relations,* February, 1951.

[25] Gooding, *op. cit.,* p. 162.

[26] J. Galbraith, *The New Industrial State,* Houghton Mifflin, Boston, 1967.

[27] Howard F. Van Zandt, "How to Negotiate in Japan," *Harvard Business Review,* November–December, 1970, p. 47.

The manager who uses groups to advantage relies on two varieties of motivation which relate to the individual's behavior in group situations —*identification* and *adaptation.* Identification relates to the "feeling of well-being" associated with having adopted a group's goals as one's own. Adaptation is a complementary motivation which involves the hope that one can affect the group's goals so that they may become more closely aligned with his own.[28]

Some ways in which the manager can utilize identification and adaptation are straightforward. He can create formal groups with well-defined goals. He can provide a feedback system so that the group is aware of its progress, and he can structure groups so that they can see the role that they play in the development of an "end product." Moreover, if the groups are "open" and structured so that everyone—regardless of rank—can participate and feel that he is contributing, the adaptation motive is addressed.

THE CONTROLLABLES OF GROUP BEHAVIOR The question of what elements can be manipulated in order to develop effective groups and to induce desired behavioral patterns in groups is even more difficult than that of operationally motivating individual behavior. One approach to the delineation of the controllables of group behavior is a proposal made by March and Simon concerning five basic hypotheses related to the identification and adaptation motives.[29] They propose that the individual has a stronger propensity to identify with the group according to:

1. *The greater the perceived prestige of the group*
2. *The greater the extent to which goals are perceived as shared by members of a group*
3. *The more frequent the interaction between the individual and the members of a group*
4. *The greater the number of individual needs satisfied in the group*
5. *The less amount of competition between the members of a group and an individual*[30]

AN OPERATIONAL APPROACH March and Simon have gone on to propose some "factors affecting group identification" which describe those elements that the manager can utilize in effectively structuring groups. Figure 15-3 shows the factors relating to the "perceived prestige" variable.

[28] Galbraith, *op. cit.*

[29] J. March and H. Simon, *Organizations,* John Wiley & Sons, Inc., New York, 1967, pp. 65–66.

[30] *Ibid.,* pp. 67–70.

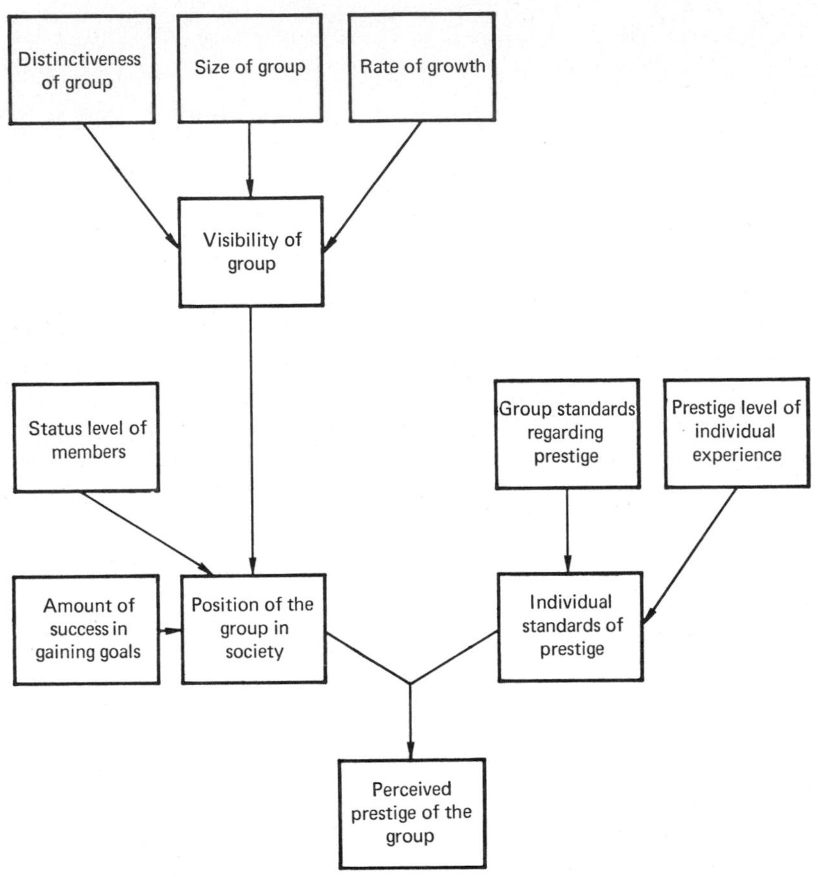

FIGURE 15-3 *Factors affecting the perceived prestige of the group. Adapted from J. March and H. Simon, Organizations, John Wiley & Sons, Inc., New York, 1967, p. 69. Used with permission.*

This figure shows that prestige varies with the "position of the group in society" *and* with "individual standards." The position in society is a function of the "cultural symbols of success," such as success in achieving group goals, the status level of participants, and group visibility. Visibility, in turn, depends on the distinctiveness of the group—in terms of goals, membership or practices, and the group's size and growth rate. Individual standards of prestige are related both to group standards and to the elaboration brought about through individual experience.

Figure 15-4 shows a similar diagram for the other controllable elements. Greater exposure to contact enhances the frequency of interaction, as do greater cultural pressure to participate and a greater homogeneity in background. The greater the size of the community, the less frequent is the interaction between the group and the individual. Homogeneity of background also affects the extent to which goals are perceived as being shared, and such goal sharing is also enhanced by similarities of *present* position in the organization.

A group which is permissive toward individual goal achievement will tend to satisfy more individual needs than one which is not, since in ". . . our culture, conformity tends to be considered a 'cost' of group membership rather than a positive advantage."[31]

Finally, the greater the independence of individual rewards, the less will be the competition among group members. If, for instance, the group were to be paid as an entity and individual remuneration was to be determined by the group, the competitive level would be very high.

Of course, these controllable elements are only partially operationalized. They provide guidance to the manager concerning which button

[31] *Ibid.*, p. 70.

FIGURE 15-4 *Factors affecting frequency of interaction, the extent to which goals are perceived as shared, the number of individual needs satisfied in the group, and the amount of competition. Adapted from J. March and H. Simon, Organizations, John Wiley & Sons, Inc., New York, 1967, p. 71. Used with permission.*

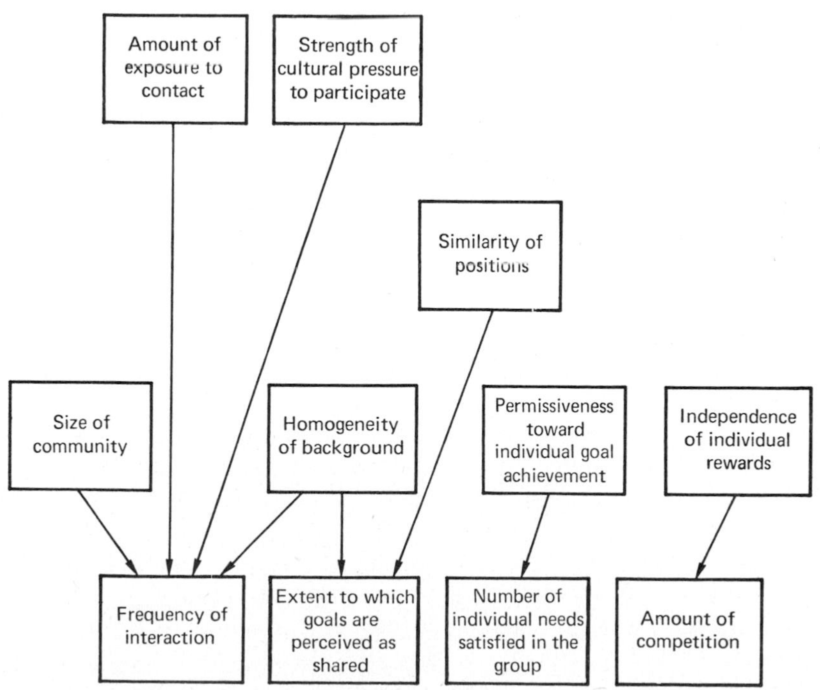

to push, but they do not tell him how hard to push, or how often, or when. In the next section, we address the question of human subsystem design to get further into those specific considerations which the manager must face in structuring an organization which makes effective use of our knowledge of individual and group behavior.

DESIGNING THE HUMAN SUBSYSTEM

One of the best ways to understand any system is to face the problem of designing it. What if no human subsystem existed in our organization? How would we design one? What considerations would be relevant?

Of course, the human subsystem designer cannot "design" the behavior, attitudes, motivation, etc., of people, nor can he program people as we would program a computer. Each individual is a unique personality made up of prejudices, attitudes, fears, background, etc., which taken together set each individual apart from other individuals. What the designer can do, however, is to consider the characteristics and features of the organization—both formal and informal—which will contribute to getting the most out of individuals and groups. He can design a *facilitating* organization—one that will serve as a catalyst for productive effort and one that will serve as a fertile environment for new ideas.

If it were possible to design a human subsystem that was rational and predictable, the job of management would be much easier than the reality of managing a human subsystem that may be emotional, irrational, and unpredictable. If we take two algebraic terms $(A + B)$ and $(A - B)$ and combine them, the outcome, based on our definitions, theory, and experience of previous behavior, is predictable, viz., $A^2 - B^2$. In contrast, take two people, person A and person B, and combine them in an organized activity. The outcome is usually uncertain. For example, we have difficulty in predicting:

1. Which person will dominate.
2. If they will establish interpersonal rapport.
3. If there will be any friction, disagreement, or hostility.
4. If they will continue to work together with social satisfaction.
5. If they will accomplish their individual and organizational objectives.

Unfortunately, these and many similar questions cannot be answered before the people are combined in an organized activity—in fact, even after they have worked together, it may be very difficult to find answers to these questions.

The General Controllables of Human Subsystem Design

Just as we previously gave attention to the controllable elements which the manager could use to affect the performance of an individual, we now turn our attention to those aspects of the human subsystem which can be controlled. In other words, if we are to design a human subsystem, to what features of it should we give our attention?

THE GENERAL ROLE OF THE MANAGER One of the elements which can readily be prescribed for the human subsystem is the general role that the manager plays in it. If we are to accept the Theory Y-based approach which we discussed earlier, the manager's role should be built on a recognition that a great deal of the organization's planning, organizing, directing, and controlling can be accomplished by subordinates, *each of whom is a manager, if only of his own time.* The degree to which the subordinate carries out his own organizational life is reflected in his perception of: (1) what the supervisor desires and (2) what the supervisor will tolerate. Our concept of leadership is one in which the supervisor is a provider and a facilitator. The supervisor provides the necessary climate and facilitates the work of the subordinates. He is less of an authority-oriented person and more of a provider of work conditions, a helper in problem solving and goal accomplishment. He is a leader who provides an environment in which people have a sense of working for themselves; he helps give visibility to organizational goals; he provides resources; and he mediates conflicts—but most of all he keeps out of the way and permits people to manage their work to the maximum degree possible.

We should emphasize at this point that this prescription of a management role is not just an attempt at achieving "nice" management or "good human relations"; it is concerned with gaining the maximum in human effectiveness through prescribing:

1. A climate of interpersonal confidence
2. Meaningful and understandable goals
3. A management system which people understand and in which they can participate

In this respect it might appear that the manager is to attempt to motivate people because he likes them; that is immaterial. We must recognize the individuality of the human being and realize that people who do not achieve their personal goals will find difficulty in identifying with organizational goals. Conversely, if they do not achieve organizational goals, they may not satisfy their personal goals. *Motivation is not solely something that we do to people, rather it is the consequence of organizational climate, goals, management systems, and the attitudes about participation in reaching meaningful objectives.*

METHODS OF THE MANAGER One of the many management methods in modern organizations is *participation.* By "participation" we mean the willingness of a manager to consult with his followers and acquaint them with the organizational problems and the decisions of the group so that the group functions as a social unit as well as a work unit. The participative manager:

> *. . . is not an autocrat, but neither is he a free-rein manager who abandons his management responsibility. The participative manager is still a responsible management representative, who retains ultimate responsibility for the operation of his unit, but he has learned to share operating responsibility with those who perform the work.*[32]

Increased employee participation in the organization tends to produce increased satisfaction and productivity and is desirable for the future success of the enterprise. The value system implicit in classical management and organization concepts emerged from the case where the person is rewarded for looking to his own job and the job of his department. Modern concepts of organization recognize that management must assume more responsibility toward the community and society in which the business competes and also in the local (departmental) society in which the individual lives. Thus, the needs of the individual must be integrated with the needs of the business; meeting the needs of people through affording them the opportunity to participate in the things that affect their work will contribute to their needs, human dignity, recognition, and self-fulfillment.

The drive for increased participation by all the clientele of an organization is a contemporary phenomenon. Max Ways speaks of the drive for increased participation by stating: "In a society where great organizations—government, business, educational—are central to the mode of life, a thrust for greater participation replaces the old drive for independence. To the threat of autocratic power, independence responded with the plea: 'Let me alone! Don't tread on me!' Participation, raising a bolder challenge, cries: 'Listen to me!'"[33] Participation does more than just provide people a sense of well-being and of contributing to the organization. In fact, there are many good ideas in an organization that start

[32] Davis, *op. cit.*, p. 128.

[33] Max Ways, "More Power to Everybody," *Fortune*, May, 1970, p. 173.

somewhere down the organizational chain of command. An interesting lesson in the value of participation is provided by looking at the Japanese industrial society. According to Van Zandt, "In comparing American and Japanese decision-making processes, one soon learns that, whereas in the United States a considerable proportion of management ideas are conceived in the executive suite and imposed from the top down, in Japan the reverse is true. More often than not proposals start from somewhere down the line."[34]

Participation increases the opportunity for the individual to develop skills, to understand himself and others, and to widen his horizons and improves his outlook on life, his team, and his job. Participation means the removal of those conditions which attempt to set people and groups apart from the organization per se, and aims to remove the conditions whereby people and groups tend to operate as if they existed independent of their organizational system.

Participation is apparent today in all of our institutions. Students participate in the university decisions which affect their lives and well-being. Even in those decision areas where students do not actually participate in the sense of casting a vote, they are kept informed of the decision problem, the alternatives, the method of resolution, and the outcome by the university administrators. In this way, they can offer ideas and feel a part of the process, even if their formal role is small.

ORGANIZATIONAL STRUCTURE The organizational structure, and how managers and workers view that structure, will affect people's attitudes on participation in the pursuit of the organization's goals. A structure that is based on authoritative management will develop attitudes hostile to organizational goals and will produce response patterns which depend on direction in detail from higher levels of the organization. If we view authority as a gravitational force that flows from the top of the organization down, then one's image of an organization will be something like Figure 15-5, in which control comes principally from outside the individual, *from the organizational boss.*

By contrast, the participative process, which builds and develops attitudes that are favorable to the integration of individual, group, and organization needs, imagines an organization something like Figure 15-6. where the participation will be supportive of organizational goals. Participation in the organization provides the means whereby the interaction and mutual influence of individuals provide a basic point of departure for the operation of the organization.

[34] Van Zandt, *op. cit.,* p. 48.

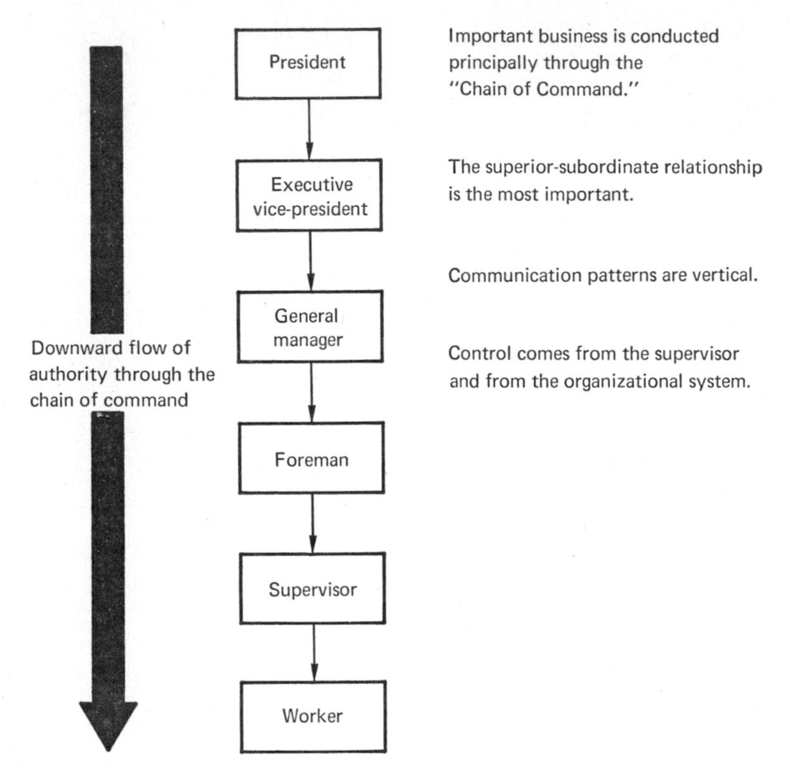

FIGURE 15-5 *An organization in which authority flows downward. Descriptive material in Figure 15-5 is paraphrased in part from William H. Read, "The Decline of the Hierarchy in Industrial Organizations," Business Horizons, Fall, 1965, p. 74.*

SUMMARY

Participatory management involves a change in the manager's perception of his job. He is not so much a manager in the traditional sense of planning, organizing, directing, and controlling his subordinates; much of these functions will be taken over by the social group itself as it exerts group influence on its members to conform to group standards. The manager's job requires that he concentrate on providing the environment to ensure that people are willing and able to participate in the group efforts. This requires that the manager provide education, training, guidance, services, arbitration, and such supportive matters, so that decisions are made and executed in a timely manner. The supervisor's role shifts from a "father" or "policeman" image to a "coordinator" of resources (human and material) and ideas.

Participative management is not a panacea—in fact, there is evi-

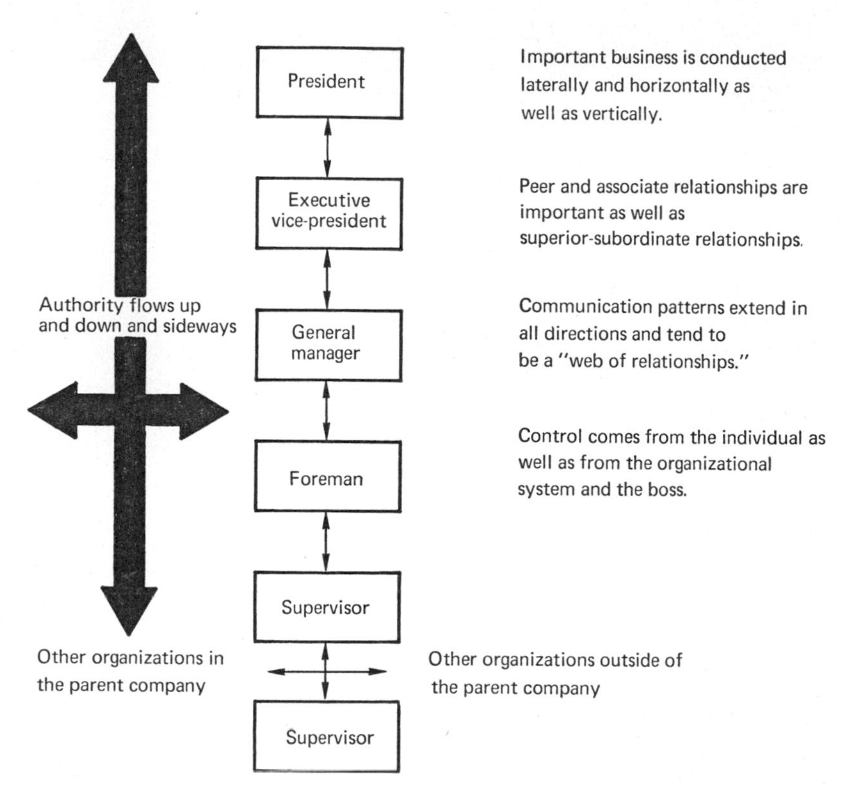

FIGURE 15-6 *An organization in which authority flows up, down, and laterally. Descriptive material in Figure 15-6 is paraphrased in part from William H. Read, "The Decline of the Hierarchy in Industrial Organizations," Business Horizons, Fall, 1965, pp. 71–75.*

dence that it does not work for everybody and in every organization. Research has shown that workers who participate are sometimes more productive, but at times there is no significant increase in production—they are simply happier. Certain people are not able to participate because of intellectual limitations or an emotional predisposition against getting involved in the negotiation and conflict that can occur in the participation process. Some supervisors are unable to *delegate,* a necessary prerequisite to participation. Then, too, in a period of emergency, the only way to get things under control is through fast, central, and sometimes tough management. It would be difficult to imagine an extensive program of participation in a military organization in a combat area. In a profit squeeze, the only way to cut costs may be to take quick action to reduce the overhead force in order to remain competitive.

Another potential difficulty with this new form of management is the proposition that it naïvely assumes that everyone wishes to participate. Some behavioral scientists are questioning the wisdom of applying participative management across the board. There is the problem of having *objective* participation by the members of the organization; personal emotions and prejudices can often get in the way. Some of the participants may enter the participation process with the attitude that their interests are mutually exclusive and any negotiation leading to compromise is a personal defeat. Under these conditions there can develop an attitude of self-protectiveness and defensiveness which does not contribute to organizational productivity.

Participative management assumes that an environment can be created and maintained so that people will enjoy being *ego-involved* in affecting the conditions of their job. It also assumes that people enjoy satisfaction in furthering the goals of the organization through self-control, rather than through reliance on organizational control by a system of detailed policies and procedures. *If a person has some choice and the opportunity to influence how he does his job, he will be more challenged than if he is given detailed, rigid specifications for his job.* Under such conditions the supervisor is a helper, not an authoritative boss; he is a teacher, consultant, chaplain, or peer, but only rarely a manager who "bosses" other people.

Thus, there are not definite rules for the modern manager to follow in dealing with the human subsystem. At best, he can develop a set of beliefs and attempt to apply them. We close the chapter with such a proposed set of beliefs:

1. People's behavior is motivated to a large degree by habit and emotion, rather than by reason.

2. People expect to be treated as individuals, with respect and understanding.

3. The best in people is brought out:

   a. When they trust and respect their supervisor; they want to know what he will permit in the work situation.

   b. When they believe their supervisor is genuinely interested in them, shows his interest in their personal lives, and enjoys taking some time to talk to them.

   c. When they know where they stand through issues of simple, understandable instructions; they know then what the supervisor expects and what he wants done.

   d. When they feel they are an essential part of the organization; they want to become and remain a fully participating member of the organization.

4. People are motivated more by being made to feel important than by fear; they enjoy receiving credit when it is deserved.
5. People tend to resist change, particularly sudden change; they will accept change more gracefully if they are given an opportunity to prepare for it.
6. People will honestly try to live up to the expectations of their superiors and peers; subordinates typically try to emulate their superior's way of doing things.
7. People expect to be judged and rewarded by a system of discipline. Subordinates expect the boss to establish patterns of behavior and effectiveness for the organization.
8. People like to be commended publicly; they resent public criticism and unfavorable comparison with others in the group situation. People do not like to lose face.
9. People want to know what is going on in the organization—what is the objective of the team. They produce more when there is some incentive present.
10. People want to feel secure in their job and in their social group. They want to be able to use the full extent of their capacities and to know they will receive the necessary challenges to do so.
11. People expect to be corrected when they are doing something in the wrong manner; they want to be able to correct their deficiencies within the limits of their potential.
12. People want to know where they stand in the organization.

EXERCISES

1. What are the controllables of the simple job performance model of Figure 15-1?
2. Discuss Theory X and Theory Y. Which theory is the basis for the operation of your family?
3. Are most labor negotiations conducted as though Theory X or Theory Y were believed to hold? Explain.
4. Which of the two theories—X or Y—makes it easier for the manager to operate? Explain. However, does the manager have a choice?
5. What is the significance of the hierarchy aspect of Maslow's hierarchy of needs? In other words, why are there not just a number of different kinds of needs rather than a *hierarchy* of needs?
6. What is meant by a "satisficing value" for a need in the hierarchy of needs? What does this imply about an appropriate need-fulfillment model?
7. Keith Davis is quoted as saying that many wage disputes are not really wage disputes. What does he mean?
8. How is need fulfillment related to individual motivation?
9. "Blue-collar workers must be managed by focusing on low-level needs;

professionals are more concerned with higher-level needs." Comment on this statement.

10. Of the controllables of human behavior listed on page 370, which would you expect would be ranked as very important by *(a)* top managers, *(b)* middle managers, *(c)* workers?
11. Define "status" and "role" as related to an individual. How can the manager influence each of these aspects of an individual under his authority?
12. What is "job enrichment"? What does it have to do with motivators and hygiene factors?
13. What is the "risky shift" in group decision making. Hypothesize its cause.
14. Discuss the question of developing a group preference function in theoretical terms. How do practicing groups avoid the problems associated with a group preference function?
15. Contrast the production line type job with a team job. Which is more satisfying to the individual? How can the lesser satisfying one be made more like the more satisfying one?
16. What are identification and adaptation? How do they relate to Herzberg's motivators?
17. Discuss the controllables of group behavior.
18. What is participative management?
19. The authors make the statement that the manager is a person who accomplishes results through working with others in the group situation; he makes things happen through the efforts of other people. How would you explain this position?
20. Why is it necessary for the manager to have an understanding of both individual and group behavior?
21. Individual performance is a function of two primary elements. Identify and demonstrate by example that you understand what these two elements are and what they mean in organized society.
22. Peer-to-peer relationships in an organizational setting may be just as important as superior/subordinate relationships. Why? Defend your position.
23. What evidence do we see in contemporary society that indicates that more "team" and "task force" organizational approaches will be used?
24. A scholar once stated that it was evident to him that the camel was designed by a committee. Why do you think the scholar made this statement?
25. How could we set about designing the human subsystem in an organization?
26. In the final analysis, we might define the manager's responsibility as facilitating an environment wherein people are able to work together with economic and social satisfaction in the pursuit of a common goal. Defend or refute this statement.
27. The authors take the position that each of us plays the role of a manager in today's society. Why is this so?
28. The best way to motivate people is to concentrate on the people themselves. Defend or refute this statement.

29. "We know enough about participative management today to recommend that participatory techniques could be used in all organizations and under all conditions of organizational environment." Defend or refute this statement. Support your position by example.
30. The authors close the chapter on the human subsystem of an organization by offering a proposed set of beliefs about people. Do you agree with this set of beliefs?
31. A procedure of "staggered work time" has been introduced in a number of industrial firms in the United States and other nations.[35] Under this arrangement, workers choose their own working hours within prescribed limits. Make your own predictions on how such a system would work in terms of *(a)* its effect on absenteeism, *(b)* its effect on productivity, *(c)* different approaches to work hours by blue-collar and white-collar workers.

REFERENCES

Albrook, Robert C., "Participative Management: Time for a Second Look," *Fortune,* May, 1967.

Catton, Bruce, "The Other Side Camp," *American Heritage,* October, 1967.

Cyert, Richard M., and James E. March, *A Behavioral Theory of the Firm,* Prentice-Hall, Inc., Englewood Cliffs, N.J., 1963.

Davis, Keith, *Human Relations at Work,* McGraw-Hill Book Company, New York, 1967.

Dubin, Robert, *Human Relations in Administration,* Prentice-Hall, Inc., Englewood Cliffs, N.J., 1961.

Filley, A. C., "Committee Management: Guidelines from Social Science Research," *California Management Review,* Fall, 1970.

Galbraith, J., *The New Industrial State,* Houghton Mifflin Company, Boston, 1967.

Gooding, J., "It Pays to Wake Up the Blue Collar Worker," *Fortune,* September, 1970.

Hekiman, James S., and Curtis H. Jones, "Put People on Your Balance Sheet," *Harvard Business Review,* January–February, 1967.

Herzberg, F., *Work and the Nature of Man,* The World Publishing Company, Cleveland, 1966.

Hodgetts, Richard M., "Leadership Techniques in the Project Organization," *Academy of Management Journal,* June, 1968.

"Job Motivation: Money Isn't Everything," *Newsweek,* April 22, 1968.

Kelly, Joe, "Making Conflict Work for You," *Harvard Business Review,* July–August, 1970.

Kurlioff, Arthur H., "An Experiment in Management—Putting Thirty Whys to the Test," *Personnel,* November–December, 1963.

[35] "Pick Your Hours," *Time,* July 19, 1971, p. 68.

Lawrence, Paul R., and J. W. Loorsch, "New Management Job: The Integrator," *Harvard Business Review,* November–December, 1967.

Leavitt, Harold J., "Unhuman Organizations," *Harvard Business Review,* July–August, 1962.

"Management: Making a Job More Than a Job," *Business Week,* April 19, 1969.

March, J., and H. Simon, *Organizations,* John Wiley & Sons, Inc., New York, 1967.

Maslow, A. H., *Eupsychian Management,* Richard D. Irwin, Inc., Homewood, Ill., 1965.

———, *Motivation and Personality,* Harper & Row, Publishers, Incorporated, New York, 1954.

McGregor, Douglas, *The Human Side of Enterprise,* McGraw-Hill Book Company, New York, 1960.

Prince, George M., "How To Be a Better Meeting Chairman," *Harvard Business Review,* January–February, 1969.

Reeser, Clayton, "Some Potential Human Problems of the Proiect Form of Organization," *Academy of Management Journal,* December, 1969.

Smith, Henry Clay, *Psychology of Industrial Behavior,* 2d ed., McGraw-Hill Book Company, New York, 1964.

Tierman, Jr., R., "Problems in Review! Committees on Trial," *Harvard Business Review,* May–June, 1960.

Van Zandt, Howard F., "How to Negotiate in Japan," *Harvard Business Review,* November–December, 1970.

Williams, Lawrence K., "The Human Side of a Systems Change," *Systems and Procedures Journal,* July–August, 1964.

16

management control

The "final" phase of the execution function of management is *control.* Control is the constraining of activities and resources to conform to a plan of action. It is usually brought about through the operation of a *control system*—a management subsystem within the organization.

In complex organizations, there are at least two "kinds" of control. Anthony et al., refer to these as "management control" and "operational control."[1] *Management control* is the process by which it is ensured that activities and resources are directed toward the efficient and effective accomplishment of objectives as specified in the planning process. *Operational control,* on the other hand, is ". . . the process of assuring that specific tasks are carried out effectively and efficiently."[2]

BASIC CONTROL CONCEPTS

Control and control systems play an important role in our daily lives. For instance, heating and air conditioning systems have automatic temperature control subsystems. Our body is replete with automatic control systems. The eye fixation system operates when one girl watches (or boy watches, as the case may be). When the image being observed becomes displaced on the retina, the brain detects the displacement and commands the eye muscles to move in such a way as to keep the "desirable" image centered on the retina. Thus, a control or feedback loop is completed.

Control Models

In Chapter 2, we discussed various models of feedback control systems. Figure 2-5 (page 38) describes the simplest such system. Figure 2-6 (page

[1] R. N. Anthony, J. Dearden, and R. F. Vancil, *Management Control Systems,* Richard D. Irwin, Inc., Homewood, Ill., 1965, chap. 1.

[2] *Ibid.,* p. 7.

39) is a model of a more sophisticated feedback system with memory. Such abstract models can readily be applied to a wide variety of economic and organizational systems.

Figure 16-1 illustrates an economic system described in Keynesian terms: total income is the sum of incomes from capital and consumer goods, each of which, in turn, is dependent on total income via the feedback loop.

Similarly, Figure 16-2 illustrates the business organization in abstract input-output and feedback terms. The organization is viewed as a "processor" which converts labor, material, and other resource inputs into product or service outputs. The processor is controlled through the comparison of feedback information with standards represented by long-range goals, objectives, policies, etc.

Planning and Control

A plan prescribes a way in which something is to be done. Control is that managerial activity that tells you how well you are doing that which you planned. All matters within an organization are grounded in some plan of action. Day-to-day matters are based on an operational plan. Activities

FIGURE 16-1 *Economic feedback cycle. Reprinted with permission of the Macmillan Company from Introduction to Continuous and Digital Control Systems by Roberto Saucedo and E. E. Schiring. Copyright © The Macmillan Company, 1968.*

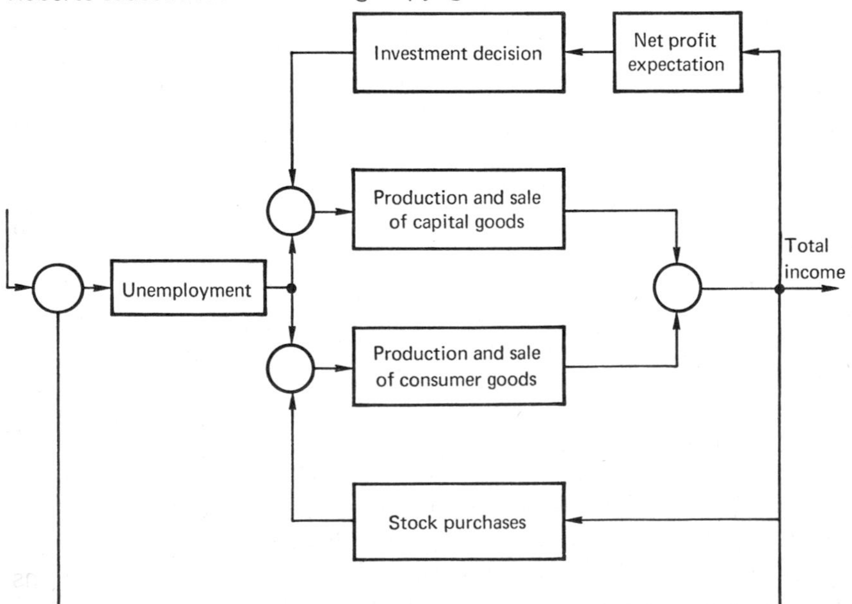

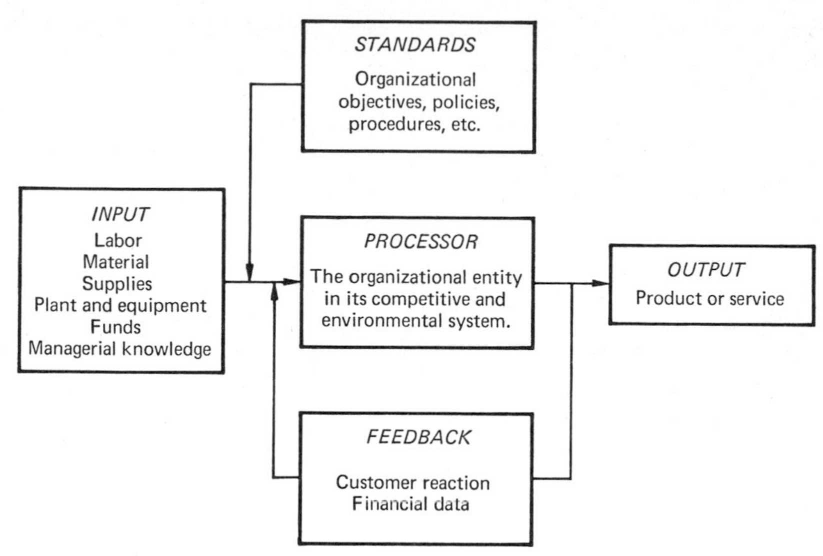

FIGURE 16-2 *The business organization as a system.*

having a long-range impact should be based on a long-range strategic plan. Organizational objectives, policies, procedures, and budgets and executive example also serve to provide standards which can be used for control purposes. Operation of the organization must stay within the constraints set by these standards. If operations do not stay within these constraints, the manager must do whatever is necessary to bring the performance back into line and must evaluate existing plans to see if they should be adjusted. This process, which is inherent in executive responsibility, is called "controlling." Control, then, has two aspects: used as a noun, it is the means and method of keeping track of progress; used as a verb, it means the active surveillance of an operation to keep it within the limits of the defined task. Control may be thought of in terms of a cycle, as portrayed in Figure 16-3.

The control cycle is therefore an endless sequence of establishing standards, observing performance, comparing performance with standards, and taking corrective action to increase the likelihood of achieving performance which "measures up" *or* to revise plans and standards in the light of actual performance.

The latter possibility should be emphasized. Control does not deal solely with performance below that of the standard, nor does it deal strictly with taking action to affect performance. Performance may be better than the standard, in which case consideration may be given to changing the standard. Or, if performance is below standard and corrective actions have no effect or are too costly, perhaps the standard is too high.

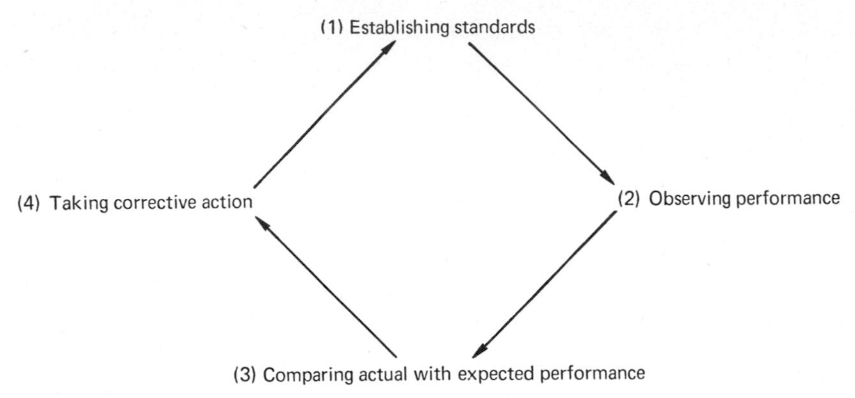

FIGURE 16-3 *The control cycle.*

Thus, planning and control are an interrelated set of activities, each of which affects the other. The simplistic view held by many that planning involves the establishment of inflexible standards which then must be sought through a control process leaves much to be desired. The best use of planning and control involves a high degree of interdependence between the two phases.

PROJECT PLANNING AND CONTROL Perhaps this is best illustrated operationally by some of the "planning methodologies" which can be used to facilitate control. In Chapter 12, we discussed project planning using network analysis. The project network, with its associated time estimates for each of the activities which make up the project, can also be a control tool.

To illustrate this, consider the network of Figure 16-4 (which is the same as that previously used to illustrate the planning uses of networks as Figure 12-7). The critical path is determined to be the lower one (events

FIGURE 16-4 *Project network with time estimates.*

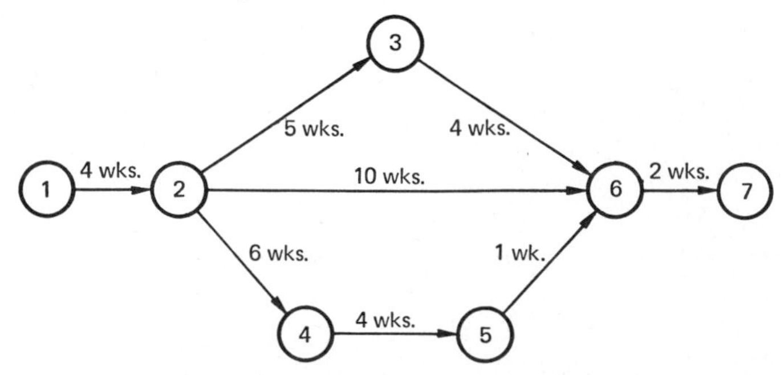

1-2-4-5-6-7), which has a planned duration of 17 weeks. The other two paths have planned durations of 15 weeks (1-2-3-6-7) and 16 weeks (1-2-6-7). Thus, the overall project has a planned duration equal to that of the critical path—17 weeks.

Now, suppose that the project begins and after 10 weeks have elapsed the project manager receives the following status report:

Activity 1–2—Completed in 4.5 weeks
Activity 2–3—Not yet complete—2 more weeks estimated for completion
Activity 2–4—Completed in 5.5 weeks

Therefore, after 10 weeks, event #4 on the critical path has been reached as planned, although activity 1–2 took longer than planned and activity 2–4 was completed ahead of time. However, activity 2–3 has already consumed 5.5 weeks (10 minus 4.5 for activity 1–2), and 2 additional weeks are estimated to be required. Thus, event #3, which was supposed to be reached after 9 weeks, is not expected to be reached until a total of 12 weeks have elapsed.

The activities on the top path in the network appear to have therefore "slipped" by a total of 3 weeks—½ week for activity 1–2 and 2.5 weeks for activity 2–3.

Now, assuming that all other estimates for incomplete activities remain the same, *the lower path is no longer critical.* It still has a planned duration of 17 weeks, but the top path now has a planned duration of 18 weeks. It has slipped 3 weeks from its original planned duration of 15 weeks.

Since the critical path is now 18 weeks long, the overall project will be expected to require 18 weeks. All schedules, delivery commitments, etc., should therefore be revised. This is therefore one variety of control action that the manager can take.

A better control approach by the manager might involve a reallocation of resources to increase the likelihood that the original 17-week schedule can be met. He could, for instance, apply additional resources to the activities on the critical path in the hope that this will result in faster completion of those activities. More personnel might be assigned to activity 2–3 in the hope of completing it in less than the anticipated 2 additional weeks. Or, additional funds might be allocated to activities 3–6 or 6–7 in the hope of completing them in less than the planned time. *If in any way a single week can be made up, the project can still be completed in an overall 17 weeks.*

But where would these resources be found? Of course, it is possible that they can come from outside the project budget, but in most organizations this would require special approval at high levels. The activity which is solely on the middle path is the most likely candidate. The middle path is now expected to require 16.5 weeks—the original 16-week estimate plus the ½-week slippage in activity 1–2. Thus, it could be allowed to slip another ½ week without endangering the completion of the project in the originally planned 17 weeks. Some personnel or funds might well be taken from activity 2–6 and reassigned to activities on the upper path to enhance the likelihood of it being completed in 17 weeks.

In performing this control action, the manager is comparing actual progress with the standard provided by the plan and is taking action to increase the chances of the plan being achieved. He should be aware of the latest date for changing delivery commitments so that, if it becomes likely that the plan cannot be met, *he can act to change the plan* by rescheduling deliveries.

Standards

Standards are the criteria by which we evaluate past or present performance. The network of standards that apply to a given organization vary depending on the particular objectives, environment, and *modus operandi* of that organization. A control system used by the commander of a military organization in a combat theater would differ from the system used by the president of an industrial firm. The theory of control is essentially the same for any organization, e.g., a military, ecclesiastical, or business organization. From the standpoint of control in general, certain broad standards can be offered:

1. *Ethical and moral standards,* i.e., the expected behavior patterns of people. This includes a wide range of criteria. Some examples would include: the Ten Commandments; Judaeo-Christian ethical principles; military code of conduct; oath of office; and the special requirements laid down by particular organizational groups.
2. *Standard costs.* Normative costs that have been established for the performance of functions or phases of operations.
3. *Ratios.* This includes a wide range of financial ratios, such as the current ratio—the relationship of current assets to current liabilities. Desired ratio patterns are developed from past experience, industry practices, competitor standards, etc.
4. *Budgets.* The cash budget, or cash flow forecast, used as a technique to predict future cash requirements, serves as a standard when it is approved.
5. *Return on investment.* Often used as a standard when evaluating product development effort.

6. *Miscellaneous criteria.* Standards that can be used to evaluate the overall performance of an organization, such as: *(a)* philosophy and quality of management; *(b)* market position; *(c)* social expectations, as in ecology or community service; *(d)* morale, employee turnover, and grievances; *(e)* customer and public relations; *(f)* personnel development; *(g)* innovation and research; *(h)* conservation of assets.

Management and "Nonmanagement" Control

The processes of management control in an organization relate to formal standards—budgets, policies, etc. However, informal control processes are simultaneously at work.

Much of the "real" control which exists in an organization, or, indeed, in a society, is created through peer-group standards rather than formal organizational standards. Just as a society depends on a "social compact" —an informal agreement among its members—so, too, does the organization rely on standards developed by the informal organization. The "informal organization" often sets standards of behavior for its members, for example, production quotas, criteria for belonging to a certain clique, behavior toward the boss, resistance to change, job status, and such related matters. In speaking of conformity in the informal organization, Keith Davis has said: "Nonconformers may be harassed or ostracized from the group until they capitulate or leave. . . . In one instance, a worker's associate refused to speak even one word to him for weeks because he did not conform."[3] Control exercised by the informal group can be both distracting and reinforcing to the manager's job. Informal social standards within such groups can blend with formal control systems to produce a viable system. According to Dubin, this blending of formal and informal control systems protects the organization ". . . from the self-destruction that would result from literal obedience to the formal policies, rules, regulations, and procedures."[4]

Even the most comprehensive of formal control systems cannot substitute for the individual responsibility and willingness to conform to a pattern of behavior expected in the organization. The fallibility of human beings and the vagaries of forces that affect an organization require a manager to exercise some form of formal control to keep his organization on the planned course.

However, even the best formal control system is itself usually not

[3] Keith Davis, *Human Relations at Work,* McGraw-Hill Book Company, New York, 1967, pp. 217–218.

[4] Robert Dubin, *Human Relations in Administration,* Prentice-Hall, Inc., Englewood Cliffs, N.J., 1951, p. 68.

adequate to accomplish *all* of the control which is necessary in order for the organization to function. If formal controls are integrated with informal ones, the result *can* be a comprehensive control system. If formal and informal controls are at odds, the result can be chaos.

Exception-oriented Control

Formal control often takes the form of prescribed checks on key operations and processes which flash a warning signal if a mistake or deviation from a charted course of action is detected. This, coupled with a system for positive intervention in the system to correct the error or to reorient the process toward the goal, is the essence of an *exception-oriented control system.*

The variety of management action which is consistent with this sort of system is referred to as *management by exception.* As a management philosophy the management by exception principle says that the manager should focus his attention on exceptions or deviations from established norms or patterns. He is thereby able to devote his scarce resources (his time, for instance) to those things which are "out of control" in either a positive or a negative way. Those deviations which have negative implications can have corrective actions applied. Deviations with positive implications serve as a basis for learning for the manager. He can ask: "Why did we do better than we had planned? What can we do to continue doing better? How can we use this information to establish new standards?"

THE CONTROL CHART A management technique which readily illustrates the exception principle is the *control chart.* First utilized in the area of statistical quality control,[5] it is potentially applicable to a wide variety of management situations.

We shall illustrate the control chart in its quality control context, since that is the easiest to understand. Suppose that items come off a production line in large batches. Since it would be too costly to inspect each item for a critical dimension, a *random sample* is taken from the batch and inspected. The quality control manager's problem is to infer something about the overall batch using only measurements made on items in the sample, which usually represents a relatively small proportion of the total number of items in the batch.

To do this, the quality control manager plots a control chart. Such a chart as that shown in Figure 16-5 is centered on the value of the dimen-

[5] Eugene L. Grant, *Statistical Quality Control,* 3d ed, McGraw-Hill Book Company, New York, 1964.

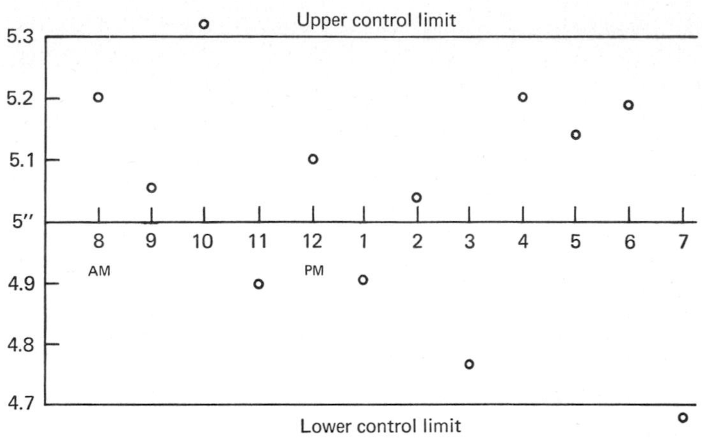

FIGURE 16-5 *Control chart.*

sion which the production line is supposed to be producing. In this case it is assured that the critical dimension is supposed to be 5 inches. Each circle on the chart represents the *average* value of the dimension *for those items in one sample.* For instance, the first circle shows that the average value in the first sample was very nearly 5.2 inches.

The manager knows that he cannot expect every item produced to have exactly the *standard* 5-inch dimension. Random variation will result in some items being smaller and some larger. Moreover, since he is plotting the average value of the dimension in a sample, his measurement is subject to *sampling error*—the error due to his observing a sample rather than the entire batch.

However, he does expect that, *unless some problem exists in the production line,* the average dimension will not be too far away from the desired 5 inches. He assesses what is "too far away" on the basis of a statistical theorem known as the "central limit theorem."[6] It tells him how to determine an *upper control limit* and a *lower control limit. These limits will be exceeded by a sample average only 5 percent of the time if the process is in control*—i.e., if the production line is actually producing at the desired 5-inch dimension. In Figure 16-5, these limits are at 4.7 inches and 5.3 inches, as indicated.

If a sample value falls outside the control limits, as do the 10 A.M. value and the 7 P.M. value, the manager should be aware that a significant deviation from the standard has occurred—the process may be *out of control.* He should focus on this exception to see if it can be explained and, consequently, to determine what control action should be taken.

[6] See W. R. King, *Probability for Management Decisions,* John Wiley & Sons, Inc., New York, 1968, pp. 297–299.

There are a number of possible explanations for an "out of control" point. First, it is possible, although not very likely, that this point represents normal random variation. The control chart is based on the concept that out-of-control points will occur about 5 percent of the time even if the process is in control. Hence, this point may be one of that 5 percent.

Alternatively, the point may be explainable through an "assignable cause." For instance, a new man may have made the 10 A.M. measurement while the regular man was on a coffee break, and his lack of familiarity with the measuring device may have introduced a bias into his measurement.

A third possibility is that the process is truly out of control—i.e., that the production line is not producing at the 5-inch standard. If this is the case, the manager must act on this feedback information to bring about control.

The control chart approach is obviously exception-oriented. Those points which are within the control limits are not given attention. When an exception occurs (an out-of-control point), the manager acts to find out why. Is there an assignable cause? If not, is the process out of control? If so, what can be done to get it back into control?

The control chart approach is applicable to a variety of circumstances. It has been used in straightforward ways to assess such quantities as worker productivity and absenteeism rates. In more sophisticated applications, it has been used for assessing the market performance of newly introduced products[7] and for determining the adequacy of the casualty insurance reserves held by an insurance company.[8]

PROJECT CONTROL AS EXCEPTION MANAGEMENT The network approaches previously described for project control represent an application of exception-oriented management. In the case of a complex project network, the manager cannot spread his attention equally over every activity. What he does, therefore, is to focus on *critical activities*—those on the critical path—and on activities where slippage has occurred.

Before a project begins, the manager treats the activities on the critical path as exceptions, since they are the ones in which delays will cause delays in the overall project. He tries to reallocate resources to these critical activities to ensure their prompt completion. Later, as the project progresses, other activities may become critical because of slip-

[7] W. R. King, *Quantitative Analysis for Marketing Management,* McGraw-Hill Book Company, New York, 1967, pp. 190–200.

[8] W. R. King, "A Control Approach to Insurance Loss Reserving," *Management Science,* February, 1971, pp. 364–370.

page. These then become the exceptions, and the manager shifts his attention to them. In each case, the "nonexceptional" activities get little attention other than monitoring, since they are on time and not likely to have significant impact on the project's completion.

CONTROL AND THE MANAGEMENT PROCESS

The relationship of control to the overall management process is depicted in Figure 16-6. This management process is common to any organized activity. Explanations of the various phases outlined in the figure follow to provide a review of the management process and to emphasize its control orientation. Greater detail is given to control-oriented functions because other functions have already been discussed extensively.

FIGURE 16-6 *Management process.*

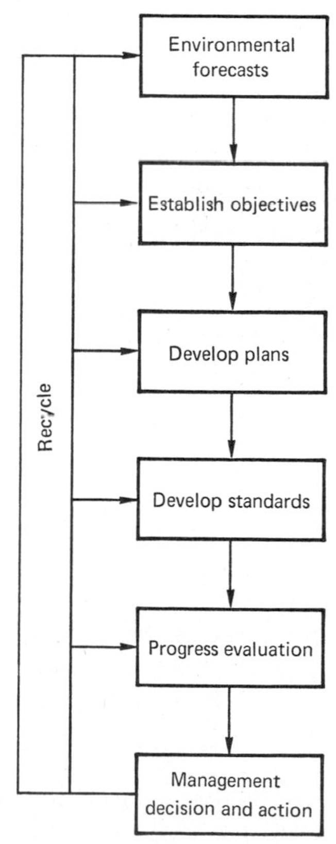

ENVIRONMENTAL FORECASTS Establish intelligence about the environment in which the organization is expected to compete.

ESTABLISH OBJECTIVES Establish the principal and supporting objectives for the organization. Delineate their relationships and prescribe the strategic means for accomplishing these objectives.

DEVELOP PLANS Prepare the necessary strategic, development, operational, and project plans that identify sequential tasks and their input/output relationships. Policies, procedures, work and information flow, budgets, and such criteria are reflected in organizational documentation, which tells what the organization is to be in the future.

DEVELOP STANDARDS Determine the paradigms for organizational and personal performance. The plans developed in the previous step can be visualized as a description of what needs to be done in the judgment of the managers based on the availability of resources. The next step in developing standards is to obtain firm measurable commitments for what will be done from all the managers concerned and for the organization as a subsystem in a larger system.

PROGRESS EVALUATION Measure current performance against commitments and evaluate the meaning of commitment deviations. Commitments for an organization are ultimately expressed as a function of some combination of time, cost, and performance. Commitment targets change as the organization changes and as different projects and programs are undertaken in the organization. Consider, for example, the organizational changes that occur as the organization, and projects within the organization, go through their life cycle. Thus, the manager must view the evolving standards against a changing scenery and across the complete span of interfacing systems. At each higher level of organization involvement, the broadening of the spans of influence increases the manager's span of products/services *and* functions. Consequently, the span of decisions, information, and communications is increased. Thus, the manager must evaluate the progress of a myriad of projects and activities in terms of dynamic standards. Having done so, he must hypothesize the cause of deviations and investigate his hypotheses to determine their validity.

MANAGEMENT DECISION AND ACTION Take necessary action to minimize commitment deviations from standards. Managers generally have considerable discretion in the decisions and actions they take for work completely under their cognizance. The problems they face in making such decisions are as numerous, complex, and different as are the technologies they manage. The manager is constantly confronted with the question: Given that things have not gone as we planned, how do we bring them back into line?

RECYCLING Recycle actions and decisions to the necessary organizational decision centers. As the manager attempts to operate a control system, he finds himself constrained by the commitments that have been made, the plans with which commitments are interlocked, the objectives that the plans were developed to satisfy, and the policies that describe the approved methods. To satisfy these constraints, the manager must evaluate the impact of proposed actions on each constraint and must recycle proposed actions of the decision centers that work with him in redetermining performance requirements, schedules, and costs, redeveloping plans, reestablishing objectives, and reformulating policies. The recycling process enables the manager to preserve a "commitment perspective" designed to effect control with evolving and changing commitments, plans, objectives, and policies.

Figure 16-6 shows that the entire management process can be thought of as facilitating the control function. Just as any person can become so parochial in his own activity or function that he believes the rest of the organization and world to exist only to facilitate his activities, so, too, can all of management be viewed as facilitating control. This viewpoint is too narrow, but it does illustrate the complementarity and interrelatedness of the various management functions.

According to this view, control requires that standards be established; consistent standards, in turn, require an overall framework, so plans must be made; plans are based on goals, so objectives must be prescribed; objectives must be based on some view of the world, so that environmental forecasts must be made. From this view, management planning, organizing, and motivating are all directed toward achieving control. So, too, is the recycling which the feedback loop entails.

The Control Matrix

Perhaps the best way of operationalizing the role of control in the management process is through a *control matrix* which focuses on management functions and performance factors. Figure 16-7 shows the management functions of the previous figure as describing rows of a matrix. The columns are defined by various pervasive performance measures—cost, time,

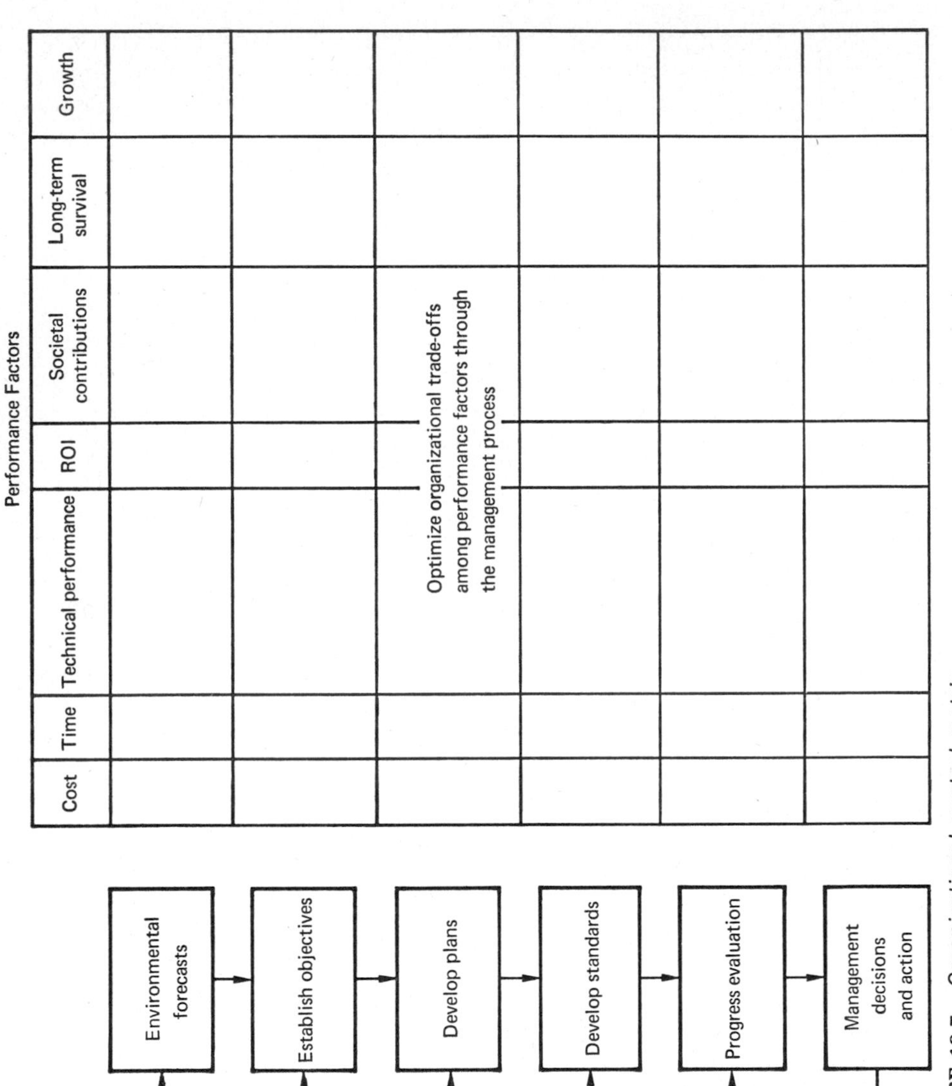

FIGURE 16-7 *Organizational control matrix.*

technical performance, return on investment (ROI), societal contributions, long-term survival, growth, etc.

Each of these performance factors must be considered in each of the phases of the management process. Some elements of the matrix are more important than others, but the entire matrix serves to illustrate the interrelationships and to ensure that none of the possible combinations go unnoticed.

Thus, the first row of the matrix says that environmental forecasts must be made in terms which will permit the assessment of the impact of environmental forces on each of the salient performance factors designated. The fourth row says that standards must be developed for each performance factor, and the fifth row says that progress in each performance area must be assessed.

SUMMARY

Control is the process of constraining activities and resources to conform to a plan. Management control is oriented toward the achievement of objectives through the control of activities and resources. Operational control focuses on the efficient and effective execution of specific tasks.

Feedback system models are commonly used to describe control systems. The feedback loop in such a model represents the flow of information concerning actual performance which is being "fed back" for comparison with a standard. Standards are an important aspect of control. Standards—be they of the financial, ethical, or cultural variety—represent the "bench marks" by which progress is gauged.

In organizations, control is achieved through both formal and informal channels. Formal controls are usually inadequate in and of themselves. When they are complemented by informal controls, such as those exerted by peer groups, a true control *system* exists. Even though such a system may be difficult to describe or assess, it must itself be kept in control in order to assure the progress of the organization.

"Management by exception" is a management philosophy which rests heavily on control concepts. This philosophy directs the manager to focus his attention on exceptions, rather than on those elements and activities which are progressing normally. In doing so, he directs his attention on those places where control action may be required, and by omitting other areas from his consideration, he is able to devote the time and energy to the exceptions which they may well require.

Thus, planning and control are intrinsically interlinked aspects of the manager's job. If he does one without the other, he has not fulfilled his duty and the organization is unlikely to achieve its objectives.

EXERCISES

1. Distinguish between management control and operational control. Relate these to the system of plans.
2. What is control? How does it relate to the management function of planning?

3. What is the role of standards in control? If you were about to undertake a research project, what standards might you establish to ensure control as the project progresses?
4. Relate the problem of reviewing and resetting standards to the feedback models of Chapter 2.
5. Discuss *management by exception.* How is it related to *(a)* control, *(b)* the use of critical path network analysis?
6. What are some of the standards frequently applied in business? What kinds of standards are becoming more important to business managers than they were previously?
7. Consider the project network shown below (estimated time in weeks on activity arrows).

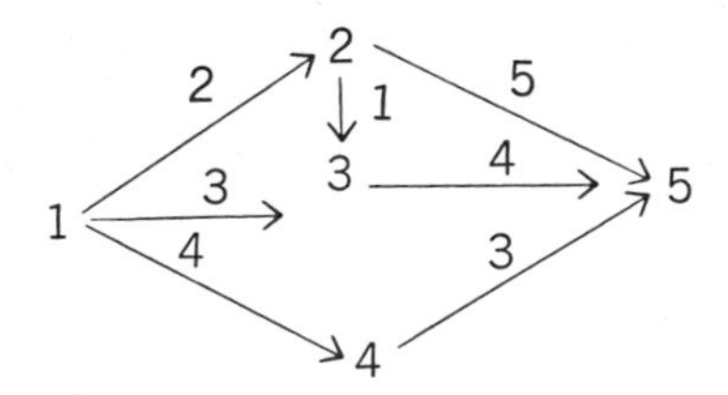

a. Determine the critical path.
b. After 4 weeks have elapsed, the following progress report is given to you as project manager:

Activity 1–2: Completed in 2 weeks
Activity 2–3: 2 weeks utilized; another week required
Activity 1–3: Completed in 4 weeks
Activity 1–4: Completed in 4 weeks

What actions might you consider?
8. What is meant by informal control? Which sort of control—formal or informal—is harder for the individual to escape?
9. What does Dubin mean when he says that the blending of informal and formal controls protects the organization from self-destruction? What are the implications of this to the overall management process?
10. How can the control function be conceptualized to form the central core of the manager's job? In this conceptualization, relate other management functions to the control function.

REFERENCES

Archibald, R. D., and R. L. Villoria, *Network-Based Management Systems,* (PERT/CPM), John Wiley & Sons, Inc., New York, 1967.

Arrow, Kenneth J., "Control in Large Organizations," *Management Science,* April, 1964.

Berkwitt, George J., "How Good Are the Management Sciences," *Dun's Review,* July, 1968.

Bigelow, C. G., "Bibliography on Project Planning and Control by Network Analysis: 1959–1961," *Operations Research,* September–October, 1962.

Davis, Keith, "Mutuality in Understanding of the Program Manager's Management Role," *IEEE Transactions on Engineering Management,* December, 1965.

Einstein, Hans A., "Systems Philosophy and the Professional Manager," Air University Library, Maxwell AFB, Alabama, June, 1965.

Frederick, William C., "The Next Development in Management Science: A General Theory," *Journal of the Academy of Management,* September, 1963.

Flaks, Marvin, and Russell D. Archibald, "The EE's Guide to Project Management—I—The Concepts Behind the Modern 'Systems' Approach to Controlling Complex Operations and Programs," *Electronic Engineer,* April, 1968.

———, "The EE's Guide to Project Management—II—The Project Manager: Who He Is and How He Goes About Organizing a Project," *Electronic Engineer,* May, 1968.

Gibson, R. E., "A Systems Approach to Research Management, Part I," *Research Management,* vol. V, no. 4, 1962.

———, "A Systems Approach to Research Management, Part II," *Research Management,* vol. V, no. 6, 1962.

Goodman, Richard A., "Ambiguous Authority Definition in Project Management," *Academy of Management Journal,* December, 1967.

Granger, James I., "Analysis of Management Control Systems," Air University Library, Maxwell AFB, Alabama, June, 1965.

Handy, Henry Wesley, *Network Analysis for Educational Management,* Prentice-Hall, Inc., Englewood Cliffs, N.J., 1967.

Johnson, Richard A., et al., "Systems Theory and Management," *Management Science,* January, 1964.

Lewis, Leonard J., *The Management of Education; A Guide for Teachers to the Problems in New and Developing Systems,* Frederick A. Praeger, Inc., New York, 1965.

Martino, R. L., "Finding the Critical Path," *Project Management and Control,* vol. 1, American Management Association, New York, 1964.

———, "Applied Operational Planning," *Project Management and Control,* vol. 2, American Management Association, New York, 1964.

———, "Allocating and Scheduling Resources," *Project Management and Control,* vol. 3, American Management Association, New York, 1964.

Souder, William E., "Experiences with an R&D Project Control Model," *IEEE Transactions on Engineering Management,* March, 1968.

17

management information systems

Information is the essence of any management control system. Indeed, often the terms "management control system" and "management information system" are used interchangeably. However, in this chapter we shall analyze information systems for both planning and control, since, ideally, the same comprehensive information system can provide information to accommodate all phases of the management process.

The organization's management information system (MIS), as we shall conceive of it, provides managers with the information which they need to intelligently make and execute decisions. It provides the basic information which is necessary for all aspects of organizational management—from the highest level of establishing the organization's basic goals and policies to the lowest operational level.

Few organizations currently possess *comprehensive* computerized management information systems of the variety which we will discuss in this chapter. However, many large organizations, such as Ford, Pillsbury, General Foods, Boeing, and Control Data,[1] have very sophisticated systems, most have advanced computer subsystems, and virtually all large, complex organizations are working aggressively toward the development of comprehensive systems.

"Information," as defined by Webster, is "the communication of knowledge or intelligence; . . . knowledge derived from reading, observation or instruction; especially unorganized or unrelated facts or data."[2] In an organization, information is communicated by mail, telephone, telegraph, teletype, personal visits, periodicals, books, newspapers, memoranda, and a variety of other internal printed documents such as policy statements and budgets.

[1] "Litton's Electronic Information Machine," *Business Week,* Mar. 28, 1970, pp. 158–164.

[2] *Webster's New Collegiate Dictionary,* G. & C. Merriam Company, Springfield, Mass., 1961, p. 430.

Every organization has its *formal information system*—a communications process and reports which are generated by individuals or departments for distribution to other individuals or departments. This system is usually mechanized to some degree; thus, it often involves data processing equipment and mass distribution of duplicated memos and reports. Some formal information systems are very extensive and sophisticated. For instance, Chrysler Corporation is reported[3] to have instituted an extensive system designed to keep top management informed of environmental occurrences which can affect the company. It involves a "news center" which is not unlike that of a radio or TV station and the submission of news analyses four times each day to the firm's top managers.

Comprehensive organizational information systems usually have both internal and external sources of data. For instance, Figure 17-1 is a

[3]"Alerting Top Brass," *Saturday Review*, Aug. 10, 1968, p. 59.

FIGURE 17-1 *Litton's management information system. From "Litton's Electronic Information Machine," Business Week, Mar. 28, 1970, p. 158. Used with permission.*

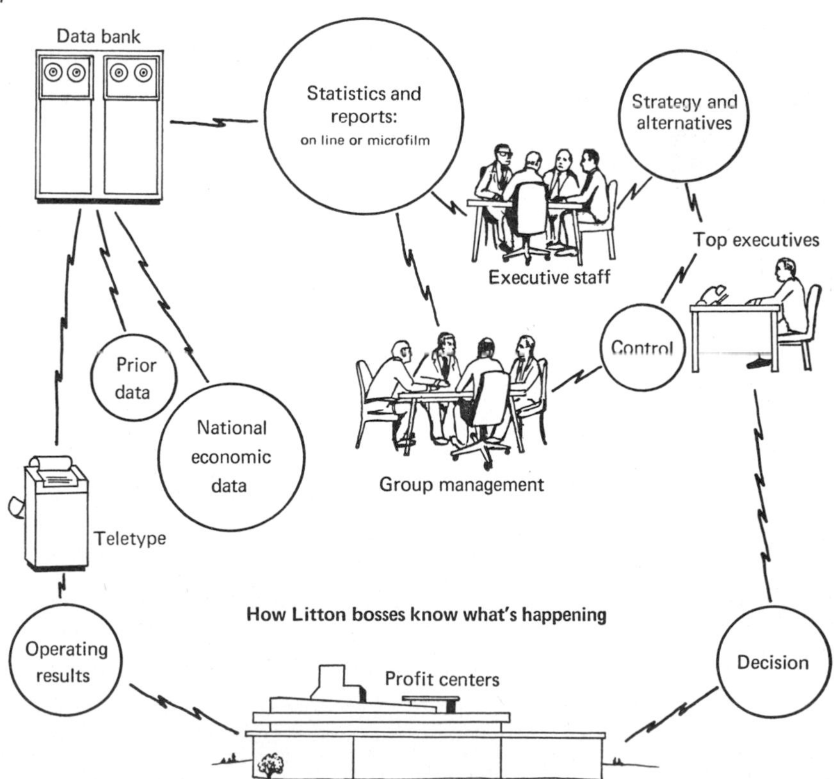

pictorial illustration of Litton Industries, Inc.,'s MIS. It shows operating results combined with historical internal information and externally generated economic information in a data bank. Statistics and reports are regularly produced for and irregularly requested by the executive staff (for the development of plans) and by group management (for control purposes). Top management acts on the recommendations made by these groups, using, if necessary, additional information obtained from the system.

All of these aspects described in Figure 17-1—data gathering, communications, data bank, reports, meetings, recommendations, etc.—constitute parts of a formal information system.

The *semiformal* aspects of the total information system consist of all memos, conversations, and other communications which are carried on from day to day within the structure of the normal "chain of command." Thus, a conversation with one's boss concerning the impact of a pending reorganization, or a phone call from a branch office to expedite an order, is a link in the semiformal system. Similarly, an executive who considers it a function of his position to maintain an awareness of the current business and financial situation by a daily perusal of the *Wall Street Journal* is performing a semiformal information collection function. A similar function might well be performed on a formal basis by a business librarian who is charged with scanning a variety of publications each day to identify those items of interest to the firm.

The *informal information system,* or "grapevine," consists of all of those interpersonal and interdepartmental linkages which are not prescribed by policy or tradition. A secretary whose roommate is a secretary in another department may deliver some "gossip" concerning goings-on in that department which are very pertinent to business decisions. Thus, the grapevine comes about because of the social interaction of people as opposed to their formal organizational interaction. According to Davis, ". . . it [the grapevine] is as fickle, dynamic, and varied as people are. It is the expression of their natural motivation to communicate. It is the exercise of their freedom of speech and is a natural, normal activity."[4] The grapevine cannot be abolished by executive decree—in fact, attempting to abolish it may in fact strengthen it and cause the informal organization to work to the detriment of the formal organization.

The organization's overall information system consists of all three of these varieties—formal, semiformal, and informal. There is little serious attempt on the part of managers to exert anything but meager control over the informal system. However, both the formal and the semiformal systems

[4] Keith Davis, *Human Relations at Work,* McGraw-Hill Book Company, New York, 1967, p. 222.

can be designed and maintained in a way which maximizes their effectiveness in terms of management decision making.

The *management information system* (MIS) is that combination of formal and semiformal "systems" which provides information to managers. The marketing department continuously needs information about the demand for the company's products and the action of the competitors producing similar products. The finance department requires information to plan and control the funds used in the business. The production people have to receive information on which to base the design and operation of the production system, so that the product is produced when required and in the form desired by the customer. Research and development managers must have information to provide them with a basis for predicting what the market of the future will demand in the way of new products and services. Thus, every manager in an organization requires information and an organized system for providing it to him—a *management information system.*

BASIC INFORMATION CONCEPTS

Several basic concepts are essential to an understanding of management information systems.

Data and Information

Two of the words in the three-word phrase, "management information systems," have already been defined and extensively discussed in earlier chapters. The only word requiring further definition here is "information." We shall go beyond Webster's definition to offer one which is more operationally useful.

Most of us think of information in terms of "facts" or "data." Indeed, data and facts are basic to the concept of "information." *Data are the unevaluated facts and messages which exist in the environment.*[5] Examples of data are statistics describing the demographic characteristics of markets, sales statistics, and records of phone conversations and telegrams. We normally think of data in recorded form—as printed in government

[5]Another view of data is in terms of "organization" rather than "evaluation." For instance, Bower et al., view data as "unorganized facts." See J. B. Bower, R. E. Schlosser, and C. T. Zlatkovich, *Financial Information Systems: Theory and Practice,* Allyn and Bacon, Inc., Boston, 1968, p. 195. Still another view is in terms of "interpretation." Blumenthal says that data are "uninterpreted raw statements of fact." See S. C. Blumenthal, *Management Information Systems,* Prentice-Hall, Inc., Englewood Cliffs, N.J., 1969, p. 30.

reports, sales summaries, and magazine articles, or as magnetic impulses on tapes or in computer memory cores. And, although we often think of data in numerical form, much of the important data which are used by modern organizations are qualitative rather than quantitative. The data which are collected, stored, and perhaps even used by the typical organization are voluminous. Organizational data are collected from each of the organizational "claimants" and stored in original or summary form on punched cards, file cards, and magnetic tapes, in photographs, and in as many other media as human ingenuity can conceive. Often data are maintained for decades—sometimes because no formal procedure has been developed for updating them or for reevaluating their utility and sometimes as the result of conscious decisions on the part of management. A recent "clean-up" campaign experienced by one of the authors is a reflection of this phenomenon in a university context. During the cleanup, data used in research projects conducted by now-departed professors over 10 years ago were unearthed and discarded to make space for newer data.

However, the maintenance of data for a long period is sometimes necessary. For instance, one major insurance firm files original life insurance applications for the lifetime of the policyholder in order to preserve an original legal record of the applicant's declarations concerning his health history. This storage requires several floors of a city-block-sized vault.

A manager may require or desire any of the organization's data in order to help him make a decision or deal with a situation. Thus, when a decision is to be made on the location of a new plant, the manager will desire data on land costs and labor costs for each of the locations under consideration. He will also wish to know about existing transportation facilities, transportation charges, the nature and dispositions of local governments, and a host of other things. So, too, when the manager is faced with a "simple" situation, such as welcoming a visiting dignitary, he will wish to have data on the visitor's background and interests to permit him to discuss and display those things which will be of greatest interest.

The essence of "information"—as opposed to data—is that *information is data which have been evaluated for use in a particular situation or class of situations.* For example, the numbers which represent records of the past monthly sales of a product in various markets represent data. If it is determined that these numbers are to be used to forecast future sales, which forecasts will in turn serve as a basis for scheduling production of the product, the data become information; they have been evaluated in terms of their utility in a management decision situation. So, too, do the filed original insurance applications encompass information, since the legal necessity for the maintenance of original documents has been established. Thus, while the original documents are not directly related to a specific decision problem, their general need in a class of important situations has been identified.

Information in Management Decisions

A simple decision feedback loop such as that in Figure 17-2 demonstrates the role of information in management decision making. The decision-making process involves information in two essential forms—input and feedback.

Input information is that provided to the decision maker from outside the context of the decision problem in question. He uses this input information in his decision-making "model"[6] and chooses an action. This action impacts on "the world" and produces some consequences. *Feedback information* describing these consequences is then provided to the decision maker's model for his use in reevaluating the decision. For instance, government statistics on population might serve as input information to a decision maker who is faced with decisions concerning a new product. He may use these data to make sales predictions. Then after the product is marketed, the manager receives feedback information on *actual sales.* He may use this feedback information to revise his predictions and his decisions concerning the product. If sales are not up to expectations, he may decide that control action is necessary—e.g., that more advertising is needed to enhance the initial appeal of the product in the consumer's mind. If sales are adequate, he may decide to try some small cuts in his advertising expenditures. In any case, he uses the input information for his initial appraisal of the decision problem and *both* the input *and* feedback information to make reevaluations.

Every decision can be thought of as a way of integrating input and

[6]The term "model" is being used here in a general sense to mean any thinking process (objective or subjective) which is used by a manager to characterize a decision situation. Thus, both "mental models" and formal mathematical ones fit this use of the term.

FIGURE 17-2 *The role of information in decision making.*

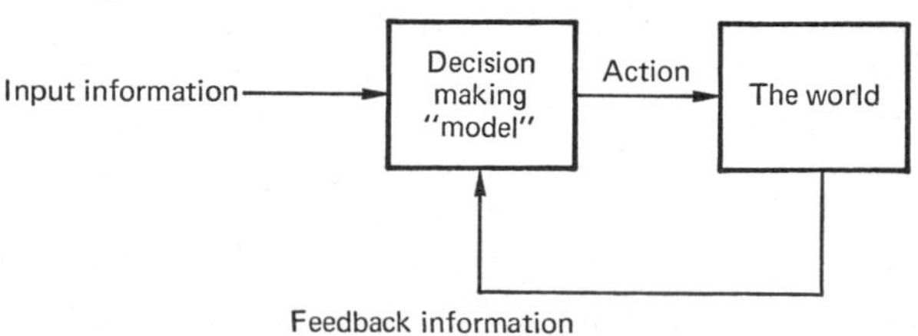

feedback information into action-oriented directives. Without information, the manager would be required to solve his problems totally on the basis of his "hunches" and intuition. In such a case, he would be operating apart from the real world in a fashion which would virtually ensure his demise.

However, the significant problem in modern organizations is not the realization of this caricature of the manager who has no information on which to base his judgments, but rather the manager who has too little *relevant* information. We shall stress this point more strongly later. For now, consider the manager whose desk is laden with reports and computer printouts, while at the same time he searches for the one crucial bit of information which would enable him to arrive at a reasoned decision. This situation is more realistic than that of the manager who is cut off from the world around him. Indeed, it is believable in part because it is so true.

In a sense, many managers are blessed with too much information, or at least, with too much data.[7] Modern computers have given organizations the ability to collect and process voluminous data. In many ways this facility has opened whole new vistas for management. For instance, a large manufacturer of household and cosmetic products has recently for the first time been able to provide managers with breakdowns of weekly sales of their numerous brands by market area and package size. This has opened up a whole range of opportunities for special product promotions and for analyses of the impact of price "deals" and promotions which were never before feasible.

However valuable may be such information, the proliferation of computerized information often results in clogged communications channels. The manager who cannot possibly read and digest the myriad reports which come to him can be rendered ineffective by a lack of *relevant* information. Unless information is presented to the manager in a way which enables him to readily cull out that which is crucial from that which is routine, the manager is being managed by a creature of his own making—the MIS.

What Is an MIS?

The role of an MIS in an organization is to provide managers with the information which they need in order to manage. This simple statement of purpose is not nearly so easy to accomplish as it may appear, however.

[7] See Russell L. Ackoff, "Management Misinformation Systems," *Management Science*, December, 1967, pp. 147–156.

Any useful MIS must accomplish this goal with reasonable cost, accuracy, and speed.

Although there is no reason why an efficient and effective MIS could not be developed without involving computers as elements of the system, the need for cost efficiency, speed, and accuracy often dictates that computers be incorporated as an important part of such systems.

One way of making clear exactly what sort of information system constitutes an MIS is to illustrate systems which *do not* qualify. For instance, most modern organizations have developed data processing subsystems which handle portions of the data necessary for the organization to operate. Often, some of these subsystems will be manual operations and others will involve electronic equipment. Sometimes, the recognition of the need for a comprehensive MIS comes about because the various existing subsystems are not integrated (or because they are inefficiently integrated).

An illustration of this is a major bank that has many separate computer-based subsystems which handle the accounting and record keeping for its various services. These subsystems were independently designed in an era when it had become clear that the old manual accounting systems simply could not keep up with the demands of a rapidly expanding business. They were implemented and installed and they did their accounting functions well. Now the managers of the bank find that they need certain integrated elements of information—parts of which are contained in the various subsystems. For instance, the marketing manager would like to have a listing of the various services provided to each of the bank's customers. If he knew, for instance, that a particular customer had a checking account *and* a loan account, he might consider this customer a prime prospect to also open a savings account. After all, the patron is already a satisfied customer of the bank and he should therefore be easier to "sell" than would be someone who has had no contact with the bank. However, this simple bit of information is not available (except at great cost), because each of the bank's services has its separate computerized subsystem, and each subsystem has its own "rules" and modes of operation. Thus, it would take a special "run" of every subsystem to obtain this sort of information—a very costly procedure which would be required on a continuing basis, since the marketing manager would reasonably require current information and he could act on only a small number of accounts at a time.

But the problem is even more complex than this, because the "pieces" of information provided on each customer by each service subsystem are not necessarily compatible, or even identifiable. For instance, the various accounts of one of the authors might be labeled as follows:

William R. King
W. R. King
Dr. William King
William King
William Richard King
King, W. R.
.
.
etc.

The process of "matching" such divergent labels to ensure that each customer's various accounts are consolidated under one label is a difficult one. To perform such an operation on a large scale is costly.

These problems clearly indicate that one essential characteristic of a true MIS is that it be *integrated.* This integration must exist in at least two senses. First, the various subsystems of the system must be compatible and interlinked. A set of incompatible or independent computerized data processors clearly do not constitute an MIS. The MIS, then, must be built on an *integrated data base.*

The Integrated Data Base

Basic to the idea of a comprehensive MIS is the concept of an *integrated data base*—a common pool of data in standardized form which can be drawn on by different subsystems and different managers in various functional elements of the organization. Thus, the marketing manager who needs to have information on the history of sales made to a particular customer should be able to obtain it, and the sales manager who wishes to have sales broken down by territory should be able to secure them.

Each of these requests for information requires the use of some common data elements—the quantity sold to a particular customer on a particular date—but each also requires some auxiliary data, such as the identification of the sales territory in which the customer is located. The idea of an integrated data base involves the recording of information in *primitive,* or basic, terms so that it can be made use of in the widest possible variety of ways. Thus, the data base would not have monthly sales summaries, since that would make it impossible to obtain certain "decomposed" information elements such as that regarding a particular order. Information would be recorded in *indecomposable* form in terms of each product ordered by each customer. If a given order number involved several different products, each product would be recorded separately, along with appropriate auxiliary information, such as the customer's industrial categorization,[8] the market area in which he is located, the salesman, etc.

Of course, the two critical aspects of the development of a data base are the *identification of the indecomposable data elements* and the *determination of the appropriate auxiliary information.* For instance, such items as the weather conditions on the day of the sale and the amount of time spent by the salesman in obtaining the order could conceivably be considered as appropriate auxiliary information. Someone might hypothesize that most sales are made on warm days and might wish to test that hypothesis. Or, an accountant might wish to do a cost study relating the time spent in completing each sale to the size of the order obtained. In either case, the weather and "sales time" auxiliary information would make it feasible to do so, and, if the information were not recorded, accomplishing it would be difficult.

The salient question, then, in the development of a data base revolves about the relative *costs and benefits* of the information to be incorporated into the data base. Each item of data must be evaluated in terms of its informational content—the benefits which will accrue to the organization from having it—and its cost—the resources which must be devoted to incorporating it into the data base and updating it. Thus, if the customer's market area designation is being evaluated for possible incorporation into an MIS data base, it must be considered in terms of the decisions and situations in which it will (or may) prove useful, how crucial it may be to those situations, how important the situations themselves are to the achievement of overall organizational objectives, and the total cost of incorporating and maintaining the information in the data base.

Of course, the actual evaluation of every data element on such a detailed and stringent cost benefit basis is neither easy nor always practical. However, the concept is valid. Moreover, the failure to apply the concept, at least in approximate form, in the design and development of an MIS can lead, and has led, to a proliferation of data that makes the system unmanageable.

Unless the data base is evaluated in these terms, one of two extreme things is likely to take place. In the first case, everything of conceivable relevance is incorporated into the data base, so that the system becomes very costly and inefficient. Alternately, a cost-conscious designer imposes arbitrary limits on the amount of data which can be recorded, and managers find that the sparse data available are not useful. Examples of the former case are legion; the latter case is illustrated by the marketing

[8] The SIC (Standard Industrial Classification) system is most commonly used for this purpose. It assigns to each industrial firm a number which identifies the industry and subindustries to which the firm belongs.

manager who was told that economics dictated that he could have only a three-digit customer SIC code[9] included. He subsequently found that a five-digit identifier was necessary for him to perform any analyses and that the three-digit code was entirely useless. In such a case, either the five-digit number should have been used or no SIC code should have been recorded at all, and the better of these two alternatives could only have been determined through the use of a cost benefit analysis.

The Role of the Computer

Although the information system concept does not necessarily involve computers, it is certainly true that most large organizations design MIS using computers as the heart of the system, and it is also true that computers have had a great impact on MIS thinking.

The earliest computers (IBM 701, Univac I) were vacuum tube models introduced in the early 1950s. This introduced the age of *electronic data processing* (EDP). Traditional organizational operations, such as payroll and accounts receivable, which had previously been performed manually or by using tabulating equipment, were readily transformed to electronic data processing. The operations were performed faster and more accurately than they had been previously, but little new information was provided to enable managers to make better decisions.

Since the introduction of solid-state computers (1958) and the third and fourth generations of high-output–low-cost computers (1964 and 1970, respectively), the feasibility of using computers to do more than automate existing functions has been greatly enhanced. For instance, the concept of "integrated data processing," in which data are processed in a continuous sequence with a minimum of transformation or change after their entry into the data base, has led to increased accuracy and better control in data processing.

It is estimated that the value of general-purpose computer installations in the United States in 1972 is $30 billion.[10] The estimated annual outlay for "software"—the programs which are the brains of the computers—in that year is $11 billion.[11] So computers are already big business in the United States.

However, the *potential* for further applications of computers in operations control and management information systems is illustrated by

[9] See previous footnote.

[10] This estimate includes peripheral equipment. See G. Burck, "The Computer Industry's Great Expectations," *Fortune*, August, 1968.

[11] See Burck, *op. cit.*

two statements which summarize portions of a computer applications survey.[12]

> *In spite of nearly ten years of company experience with computers, automated systems for operations were observed as being in the early stages of development.*
>
> *In general, computer uses in technical areas were found as facilitating operations rather than aiding in their management.*

Thus, while computers play an important role in business, industry, and government information systems, their role in true management information systems still leaves much to be desired.

THE ORGANIZATIONAL INTELLIGENCE SYSTEM[13]

In discussing management information systems, great emphasis is usually placed on the information subsystem which requires only (or largely) internal inputs. In the experience of the authors, organizations tend to defer development of the subsystem requiring principally external inputs until the internal system is operative. Often, they never get to the external part. And, although some obviously do develop such systems, less is known about them because there is less total experience in this milieu.

Organizational Intelligence

"Organizational intelligence" refers to *competitive and environmental data* which have been evaluated and found to be useful in a specific situation, project, or class of situations. However, the term is not restricted to external data, since a great deal can be discovered about competition and the environment from internal sources.

No person or organization can effectively operate in any competitive environment without a basic understanding of that environment or without up-to-date information on happenings in the environment. This point really needs no extensive logical justification. However, on a pragmatic level, one can simply point to the intelligence systems being developed by many diverse firms. Witness, for example, the *Harvard Business Review*

[12] N. C. Churchill, J. H. Kempster, and M. Uretsky, *Computer-based Information Systems for Management: A Survey,* NAA Research Study, New York, 1969.

[13] The authors are indebted to Dr. John H. Manley for his aid in developing and reviewing this section.

article, from more than a decade ago, which said that: "Commercial intelligence departments are appearing more and more on corporate organization charts."[14]

The types of information about competitors which is generally considered "fair game" for collection include market pricing, discounts, terms, specifications, total market volume for a given product, historical trends, estimates of competitor's share, competitor's trends, evaluations of competitive product quality and performance, estimates of marketing policies and plans of each competitor, and new competitive systems and/or trends.

However, it is equally important that decision makers be aware of more subjective, and potentially more valuable, information obtained through personal contact with competitors, government officials, etc. To do this effectively requires that:

1. Marketing people, executives, and others in the organization be assigned specific intelligence-gathering responsibilities as an inherent part of their position.
2. Organizational operatives be assigned to monitor certain specific activities of other organizations.
3. A formal system be developed for collecting, evaluating, and dispensing business intelligence.

The Necessity for a "System"

Illustrations of the nature of business intelligence serve to motivate the need for a formal intelligence-gathering, -processing, and -monitoring system. A useful broad delineation of the various kinds of business intelligence is given below.[15]

1. *Corporate and financial*
   Corporate structure, interlocking directors
   Management personnel changes, organizational changes
   Mergers and acquisitions
   New Financing
   News of capital investments
   Quarterly and annual financial reports

[14] D. R. Daniel, "Management Information Crisis," *Harvard Business Review,* vol. 39, no. 5, September–October, 1961, pp. 111–121.

[15] Based on material in Peter Hamilton, *Espionage and Subversion in an Industrial Society,* Hutchinson Publishing Group, Ltd., London, 1967, p. 187.

2. *Production*
   Raw materials—nature, sources, costs
   Present capacity, present output, planned capacity
   Production costs
   Processing techniques—use of new equipment and instrumentation
   Product improvements
   New plans—where, when, what, how, and why
3. *Marketing*
   Method of marketing
   Warehousing methods
   Price lists, discounts, contracts, reciprocal agreements
   Advertising plans
   Customer surveys
   Customer brochures
4. *Technical*
   R&D staff—quantity, quality, and location
   Analysis of technical recruitment advertising
   New products and processes—research findings, pilot-plant work
   Government research contracts
   Quality of products as per analysis and tests
5. *Legal*
   Trademarks used and pending
   Patents issued and pending
   Patent or trademark infringement suits

To even attempt to address such a broad diversity of business intelligence without systematization would be utter folly. No individual or group could reasonably be expected to be aware of, and much less to have evaluated, such a range of data. Therefore, any action taken without such a system is necessarily made without the full Information which could contribute to that action.

Even subjective personally derived information needs to be systematized. The "slip of the lip" will stand a much greater chance of being caught and passed on to the key decision makers if it can be evaluated, put into its proper context, and highlighted as useful intelligence.

To systematize business intelligence really means only that questions such as those that follow be answered and the answers made optional.

1. What needs to be known?
2. Where can the data be obtained?
3. Who will gather the data?

4. How will the data be gathered?
5. Who will analyze and interpret the data?
6. How will extracted information be stored most efficiently for equally efficient future retrieval?
7. How can extracted intelligence be disseminated to the proper parties at the right time for consideration?
8. How will the system be protected from "leakage" and from sabotage?

Unless the business intelligence problem is focused toward a single program or objective and can therefore be approached on a somewhat ad hoc basis, reasonable answers to these questions generally dictate that a computer system be utilized for effective storage and retrieval operations. Dissemination and display over and above periodic briefings and responses to specific requests can be facilitated by assembling user interest profiles and feeding them into the computer. Then, as additions are made to the file, match-up with these profiles can automatically trigger intelligence outputs which might have been missed by the human data interpreter. This is a common approach taken by military intelligence units which will permit a double check on the process and ensure rapid dissemination of vital information to the proper users.

An Action-oriented System

An organizational intelligence system should be action-oriented in that it informs managers *when* actions should be taken and provides some indication of the *best action* to take. If the system simply provides volumes of reports on a continuing basis, action directives are left solely to the manager; he must sift through the mire of data to evaluate their relevance and applicability.

If the user-manager is to be relieved of this task, the business intelligence system should be *exception-oriented;* that is, in addition to a continuing flow of relevant information, it should be programmed to provide "red flag" signals where action is required.

For instance, the emergence of a new line of thought in a customer's organization might signal the need for a new project. In such a case, the "project" might simply be the assignment of an individual to follow the development of the idea in the customer organization, to evaluate its potential for new business, and to report on it.

To accomplish this requires two primary things: *(a)* that the organization be monitored, and *(b)* that the business intelligence system be capable of reporting this "exception."

Some of the other actions which can be signalled by a business intelligence system are:

1. Difficulties in ongoing projects that may require control actions within the company
2. Changes in competitor's policies which may require policy reviews
3. Emerging technologies which may require study
4. Suggestion of emerging situations where the organizational image is threatened.
5. The emergence of forces that suggest the need to face a strategic decision

Intelligence exists in many different forms in a variety of places. Perhaps a summary of some intelligence-gathering methods is most descriptive of this variety:

1. Published material and public documents
2. Disclosures made by competitor's employees and obtained without subterfuge
3. Market surveys and consultant's reports
4. Financial reports and broker's research surveys
5. Trade fairs, exhibits, and competitor's brochures
6. Analysis of competitor's products
7. Reports of company salesmen and purchasing agents
8. Legitimate employment interviews with people who worked for a competitor
9. Camouflaged questioning and "drawing out" of competitor's employees at technical meetings
10. Direct observation under secret conditions
11. False job interviews with competitor's employee (i.e., where there is no real intent to hire)
12. False negotiations with competitor for license
13. Hiring a professional investigator to obtain a specific piece of information
14. Hiring an employee away from the competitor, to get specific know-how
15. Trespassing on competitor's property
16. Bribing competitor's supplier or employee
17. "Planting" your agent on competitor's payroll
18. Eavesdropping on competitors (e.g., via wire-tapping)
19. Theft of drawings, sample, documents, or similar property
20. Blackmail and extortion

This distribution of methods ranges from the obvious and ethical to the less obvious and unethical. We shall take up the question of ethics later. Here, we wish to emphasize that *the important sources of information are largely of the readily and ethically available variety.* To illustrate this, consider the fact that Admiral Ellis Zacharias, Deputy Chief of Naval Intelligence during World War II, has indicated that the U.S. Navy obtained about 95 percent of its peacetime intelligence through public sources, 4 percent from semipublic sources, and only 1 percent from secret agents.[16]

The Ethics of Intelligence

The predominantly public or semipublic basic nature of intelligence, coupled with the fact that any good intelligence system will operate with organizational personnel in visible positions in other organizations, determines that unethical or illegal methods using "secret agents" to collect competitive business information are not likely to be necessary for attaining an effective level of intelligence. This supposition is supported in the literature of the field. For example, *Industrial Management* says that ". . . the sources are so numerous . . . one need not stoop to any unethical practices."[17]

The use of unethical means of intelligence gathering can be counterproductive in the long run. If one pays an agent or bribes a competitor's agent to gain information, that person can undoubtedly be "bought" by the competitor, thus making the information possibly detrimental rather than beneficial. If one then uses a counterintelligence system to preclude such erroneous data, the costs of data collection will rapidly escalate to the point of negating the benefit of the original system. This argument does not even take into account the risk of exposure and resultant loss of customer good will and possible fines or even imprisonment. Illegal and unethical activities simply do not pay in the long run.

MIS DESIGN

Perhaps the best way to understand what an MIS is, what it should do, and what the problems are in developing an MIS which will truly serve the need of managers is to consider an idealized MIS design process.

[16]Captain E. M. Zacharias, USN, *Secret Missions, The Story of an Intelligence Officer,* G. P. Putman's Sons, New York, 1946, pp. 117–118.

[17]"Guide to Gathering Market Intelligence," *Industrial Management,* vol. 47, March, 1962, p. 84.

Idealized MIS Design Process

The process of designing an MIS might consist of six major phases, as outlined below:

1. Analysis of the objectives and goals for a management information system
2. Development of a "decision inventory"
3. Analysis of information requirements
4. Development of a data base
5. Analysis of software requirements
6. Analysis of hardware requirements

ANALYSIS OF THE OBJECTIVES FOR A MANAGEMENT INFORMATION SYSTEM The first phase of the management information systems design process involves the determination of appropriate objectives *for the system* to be developed. Since the objectives and goals for a broad-based system which will serve the informational needs of many managers will invariably be myriad and varied, this determination requires the gathering of opinion and judgment from various managers concerning the organization's needs and desires for such a system.

After these judgments concerning goals and objectives of the system have been articulated, they must be checked for compatibility and practicability. In other words, are the stated objectives in conflict with one another, and are they really operationally obtainable? While it is true that an organization may operate for a period of time without a clear explanation of compatible and operationally obtainable goals, the same is not true for an MIS. It is possible for an organization to do this because it can subjugate certain goals during certain time intervals and emphasize them during others. Thus, it can emphasize safety goals during a period immediately following a serious plant accident while paying only lip service to this goal during peak periods of profitable production. In the case of an MIS, a single system is to be designed and implemented, so its goals must be explicit, compatible, and operationally obtainable.

Perhaps the best way of assessing the goals which have been stated by various managers is in terms of a priority list of objectives which can be developed and evaluated by those managers who have made contributions. While the time and cost of making such reevaluations of goals are great, any MIS which is developed without an explicit delineation and evaluation of its proposed objectives will inevitably fail to be a true management-oriented system. Even though the process of determining systems objectives is always a tedious one, the cost of not doing so is very great.

Illustrative of the sort of operational goals which can be the output of this phase of the MIS design process are the specifications that the MIS should provide information for such activities as:

The establishment of sales and cost objectives at the corporate level
The establishment of operational objectives for branch offices
The establishment of performance standards for managerial personnel
The prediction and evaluation of competitive market actions
The prediction evaluation of changes in market structure
The evaluation of new services
The identification of potential customers

DEVELOPMENT OF A DECISION INVENTORY After the goals and objectives for the system have been defined, the specific organizational decision areas which will be served by the system must be determined and analyzed. Since the primary purpose of the MIS is to provide a basis for better decisions and better control, it is impractical to attempt to determine "information requirements" without first relating them to the decision for which they are "required." Thus, a "decision inventory" which specifies the decisions made within an organization, together with those who are responsible for the decisions, provides a basic tool for MIS design.

ANALYSIS OF INFORMATION REQUIREMENTS After the decision inventory has been compiled, the specific elements of information which are potential inputs to the system must be evaluated in terms of their relevance to the objectives. The decision inventory is a guide to the accomplishment of this phase, and it describes the roles of individual managers in each of the major decisions of the organization. Thus, if one of the objectives of the system is to provide information which will permit the relative evaluation of new products, the decision inventory prescribes those involved in the decision. Each of these managers can be consulted so that an initial specification of the "information requirements" can be made. Thus, each of the managers might specify certain items of information which he deems to be necessary to fulfill his role in the new-product decision.

Each of the elements of information which are so specified must then be subjected to some process of objective analysis. MIS technicians and systems analysts have learned that allowing managers to specify those elements of information to be included in an MIS without regard to the *economics* of the inclusion inevitably leads to costly and unworkable systems. Thus, the initial information requirements specified by management must be evaluated on a cost benefit basis. Each element must be assessed in terms of the total cost of incorporating it into the system and of continuously updating and maintaining it versus the benefits which will be achieved by the organization from having the information available.

DEVELOPMENT OF A DATA BASE The data base consists of those elements of information which have been evaluated on a cost benefit basis and assessed to warrant including them in the system. In essence, the cost benefit approach aims at the achievement of an "optimal" data base—one which provides the essential benefits to the organization while not incurring excess costs.

ANALYSIS OF SOFTWARE REQUIREMENTS Once the data base has been specified, the programs necessary to operate the systems must be outlined. Thus, a true MIS should not be exclusively oriented toward the development of reports provided on a routine basis. Rather, it should be capable of exception reporting—the reporting of unusual events on a nonroutine basis—and it should be capable of providing information for a special analysis to be used in decision making.

ANALYSIS OF HARDWARE REQUIREMENTS After the system outline has been specified and software requirements have been analyzed, consideration of the "mechanical" part of the system should begin. Evaluations must be made of each of the existing systems which has a capability of meeting the designed criteria. Again, the basis for the evaluation will be a general cost benefit one.

SUMMARY

A management information system (MIS) provides managers with the information that they need to fulfill their planning and execution functions. Most current MIS's fail to provide adequate information for top-level organizational planning. Thus, this is both the area of greatest weakness in MIS technology and the area of greatest opportunity for MIS development.

An organization's overall information system is composed of formal, semiformal, and informal subsystems. The MIS usually involves the formal and semiformal subsystems, but seldom the informal one.

In discussing management information systems, it is important to distinguish between data and information. Data are the unevaluated facts and messages in the environment. Information is data which have been evaluated for some purpose.

Computers are not a necessary part of an MIS, but most large organizations require a quantity of information and a level of analysis which makes computers economic to use. Only in the recent past has computer technology developed to the point that the development of a true MIS has become feasible.

The intelligence system of the organization should also be thought of as a part of an MIS. Usually the term "intelligence system" conjures up in our minds ideas of secret agents, clandestine meetings, etc. However, most organizational intelligence systems operate through a concerted, formalized effort to peruse readily-available information sources such as printed media. When careful analysis is performed in such a system, valuable information can result.

To design an MIS requires that system objectives be established, a decision inventory be developed, and information requirements be established. Once this is done on a cost-benefit basis, the more readily recognized aspects of analyzing software and hardware requirements can proceed.

EXERCISES

1. How is an MIS different from a data processing system?
2. Describe the elements of an MIS in an organization with which you are familiar.
3. Distinguish between the formal, semiformal, and informal information systems in an organization. Relate these systems to the processes of formal and informal control as discussed in Chapter 16.
4. Distinguish between data and information.
5. Use a decision theoretic approach to formulate the problem of determining how long data should be kept in the files of an organization.
6. "Modern managers have too much irrelevant information and too little relevant information." Discuss.
7. How is the management by exception principle related to the information that the manager should receive from an MIS?
8. What is an integrated data base? Why is it desirable?
9. One of the most crucial questions in MIS design is that of what information to include in the data base. Discuss this question and conceptualize the solution to it. What problems are associated with actually applying your conceptual solution?
10. Discuss the development of electronic computers as related to the management information capabilities of an organization.
11. What is an organizational intelligence system? How is it related to the organization's MIS?
12. What sort of information can be incorporated into an intelligence system?

13. Why is a formal intelligence system necessary?
14. How can a computerized intelligence system be designed to provide a good likelihood that information will be available to those individuals requiring it?
15. Discuss the ethics of business intelligence. In this context why is good ethics also good business?
16. Discuss the process of designing a comprehensive MIS.
17. How do the objectives for an MIS relate to the objectives of the organization?
18. What is a "decision inventory"? How can it be used in MIS development?
19. Shouldn't an organization's information requirements be determined by simply asking managers what infoımation they need? Why or why not?
20. What is the role of cost-benefit analysis in MIS design?

REFERENCES

Ackoff, R. L., "Management Misinformation Systems," *Management Science,* December, 1967.

Anthony, Robert N., John Dearden, and R. N. Vancil, *Management Control Systems,* Richard D. Irwin, Inc., Homewood, Ill., 1965.

Bauer, W. F., and R. H. Hill, "Economics of Time-Shared Computing Systems," *Datamation,* vol. 13, nos. 11, 12, November, December, 1967.

Blumenthal, Sherman C., *Management Information Systems,* Prentice-Hall, Inc., Englewood Cliffs, N.J., 1969.

Boyd, D. F., and H. S. Krasnow, "Economic Evaluation of Management Information Systems," *IBM Systems Journal,* vol. 2, pp. 2–23, March, 1963.

Brandon, D. H., *Management Planning for Data Processing,* Brandon/Systems Press, 1970.

Burck, G. "The Computer Industry's Great Expectations," *Fortune,* August, 1968.

Cohen, I. K., and R. L. Van Horn, "A Laboratory Exercise for Information System Evaluation," in J. Speigel and D. E. Walker (eds.), *Second Congress on the Information System Sciences,* Spartan Books, Inc., 1965.

Dearden, John, and I. W. McFarlan, *Management Information Systems,* Richard D. Irwin, Inc., Homewood, Ill. 1966.

Diebold, J., "Bad Decisions on Computer Use," *Harvard Business Review,* January–February, 1969.

Emery, James C., *Oranizational Planning and Control Systems,* The Macmillan Company, New York, 1969.

Greenwood, W. T., *Decision Theory and Information Systems,* South-Western Publishing Company, Incorporated, Cincannati, 1969.

Hodge, Barton, and Robert Hodgson, *Management and the Computer in Information and Control Systems,* McGraw-Hill Book Company, New York, 1969.

Head, R. V. "Management Information Systems: A Critical Appraisal," *Datamation,* May, 1967.

Joslin, E. O., and M. J. Mullin, "Cost-Value Technique for Evaluation of Computer System Proposals," *AFIPS Conference Proceedings,* vol. 25, 1964.

Kanter, J., *Management Guide to Computer System Selection and Use,* Prentice-Hall, Inc., Englewood Cliffs, N.J., 1970.

King, William R. and D. I. Cleland, "Manager-Analyst Teamwork in MIS," *Business Horizons,* April, 1970.

Marschak, J., "Economics of Inquiry, Communication, Deciding," *American Economic Review,* May, 1968, pp. 1–18.

McKinsey & Company, Inc., "Unlocking the Computer's Profit Potential," *Computers and Automation,* April, 1969, pp. 24–33.

Miller, J. C., "Conceptual Models for Determining Information Requirements," *AFIPS Conference Proceedings,* vol. 25, 1964.

Prince, Thomas R., *Information Systems for Management Planning and Control,* Richard D. Irwin, Inc., Homewood, Ill., 1966.

Rappaport, Alfred (ed.), *Information for Decision Making,* Prentice-Hall, Inc., Englewood Cliffs, N.J., 1970.

Rappaport, A., "Management Misinformation Systems—Another Perspective," *Management Science,* December, 1968.

Sharpe, W. F., *The Economics of Computers,* Columbia University Press, New York, 1969.

Will, H. J., "Some Comments on Informative Systems," *Management Science,* December, 1969.

Williams, Thomas H., and Charles H. Griffin, *Management Information, A Quantitative Accent,* Richard D. Irwin, Inc., Homewood, Ill., 1967.

index